T0332236

Cellular Biophysics

Volume 2

Thomas Fischer Weiss

Cellular Biophysics

Volume 2: Electrical Properties

A Bradford Book
The MIT Press
Cambridge, Massachusetts
London, England

This book was set in Lucida Bright by Windfall Software using ZzT_EX.

Library of Congress Cataloging-in-Publication Data
Weiss, Thomas Fischer
    Cellular biophysics / Thomas Fischer Weiss
    v. <1- > ; cm.
    Includes bibliographical references and index.
    Contents: v. 1. Transport — v. 2. Electrical properties.
    ISBN 978-0-262-23183-1 (v. 1). — ISBN 978-0-262-52957-0 (v. 2)
    1. Cell physiology.   2. Biophysics.   3. Biological transport.
    4. Electrophysiology. I. Title.
    QH631.W44   1995
    574.87'6041—dc20   95-9801
                                                                      CIP

The MIT Press is pleased to keep this title available in print by manufacturing single copies, on demand, via digital printing technology.

To Aurice B, Max, Elisa, and Eric

Contents

Contents in Detail

Preface

In scientific thought we adopt the simplest theory which will explain all the facts under consideration and enable us to predict new facts of the same kind. The catch in this criterion lies in the word 'simplest.' It is really an aesthetic canon such as we find implicit in our criticisms of poetry or painting. The layman finds such a law as $\partial x/\partial t = \kappa(\partial^2 x/\partial y^2)$ less simple than 'it oozes,' of which it is the mathematical statement. The physicist reverses this judgement, and his statement is certainly the more fruitful of the two, so far as prediction is concerned.
—Haldane, 1985

Subject and Orientation of the Book

This and the companion text (Weiss, 1996) consider two basic topics in cellular biophysics, which we pose here as questions:

- Which molecules are transported across cellular membranes, and what are the mechanisms of transport? How do cells maintain their compositions, volume, and membrane potential?

- How are potentials generated across the membranes of cells? What do these potentials do?

Although the questions posed are fundamentally biological questions, the methods for answering these questions are inherently multidisciplinary. For example, to understand the mechanism of transport of molecules across cellular membranes, it is essential to understand both the structure of membranes and the principles of mass transport through membranes. Since the transported matter may combine chemically with membrane-spanning macromolecules and/or carry an electrical charge, it is essential to understand the principles of chemical kinetics and of transport of charged molecules in an electric field.

Knowledge of transport through membranes is based on measurements. These measurements lead to physically and chemically based mathematical models that are used to test concepts based on measurements. The role of mathematical models is to express concepts precisely enough that precise conclusions can be drawn (see quote by Haldane). In connection with all the topics covered, we will consider both theory and experiment. For the student, the educational value of examining the interplay between theory and experiment transcends the value of the specific knowledge gained in the subject matter.

My aim has been to produce textbooks on cellular biophysics that clarify rather than mystify and that eschew glib development. Topics were chosen to emphasize well-established principles, but also to include some more recent and sometimes controversial material. Most topics are introduced with a brief historical perspective. Challenging problems are included to aid students in learning the material. Extensive references, much more extensive than the citations, are provided to aid readers to pursue topics of interest in further depth.

Expected Background of the Reader

It has been assumed that the background of the reader of these texts includes one year of college physics, a semester of chemistry, a semester of biology, one year of calculus, a semester of differential equations, plus a course in which differential equations are solved in some physical context (e.g., electric network theory, mechanical dynamics, chemical kinetics, etc.). Some background in probability theory is also desirable.

A Note to the Instructor

The material in this and the companion text (Weiss, 1996) has been used to teach an undergraduate course taken by juniors and seniors at the Massachusetts Institute of Technology, which is the first of a sequence of three courses in bioengineering offered in the School of Engineering. The course, called Quantitative Physiology: Cells and Tissues, consists of the following activities:

- Three lectures each week to introduce new material.

- Two recitations each week to review material, solve problems, and answer questions.

- One laboratory exercise, which requires a written report, to provide direct experience with experimental techniques and with communication of experimental results and interpretations.

- One theoretical research project on the Hodgkin-Huxley model, which requires a written report, to provide an opportunity for students to define a testable hypothesis, conduct a theoretical investigation, and communicate the results and conclusions.

- A homework assignment each week that enables students to actively assimilate the course material.

- Fifteen-minute weekly quizzes that encourage students to learn the material as it is presented and provide a regular assessment of their progress. Many of the exercises at the ends of each chapter were devised as quiz problems.

- One midterm examination and one final examination that give students an opportunity to integrate the course material and to obtain an objective evaluation of their understanding of the material. Many of the problems at the ends of each chapter were devised as examination problems.

Students are expected to devote twelve hours of time (contact hours plus preparation time) to the course each week for thirteen weeks. The two texts cover more material than is covered in the course. The list of lectures on the next page indicates the coverage of material in a one-semester course. In the course, we use computers extensively in lectures, recitations, homework assignments, and projects, although the texts do not require the use of computers. The use of computers in teaching this course is discussed elsewhere (Weiss et al., 1992).

Preparation of the Manuscript

Typesetting was done in TEX with LaTEX macros on a Macintosh computer using Textures. Spelling was checked with the LaTEX spell checker Excalibur. Theoretical calculations were done with Mathematica, Macsyma, and MATLAB. Several tools were used to reproduce data from the literature. Figures from the literature were scanned at 300 dpi using either Ofoto or FotoLook. For figures consisting of a small number of data points, the coordinates of the points

List of Lectures

were obtained from the scanned images by means of FlexiTrace. The data points were plotted by means of the charting program DeltaGraph Pro. More complex figures were processed with Adobe PhotoShop and/or Adobe Streamline. Most graphic files were imported to Adobe Illustrator for annotation and saved as encapsulated Postscript files that were included electronically in the text. Mathematical annotations were obtained by typesetting the mathematical expressions with Textures and saving the typeset version as a file that was read by Adobe Illustrator. Chemical formulas were done with ChemDraw Plus and three-dimensional models with Chem3D Plus. A database of genes of membrane proteins was obtained with DNAStar, which also allowed translation of nucleotide sequences into amino acids sequences from which properties of the protein could be displayed.

Personal Perspective

It was a shock to look over my records and to discover that, although I did not realize it at the time, I began to write these books in the spring of 1966, when I taught a course called Introduction to Neuroelectric Potentials. That spring, I started writing notes to clarify my thinking on the course matter and as a supplement to the lectures for students. At the time, there were no textbooks available on the topics I wished to cover—none that presented empirical findings as well as derivations of theoretical conceptions from first principles.

The notes grew annually and were reproduced for each fall's class. Changes in methods of reproduction of the notes reflect the changing technology. Originally, the notes were typed and reproduced, first using ditto masters—I think there is still some blue stuff under one fingernail from the last of these messy products—then by xerographic reproduction. With the availability of technical typesetting languages, the notes began to take on an improved appearance around 1984, when they approached several hundred pages of Troffed text. By 1986 the notes were converted to $\LaTeX$, but the figures were still being pasted in with rubber cement—a process that left a succession of secretaries tired but surprisingly giddy. In order to save time, to improve the quality of the figures, and to avoid tripping out my secretary, in 1990 I began the long but rewarding process of replacing the figures pasted in with glue with figures included electronically in the notes. Experimental data were digitized from original sources, photographs were scanned, and theoretical calculations were done anew. It is remarkable what one can learn about a subject through this process. Frankly, this effort was only possible because I have an obsessive personality. A colleague once told me, "You're too thorough," which I believe was meant as a mild put-down, but which I always regarded as a compliment.

For over twenty years, the notes became more and more extensive and complete without any conscious inclination on my part to publish a text. I took seriously the admonition of a colleague who said, "Never publish a book." During this period, my children grew up—two of the three have married—and my wife and I have qualified for discounts as senior citizens. The transition from notes to textbook, which was clearly occurring subconsciously, took a final serendipitous turn into my consciousness. Sometime after 1984, I learned from a graduate student who came to MIT from mainland China that the notes had been translated into Chinese. Eventually it occurred to me that the notes were receiving a highly skewed international distribution with foci in China and Cambridge, Massachusetts. To achieve a more uniform global access, I decided to make the notes into a textbook. I sought and received a sabbatical leave in the calendar year 1991 to complete the textbook. Three and one half years later, the job is finally done—and then some. The notes have become two textbooks! The final step in the preparation of the manuscripts was producing indices, a task for which I have newfound respect. During this period I learned the meaning of the term *repetitive stress disorder*. Apparently, one of my computers did as well. Four days before final completion, the Macintosh computer in my study simply quit. It just couldn't do any more. I am now off to a vacation in North Truro on Cape Cod in Massachusetts where many chapters of these books were contemplated in summers past. But not this summer!

Acknowledgments

I welcome the opportunity to acknowledge the contributions of the many colleagues, friends, and family members who have contributed directly or indirectly to this work. I have been influenced by many of my colleagues, but I would like to single out a few who have been particularly important in my intellectual growth—Nelson Kiang, Charlie Molnar, Bill Peake, Walter Rosenblith, and Bill Siebert. Members of the Department of Electrical Engineering and Computer Science and the School of Engineering have been very supportive of my efforts. These include department chairmen and deans, of whom I wish to particularly thank Dick Adler, Joel Moses, Paul Penfield, Jeff Shapiro, and Jerry Wilson.

During the latter portions of this project, I was the fortunate recipient of a chair donated to MIT by Gerd and Tom Perkins. This support has given me the freedom to pursue this writing project with additional vigor in the latter years.

A number of faculty and staff have shared in teaching the material that comprises these texts, including Dick Adler, Denny Freeman, Al Grodzinsky, Rafael Lee, Bill Peake, and Bill Siebert. All have made contributions to the development of these materials. In particular, Denny Freeman and Bill Peake have formulated problems that appear in this text. Many of the problems have been used for years, so I no longer remember who made them up. Denny Freeman and Bill Peake have also given me a great deal of critical feedback on the text. Weaknesses, from specious arguments to indefinite antecedents, have been reduced appreciably through their efforts. I also wish to thank my colleague John Wyatt, who told me that I could "continue to work on the material *after* the books are published." This remark made while waiting for an elevator relieved my angst in parting with the manuscripts, which are still somewhat short of what I have in mind. I thank my friend Helen Peake for her support and for translating a paper by J. Perrin for me. My secretaries have also been very important to the completion of the task. There have been three that I would like to single out: Sylvia Nelson, Susan Ross, and Janice Balzer. Janice, in particular, has helped in many ways.

Graduate student teaching assistants have also made important contributions to the text, particularly in troubleshooting the problems. In this regard, I would particularly like to acknowledge help from Eero Simoncelli. Many students have used the notes over the years. Their questions have been very important in helping me to determine the issues that they found difficult to comprehend. Often I found that these issues were ones on which my thinking was unclear. Students have also been a great source of proofreading over the years. A few students made heroic efforts to find errors in the notes. I especially thank Greg Allen, Andy Grumet, Chris Long, Karen Palmer, and Susan Voss for their efforts. Despite the efforts of many people, errors no doubt still exist. I would appreciate hearing from readers about errors—typographic or conceptual—either via mail or electronic mail (tfweiss@mit.edu).

I wish to thank Fiona Stevens of The MIT Press for her help throughout this process. After I had missed yet another deadline, Fiona told me that "the books that come in the latest are often the best books." That was a kind thing to say, and I appreciated it. I thank Katherine Arnoldi, senior editor at MIT Press, who always provided calm reassurance during the final stages of production of these books. The copy editors, Suzanne Schafer and Marilyn Martin, have allowed me to perpetuate the myth that I know how to use commas correctly and that I correctly distinguish between "that" and "which." Paul Anagnostopoulos of Windfall Software was my constant companion over the telephone and the Internet as the electronic manuscripts were transformed into electronic books. I appreciate his constructive suggestions, expertise, and exuberance.

Finally, and most important, my family has been my greatest source of support. My children, Max, Elisa, and Eric, were always interested in this project and felt a sense of pride in me that helped me to keep going when I occasionally became despondent about ever finishing. My children-in-law, Kelly and Nico, were also very supportive. At times when I believed that the books were a total disaster and no one would ever read them, my wife, Aurice, always told me she was convinced they would turn out well. At times when I told her that I thought these books were the greatest texts written in any discipline, she told me she believed me. She always had a positive word to say, and she listened patiently to daily accountings of the state of the chapter of the day. It must have been terribly boring, but she always led me to believe that she was interested in the latest accounting.

References

Haldane, J. B. S. (1985). Science and theology as art-forms. In Smith, J. M., ed., *On Being the Right Size and Other Essays*, 32–44. Oxford University Press, New York.

Weiss, T. F. (1996). *Cellular Biophysics,* vol. 1, *Transport.* MIT Press, Cambridge, MA.

Weiss, T. F., Trevisan, G., Doering, E. B., Shah, D. M., Huang, D., and Berkenblit, S. I. (1992). Software for teaching physiology and biophysics. *J. Sci. Ed. Tech.,* 1:259–274.

Units, Physical Constants, and Symbols

Units

In this text we use the International Systems of Units, called the SI units (Mechtly, 1973).

Base SI Units

The SI units contain seven base units from which all other units are derived. The seven base units are presented in the following table.

Quantity	Name	Symbol
Amount of substance	mole	mol
Electric current	ampere	A
Length	meter	m
Luminous intensity	candela	cd
Mass	kilogram	kg
Thermodynamic temperature	kelvin	K
Time	second	s

Derived SI Units

Many units are derived from the base units. Some are obvious, e.g., the unit for volume is m^3; others involve new terms. The derived units that are not obvious and that are used in this text are as follows.

Quantity	Name	Symbol	Units
Capacitance	farad	F	$A \cdot s \cdot V^{-1}$
Electric resistance	ohm	Ω	$V \cdot A^{-1}$
Electric charge	coulomb	C	$A \cdot s$
Electric conductance	siemens	S	$A \cdot V^{-1}$
Electric potential difference	volt	V	$N \cdot m \cdot C^{-1}$
Energy	joule	J	$N \cdot m$
Force	newton	N	$kg \cdot m \cdot s^{-2}$
Frequency	hertz	Hz	s^{-1}
Inductance	henry	H	$V \cdot s/A$
Power	watt	W	$N \cdot m \cdot s^{-1}$
Pressure	pascal	Pa	$N \cdot m^{-2}$

Decimal Multiples and Submultiples of SI Units

The following prefixes are used to express multiples or submultiples of SI units.

Prefix	Symbol	Factor	Prefix	Symbol	Factor
exa	E	10^{18}	deci	d	10^{-1}
peta	P	10^{15}	centi	c	10^{-2}
tera	T	10^{12}	milli	m	10^{-3}
giga	G	10^{9}	micro	μ	10^{-6}
mega	M	10^{6}	nano	n	10^{-9}
kilo	k	10^{3}	pico	p	10^{-12}
hecto	h	10^{2}	femto	f	10^{-15}
deca	da	10^{1}	atto	a	10^{-18}

Commonly Used Non-SI Units and Conversion Factors

The following non-SI units are commonly used.

Quantity	Name	Symbol	Units
Energy	calorie	cal	4.184 J
Energy	erg	erg	10^{-7} J

Energy	electronvolt	eV	1.602×10^{-19} J
Force	dyne	dyne	10^{-5} N
Length	angstrom	Å	10^{-10} m
Molecular weight	dalton	D	$gm \cdot mol^{-1}$
Pressure	atmosphere	atm	1.013×10^5 N $\cdot$ m^{-2}
Pressure	bar	bar	1×10^5 N $\cdot$ m^{-2}
Pressure	millimeter of mercury	mmHg	133.3 N $\cdot$ m^{-2}
Temperature	Celsius	°C	$T_C = T - 273.15$
Temperature	Fahrenheit	°F	$T_F = (9/5)T_C + 32$
Volume	liter	L	10^{-3} m^3

T, T_C, and T_F are the temperatures in kelvins, degrees Celsius, and degrees Fahrenheit.

Physical Constants

Fundamental Physical Constants

The following fundamental physical constants (Mechtly, 1973) are used in the text.

Name	Symbol	Value
Acceleration of gravity	g	9.807 m$\cdot$s^{-2}
Avogadro's number	N_A	6.022×10^{23} mol^{-1}
Boltzmann's constant	$k = R/N_A$	1.381×10^{-23} J $\cdot$ K^{-1}
Electronic charge	e	1.602×10^{-19} C
Faraday's constant	$F = N_A e$	9.648×10^4 C $\cdot$ mol^{-1}
Molar gas constant	R	8.314J $\cdot$ mol^{-1} $\cdot$ K^{-1}
Permittivity of free space	ϵ_0	8.854×10^{-12} F $\cdot$ m^{-1}
Planck's constant	h	6.626×10^{-34} J $\cdot$ s

Physical Properties of Water

The following physical properties of water (Robinson and Stokes, 1959, Eisenberg and Crothers, 1979, Lide, 1990) are given at standard pressure (100 kPa).

Property	Temperature °C	Value
Density	20	$0.998 \, \text{g} \cdot \text{cm}^{-3}$
Dielectric constant	20	80.20
Diffusion coefficient	25	$2.4 \times 10^{-5} \, \text{cm}^2 \cdot \text{s}^{-1}$
Molar volume	20	$18.05 \, \text{cm}^3 \cdot \text{mol}^{-1}$
Viscosity	20	$1.002 \, \text{mPa} \cdot \text{s}$

Atomic Numbers and Weights

The atomic numbers and weights of the elements (Ebbing, 1984) are as follows (the values in parentheses are the mass number of the isotopes of longest half-life).

Element	Symbol	Atomic Number	Atomic Weight	Element	Symbol	Atomic Number	Atomic Weight
Actinium	Ac	89	227.0278	Molybdenum	Mo	42	95.94
Aluminum	Al	13	26.98154	Neodymium	Nd	60	144.24
Americium	Am	95	(243)	Neon	Ne	10	20.179
Antimony	Sb	51	121.75	Neptunium	Np	93	237.0482
Argon	Ar	18	39.948	Nickel	Ni	28	58.69
Arsenic	As	33	74.9216	Niobium	Nb	41	92.9064
Astatine	At	85	(210)	Nitrogen	N	7	14.0067
Barium	Ba	56	137.33	Nobelium	No	102	(259)
Berkelium	Bk	97	(247)	Osmium	Os	76	190.2
Beryllium	Be	4	9.01218	Oxygen	O	8	15.9994
Bismuth	Bi	83	208.9804	Palladium	Pd	46	106.42
Boron	B	5	10.81	Phosphorus	P	15	30.97376
Bromine	Br	35	79.909	Platinum	Pt	78	195.08 ± 3
Cadmium	Cd	48	112.41	Plutonium	Pu	94	(244)
Cesium	Cs	55	132.9054	Polonium	Po	84	(209)
Calcium	Ca	20	40.08	Potassium	K	19	39.0983
Californium	Cf	98	(251)	Praseodymium	Pr	59	140.9077
Carbon	C	6	12.011	Promethium	Pm	61	(145)
Cerium	Ce	58	140.12	Protactinum	Pa	91	231.0359

Chlorine	Cl	17	35.453	Radium	Ra	88	226.0254
Chromium	Cr	24	51.996	Radon	Rn	86	(222)
Cobalt	Co	27	58.9332	Rhenium	Re	75	186.207
Copper	Cu	29	63.546	Rhodium	Rh	45	102.9055
Curium	Cm	96	(247)	Rubidium	Rb	37	85.4678
Dysprosium	Dy	66	162.50	Ruthenium	Ru	44	101.07
Einsteinium	Es	99	(252)	Samarium	Sm	62	150.36
Erbium	Er	68	167.26	Scandium	Sc	21	44.9559
Europium	Eu	63	151.96	Selenium	Se	34	78.96
Fermium	Fm	100	(257)	Silicon	Si	14	28.0855
Fluorine	F	9	18.998403	Silver	Ag	47	107.8682
Francium	Fr	87	(223)	Sodium	Na	11	22.98977
Gadolinium	Gd	64	157.25	Strontium	Sr	38	87.62
Gallium	Ga	31	69.72	Sulfur	S	16	32.06
Germanium	Ge	32	72.59	Tantalum	Ta	73	180.9479
Gold	Au	79	196.9665	Technetium	Tc	43	(98)
Hafnium	Hf	72	178.49	Tellurium	Te	52	127.60
Helium	He	2	4.00260	Terbium	Tb	65	158.9254
Holmium	Ho	67	164.9304	Thallium	Tl	81	204.383
Hydrogen	H	1	1.00794	Thorium	Th	90	232.0381
Indium	In	49	114.82	Thulium	Tm	69	168.9342
Iodine	I	53	126.9045	Tin	Sn	50	118.69
Iridium	Ir	77	192.22	Titanium	Ti	22	47.88
Iron	Fe	26	55.847	Tungsten	W	74	183.85
Krypton	Kr	36	83.80	Unnilhexium	Unh	106	(263)
Lanthanum	La	57	138.9055	Unnilpentium	Unp	105	(262)
Lawrencium	Lr	103	(260)	Unnilquadium	Unq	104	(261)
Lead	Pb	82	207.2	Uranium	U	92	238.0289
Lithium	Li	3	6.941	Vanadium	V	23	50.9415
Lutetium	Lu	71	174.967	Xenon	Xe	54	131.29
Magnesium	Mg	12	24.305	Ytterbium	Yb	70	173.04
Manganese	Mn	25	54.9380	Yttrium	Y	39	88.9059

| Mendelevium | Md | 101 | (258) | Zinc | Zn | 30 | 65.38 |
| Mercury | Hg | 80 | 200.59 | Zirconium | Zr | 40 | 91.22 |

Symbols

The principal symbols used in the text are given in the table below, which also indicates the chapters in which each symbol is used. The symbols are not all distinct, but the context should resolve ambiguities. The math accent ($\hat{\ }$) is used to indicate a normalized variable, and the tilde ($\tilde{\ }$) to distinguish between permeant and impermeant solutes. Vectors are indicated by boldface. An asterisk ($*$) in the Units column below indicates that the units depend on the context.

Name	Symbol	Units	Chapters
Absolute temperature	T	K	3–8
Area	A	m^2	3–8
Association constant	K_a	$*$	6
Avogadro's number	N_A	mol^{-1}	3
Boltzmann's constant	k	$J \cdot K^{-1}$	3, 6
Carrier density	$\mathfrak{N}$	$mol \cdot m^{-2}$	6
Charge density	ρ	$C \cdot m^{-3}$	7
Charge relaxation time	τ_r	s	7
Chemical potential	μ	$J \cdot mol^{-1}$	3–7
Conductance	$\mathcal{G}$	S	7
Convection velocity	v	$m \cdot s^{-1}$	3, 7
Current	I	A	7
Current density	J	$A \cdot m^{-2}$	7, 8
Debye length	Λ_D	m	7
Diffusion coefficient	D	$m^2 \cdot s^{-1}$	3, 4, 5, 7
Diffusive permeability	P	$m \cdot s^{-1}$	3, 4, 5, 6, 8
Dissociation constant	K	$*$	6, 7, 8
Electric conductivity	σ_e	$S \cdot m^{-1}$	3, 7
Electric field intensity	$\mathcal{E}$	$V \cdot m^{-1}$	7
Electric potential	ψ	V	7

Electric potential difference	V	V	7, 8
Electrochemical potential	$\tilde{\mu}$	J·mol^{-1}	7
Energy	E	J	6
Equilibrium time constant	τ_{eq}	s	3
Faraday's constant	F	C·mol^{-1}	7
Force on a mole of particles	f	N·mol^{-1}	3, 7
Force on a particle	f_p	N	3
Forward rate constant	α	$*$	6
Free energy	$\mathfrak{G}$	J·mol^{-1}	7
Hydraulic conductivity	$\mathcal{L}_V$	m · Pa^{-1} · s^{-1}	4, 5, 8
Hydraulic permeability	κ	m^2 · Pa^{-1} · s^{-1}	4
Hydraulic pressure	p	Pa	4, 5, 8
Mass density	ρ_m	kg·m^{-3}	3, 4
Mean free path	l	m	3
Mean free time	τ	s	3
Mean velocity	$\overline{v}$	m·s^{-1}	3
Mean-squared velocity	$\overline{v^2}$	m^2 · s^{-2}	3
Mean value	m	$*$	3
Membrane thickness	d	m	3, 4, 5
Membrane potential	V_m	V	7, 8
Molar concentration	c	mol·m^{-3}	3–8
Molar electric mobility	$\hat{u}$	m^2 · s^{-1} · V^{-1}	7
Molar flux	ϕ	mol·m^{-2}·s	3–8
Molar gas constant	R	J·mol^{-1} · K^{-1}	1, 3–8
Molar mechanical mobility	u	m·mol · s^{-1} · N^{-1}	3, 7
Molar particle density	n	mol·m^{-2}	3
Molar quantity	n	mol	5, 6, 8
Molecular weight	M	g·mol^{-1}	3
Mole fraction	x	dimensionless	4
Osmotic coefficient	χ	dimensionless	4
Osmotic permeability	$\mathcal{P}$	m·s^{-1}	4, 5
Osmotic pressure	π	Pa	4, 5

Partial molar volume	$\overline{V}$	$m^3 \cdot mol$	4, 5
Particle mass	m	kg	3
Particle mechanical mobility	u_p	$m \cdot s^{-1} \cdot N^{-1}$	3
Particle radius	a	m	3
Partition coefficient	k	dimensionless	3
Permittivity	ϵ	$C \cdot m^{-1} \cdot V^{-1}$	7
Pore density	$\mathcal{N}$	m^{-2}	3, 4, 5
Porosity	$\mathcal{P}$	dimensionless	3
Probability	W	dimensionless	3
Probability	x	dimensionless	7
Probability density	p	*	3
Q_{10}	Q_{10}	dimensionless	6
Radius—particle	a	m	3, 4
Radius—pore	r	m	3, 4, 5
Reflection coefficient	σ	dimensionless	5, 8
Resistance	$\mathcal{R}$	Ω	7
Resistivity	ρ_e	$\Omega \cdot m$	7
Resting membrane potential	V_m^o	V	7, 8
Reverse rate constant	β	*	6
Single-channel conductance	γ	S	7
Single-channel current	$\mathcal{I}$	A	7
Solute density	s	$mol \cdot m^{-2}$	6
Specific conductance	G	$S \cdot m^{-2}$	7, 8
Specific resistance	R	$\Omega \cdot m^2$	7
Standard deviation	σ	*	3
Steady-state time constant	τ_{ss}	s	3
Stoichiometric coefficient	ν	dimensionless	6, 7, 8
Surface charge density	Q_f	$C \cdot m^{-2}$	7
Tortuosity	$\mathcal{T}$	dimensionless	3
Total molar concentration	C_Σ	$mol \cdot m^{-3}$	4, 5, 6, 8
Total molar quantity	N_Σ	mol	4, 5, 8
Valence	z	dimensionless	7, 8

Viscosity	η	Pa·s	3, 4, 5
Volume	V	m^3	3, 4, 5, 6, 8
Volume flux	Φ_V	$m \cdot s^{-1}$	4, 5, 8

References

Ebbing, D. D. (1984). *General Chemistry*. Houghton Mifflin, Boston.

Eisenberg, D. and Crothers, D. (1979). *Physical Chemistry with Applications to the Life Sciences*. Benjamin-Cummings, Menlo Park, CA.

Lide, R. R. (1990). *Handbook of Chemistry and Physics,* 71st ed. CRC Press, Boston.

Mechtly, E. A. (1973). The international system of units. Technical report. NASA SP-7012, National Aeronautics and Space Administration, Washington, DC.

Robinson, R. A. and Stokes, R. H. (1959). *Electrolyte Solutions*. Butterworth, London.

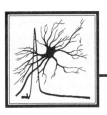

Introduction to Electrical Properties of Cells

"The most general conclusion related to the cellular morphology of nerve centers is the absence of continuity between neural, epithelial, and neuroglial processes. The neural elements are true cellular entities or neurons . . . "
—Cajal, 1894

1.1 A Brief Historical Perspective

By the end of the eighteenth century the notion that animal tissue both generated and responded to electricity was reasonably well established. The sensation experienced upon touching certain (electric) fish was explained as an electrical phenomenon. Galvani's experiments purported to demonstrate that an electric current could provoke a muscle contraction and that a muscle could produce an electric current. Galvani's interpretations of those experiments were challenged by Volta, and it took a half century to resolve the controversy. By 1850 du Bois-Reymond had demonstrated that Galvani's conclusions were essentially correct. At about the same time, improvements in optics and tissue fixation enabled observations that led to the cell doctrine of Schleiden and Schwann, namely, that living organisms are made of cells. By the end of the nineteenth century, the Spanish histologist Ramón y Cajal had demonstrated that the nervous system also consisted of individual cells called *neurons*. Thus, interpretation of the electrical properties of living tissues began to focus on the electrical properties of living cells. At the end of the nineteenth century, responses of nerve and muscle to electrical stimulation were investigated systematically. Bernstein interpreted these results in physicochemical terms. He postulated that a cell consists of a semipermeable membrane surrounding an electrolytic interior. When the cell is at rest, an electric potential difference exists across the membrane, and the electrical activity recordable from such a cell consists of a change in this potential difference resulting from changes in membrane permeability.

With the rapid development of more sensitive instruments (e.g., the string galvanometer), it was possible to measure not only electric potentials in response to electric shocks, but also potentials that occur naturally in animals; for example, the electric potential accompanying the beating of the heart (measured by the electrocardiogram or EKG) was recorded. After World War I, the vacuum tube amplifier and later the cathode-ray oscilloscope became available and allowed a thousand-fold increase in sensitivity and higher temporal resolution in the measurement of electric potentials generated by living organisms. It then became possible to measure electrical responses of sense organs in response to sensory stimuli, electrical activity of peripheral nerve fibers, electrical activity of muscles (measured by the electromyogram or EMG), and electrical activity originating in the brain (measured by the electroencephalogram or EEG). Techniques for extracellular recording of the activity of single cells led to the recognition that nerve fibers transmit information coded in the form of a temporal sequence of identical pulses called *action potentials*, and the all-or-none character of the action potential generated by electrically excitable cells was recognized.

In the decade from 1940 to 1950, techniques were developed to record intracellularly, and it became possible to measure and/or control the potential across cellular membranes. Initially this was accomplished in large cells such as the giant axons of invertebrates and in vertebrate muscle fibers. Since 1950, microelectrode recording techniques have been greatly refined to enable intracellular recording from smaller cells. A focus has been on determining the membrane mechanisms responsible for the generation of electric potentials and on determining the effects of these potentials on other cellular processes.

In the three decades beginning in 1960, microanatomical techniques, especially the development and routine use of the electron microscope, have enabled visualization of cellular and subcellular structure with much higher resolution. Microchemical techniques have revealed information about the molecular constituents of cells. Genetic engineering has enabled investigations of the relation between molecular constituents of parts of simple organisms and the behavior of whole organisms. Microelectrophysiological techniques have been developed to measure the currents flowing through individual transmembrane ionic channels. Finally, the advent of the digital computer has greatly extended the ability of experimenters to quantify both anatomical and physiological measurements, and for theoreticians to develop conceptualizations of cellular systems. The aims of electrophysiology remain the same: to explain the electrical phenomena of living cells in molecular terms, and to explain the behavior of organisms in terms of the underlying cellular electrical processes.

1.2 Cellular Electric Potentials

There is a maintained difference of electric potential across the membrane of a living cell called the resting potential (Weiss, 1996, Chapter 7). This membrane potential changes in response to changes in numerous physicochemical variables, e.g., temperature, pH, extracellular concentrations of ions, extrinsic electric current, and so on. In this section, we categorize such potential changes exhibited by cells.

1.2.1 Electric Potentials in Electrically Small Cells: Graded and Action Potentials

By definition, an *electrically small cell* is one for which the potential difference across its membrane is the same everywhere along the membrane. The notion of a small cell will be made more precise in Chapter 3, where we shall show that a cell is electrically small if its dimensions are smaller than a characteristic electrical length called the *cell space constant*. For the moment, we shall merely assume that we are dealing with such a cell. Small cells can be categorized as *electrically inexcitable* and *electrically excitable*.

1.2.1.1 Graded Potentials

If a pulse of electric current is passed through the membrane of an electrically inexcitable cell, the membrane potential response is as shown schematically in Figure 1.1. Outward current passed through the membrane causes a depolarization of the membrane, i.e., an increase in the potential across the membrane from its normal resting value. Increasing the amplitude of this current tends to increase the depolarization. Current passed inward through the membrane causes a hyperpolarization of the membrane. Such potential changes are called *graded potentials*. Graded potentials have the property that an incremental increase in the current amplitude leads to an incremental increase in the membrane potential. Note that the waveform of the graded potential that results from a current pulse is not identical to that of the membrane current; the sharp edges of the current pulse are not reproduced in

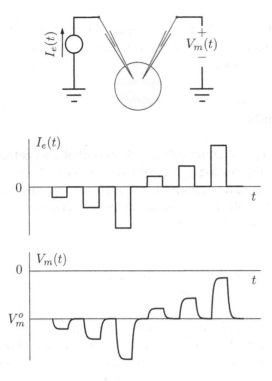

Figure 1.1 Schematic diagram of the membrane potential changes resulting from the passage of rectangular pulses of current through the membrane of an electrically inexcitable cell. When $I_e(t)$ is positive, current flows into the cell cytoplasm from an intracellular micropipette and outward through the membrane of the cell to the reference electrode (not shown), which is connected to ground. The membrane potential $V_m(t)$ is the difference in potential between that recorded by the intracellular micropipette and that recorded by an extracellular reference electrode.

the membrane potential change. We shall see in Chapter 3 that the capacitance of the membrane limits the rate of change of the membrane potential and thus filters out the sharp transitions present in the current.

Changes in other variables also generate graded potentials. For example, a change in intracellular and extracellular ion concentration leads to a change in the membrane potential (Weiss, 1996, Chapter 7). Figure 1.2 shows an example of a graded potential caused by a change in ion concentration in the bath surrounding a muscle fiber. Similarly, changes in temperature, pH, and mechanical deformation of the membrane can also cause graded potentials. Thus, all cells produce graded potentials in response to changes in a wide variety of physicochemical variables. In addition, the membrane potentials of certain specialized cells are exquisitely sensitive to specific physicochemical variables. We shall consider these sensitivities next.

Photoreceptors

The membrane potential of a photoreceptor cell in the vertebrate retina changes in response to light shining on the cell. This change in potential of a receptor cell is called a *receptor potential*. The receptor potential is a graded

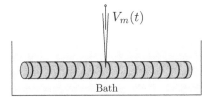

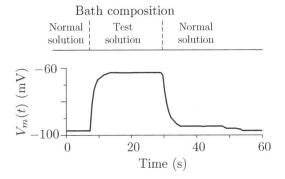

Figure 1.2 The change in membrane potential resulting from a change in external ion composition of a single (semitendinosus) muscle fiber of the frog *Rana temporaria* (adapted from Hodgkin and Horowicz, 1959, Figure 2). The initial and final extracellular solutions (normal solutions) contained the normal composition of extracellular solutions that included 2.5, 120, and 121 mmol/L of K^+, Na^+, and Cl^-, respectively. The test solution, which had the same ionic strength and osmotic pressure as the normal solution, contained 10, 83, and 30 mmol/L of K^+, Na^+, and Cl^-, respectively. The temperature was 20°C.

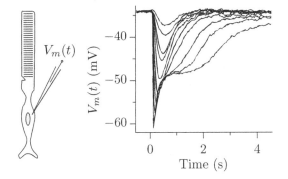

Figure 1.3 Receptor potentials of an isolated rod from the retina of the salamander *Ambystoma tigrinum* in response to brief flashes of light of different intensities (adapted from Baylor and Nunn, 1986, Figure 3). The flash duration was 11 ms. The light intensity was varied from 1.5 to 430 photons/μm^2 by factors of about 2; the responses to these flashes are superimposed.

potential whose time course and amplitude vary as the intensity, wavelength, time course, and spatial pattern of the light are varied. As shown in Figure 1.3, the response of a vertebrate rod to a brief flash of light is a receptor potential with a peak-to-peak value of as much as 25 mV, a hyperpolarizing polarity, and a duration of a few seconds. Photoreceptors are extraordinarily sensitive to light; even single photons striking the receptor cause a detectable receptor potential. The waveform of the receptor potential changes systematically as the light intensity of the flash is changed. The peak-to-peak value of the receptor potential increases as the light intensity increases. However, an in-

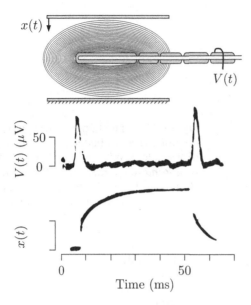

Figure 1.4 Receptor potential of a Pacinian corpuscle from the mesentery of a cat in response to a compression of the corpuscle (adapted from Loewenstein and Mendelson, 1965, Figure 2). The receptor potential is measured with an extracellular electrode on the nerve relative to a remote reference electrode (not shown) by methods that will be described in Chapter 2. The displacement of the corpuscle $x(t)$ was measured with a photodetector.

cremental change in light intensity leads to only an incremental change in this *graded* receptor potential.

Mechanoreceptors

Mechanoreceptors are sensitive to mechanical stimulation of their membranes. We shall consider two mechanoreceptors, the Pacinian corpuscle and the hair cell. Pacinian corpuscles are pressure receptors found in the skin, muscles, tendons, and joints. Each consists of a myelinated fiber with an unmyelinated nerve ending surrounded by connective tissue lamellae structured as an onion. The corpuscle has a length of about 1 mm, making it a relatively large mechanoreceptor. Using recording techniques described in more detail in Chapter 2, the response to mechanical deformation of the corpuscle can be recorded extracellularly along the nerve. As shown in Figure 1.4, compression of the corpuscle produces a graded receptor potential. This potential is graded since an incremental change in the deformation causes an incremental change in the receptor potential. The relation of the receptor potential to the deformation is said to *adapt,* since a maintained compression does not lead to a maintained receptor potential. The relation between the receptor potential and the compression of the corpuscle exhibits high-pass filtering.

Hair cells are sensory receptors found in the inner ears of all vertebrates and in the lateral line organs of aquatic vertebrates. They subserve the senses

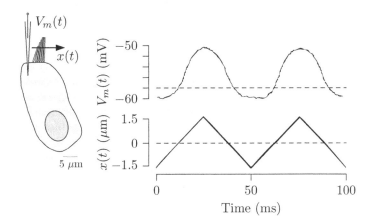

Figure 1.5 Receptor potential of a hair cell in the sacculus of the bullfrog *Rana catesbeiana* in response to a displacement of the hair bundle $x(t)$ (adapted from Hudspeth and Corey, 1977, Figure 3). A glass probe was used to produce a triangular displacement of the hair bundle, and an intracellular micropipette was used to measure the potential across the membrane of the hair cell.

of hearing, vibration, equilibrium, motion of the head, and water motion detection (in aquatic vertebrates). Sensory stimuli (e.g., sound delivered to the ear or changes in head position) are conveyed to these cells by mechanical linkages that cause displacement of the bundles of sensory hairs that project from the surfaces of these hair cells. Displacement of a hair bundle results in a graded receptor potential across the hair cell membrane as shown in Figure 1.5. Hair cells are very sensitive mechanoelectric transducers and produce graded receptor potentials that are effective in signaling the presence of hair bundle displacements of atomic dimensions.

Pacinian corpuscles and hair cells are only two types of mechanoreceptors. There are a variety of additional mechanoreceptors found in skin, in muscles (e.g., stretch receptors that detect muscle stretch), and in tendons (e.g., Golgi tendon organs that detect tension on the tendons). All these mechanoreceptors produce graded receptor potentials in response to mechanical stimuli.

Chemoreceptors

The class of specialized chemoreceptors is especially rich. These receptors subserve the special chemical senses of olfaction (smell) and gustation (taste). Receptor cells in the olfactory epithelium of the nose produce a graded receptor potential in response to an increased concentration of odorants. Taste receptors in the tongue respond with a graded receptor potential to the presence of taste stimuli (Figure 1.6).

In addition to these highly specialized chemoreceptors, most cells contain membrane patches that are chemoreceptive to specific substances in the environment called *neurotransmitters, neuromodulators,* and *hormones.*

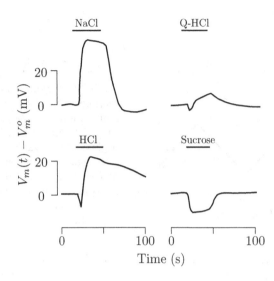

Figure 1.6 Responses of a taste receptor in the rat tongue to four basic taste sensations, salty (0.5 mmol/L NaCl), bitter (0.02 mmol/L quinine HCl, Q-HCl), sour (0.01 mmol/L HCl), and sweet (sucrose) (adapted from Sato and Beidler, 1983, Figure 1B). The solutions were applied to the tongue for the durations indicated by the horizontal lines. The difference in membrane potential from its resting value is shown.

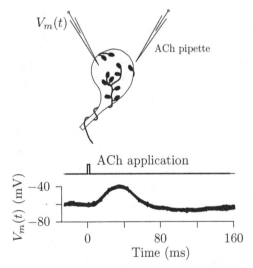

Figure 1.7 The effect of acetylcholine applied to a parasympathetic ganglion neuron in the frog heart. The upper panel shows a schematic diagram of the stimulating and recording arrangement. The ganglion cell is shown with synaptic endings (in black) from preganglionic fibers. One micropipette is used to measure the membrane potential with respect to an extracellular electrode (not shown). The other micropipette is used to apply acetylcholine (ACh) iontophoretically to the surface of the neuron. The lower panel shows the membrane potential recorded in response to the application of ACh (adapted from Dennis et al., 1971, Figure 6).

Neurons and muscle fibers have localized membrane patches, at structures called *chemical synapses,* that are sensitive to neurotransmitters. Application of neurotransmitter substances to these specialized membranes causes a graded potential called the *postsynaptic potential.* For example, Figure 1.7 shows the response of a neuron to the application of a puff of the neurotransmitter acetylcholine (ACh) to its surface. A postsynaptic potential of about 15 mV is produced and lasts for about 50 ms. This postsynaptic potential is a

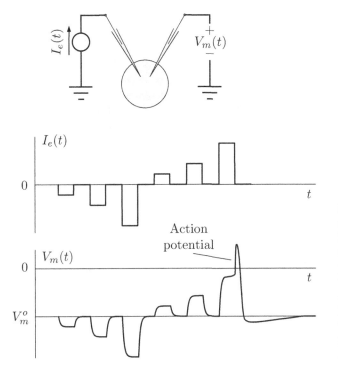

Figure 1.8 Schematic diagram of the membrane potential changes resulting from current pulses passed through the membrane of an electrically excitable cell. If the outward current through the membrane is sufficiently intense, an action potential occurs.

graded potential, since an application of an incrementally larger quantity of ACh produces an incrementally larger postsynaptic potential.

In all of these examples, the potential changes across the membrane are graded potentials. That is, an incremental change in the intensity of the physical stimulus leads to an incremental change in the membrane potential.

1.2.1.2 Action Potentials

Certain cells in living organisms (nerve cells, muscle cells, certain sensory receptor cells in animals, and some plant cells) are electrically excitable. When an electric current of sufficient strength is passed through the cellular membrane of an electrically excitable cell, a change in the membrane potential called an *action potential* is generated (Figure 1.8). If the electric current is not of sufficient strength to elicit an action potential, a smaller graded potential is generated. Thus, electrically excitable cells produce both graded and action potentials. Whereas the mechanism of generation of graded potentials shows a continuous relation between stimulus strength and membrane potential, the

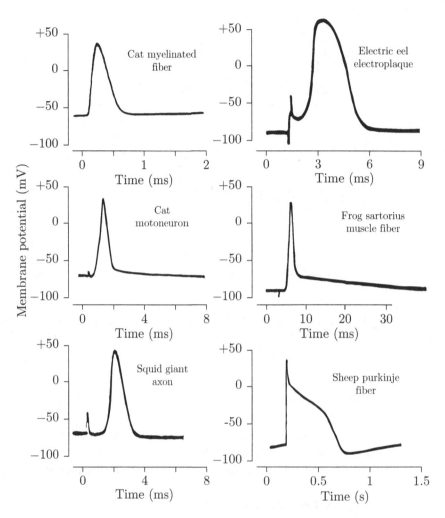

Figure 1.9 Examples of action potentials measured in a variety of cell types (adapted from Keynes and Aidley, 1991, Figure 2.4).

mechanism of generation of action potentials shows a discontinuous relation between stimulus strength and membrane potential.

Waveforms of Action Potentials

Figure 1.9 shows action potentials recorded from several different cell types. Action potentials have an amplitude that is invariably about 10^2 mV. As we shall see in Chapter 4, this small range results largely from the differences in Nernst equilibrium potentials of the permeant ions. In contrast to the small

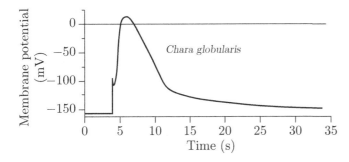

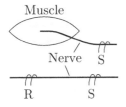

Figure 1.10 Action potential of the freshwater alga *Chara globularis* (adapted from Gaffey and Mullins, 1958, Figure 2).

Figure 1.11 Schematic diagram of two techniques for studying electrically excitable cells. In one, the nerve to a muscle is stimulated electrically with a pair of electrodes (S) and contraction of the muscle is the response variable. In the other, the electrical response of a whole nerve is measured with a pair of extracellular recording electrodes (R) in response to a current through the stimulus electrodes (S).

range of action potential amplitudes, the time course of action potentials varies greatly across cell types. The action potential duration is less than 1 ms in certain neurons, is on the order of 10 ms in skeletal muscle fibers, on the order of 0.5 s in cardiac muscle fibers, and many seconds in certain electrically excitable plant cells (Figure 1.10). The action potential time course depends strongly on temperature, so that some (relatively small) fraction of the variability seen in Figure 1.9 is due to differences in temperature of the different preparations. As is also illustrated in Figure 1.9, not only the duration, but also the waveshape of the action potentials, varies among different cell types.

Properties of Action Potentials

The generation of action potentials in electrically excitable cells has been investigated with a variety of techniques. Earlier techniques, shown schematically in Figure 1.11, were indirect but nevertheless enabled the recognition of key electrical properties. These methods included dissection of a muscle and its innervating nerve and placing these in a chamber that allowed for electrical stimulation of the nerve. The generation of action potentials in the nerve was signaled by a contraction of the muscle. A somewhat more direct method was to dissect a peripheral nerve, mount it in a chamber, and stimulate the nerve electrically. The electrical response of the nerve was recorded with extracellular electrodes. Suprathreshold electrical stimulation of the nerve pro-

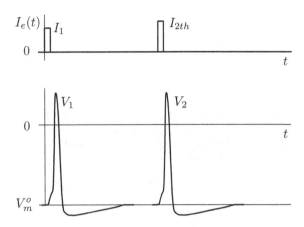

Figure 1.12 Schematic diagram showing the responses to two pulses of current. The amplitude of the first pulse is suprathreshold, whereas the amplitude of the second pulse is adjusted to be the minimum required to elicit an action potential.

duced a *compound action potential,* which is the superposition of the extracellular potentials produced by the nerve fibers in the nerve. Beginning in the 1930s, methods were developed to record electrical responses from single cells using extracellular electrodes. These will be described in Chapter 2. The most direct method by which to investigate electrically excitable cells is with intracellular recording from single cells (Weiss, 1996, Figure 7.2). Properties of electrically excitable cells have been discerned with the use of all these techniques.

Threshold

As suggested in Figure 1.8, there is a threshold for eliciting an action potential in response to a brief pulse of current. This threshold is exceedingly sharp, so that a 0.1% change in current amplitude can separate subthreshold from suprathreshold currents. Electric currents greater than this threshold may elicit additional action potentials, but each action potential has a similar waveform. Hence, above this threshold value the intensity of the current stimulus determines the temporal pattern of action potentials, but not the waveform of the action potentials. For this reason, the action potential is said to be an *all-or-none* potential.

Refractoriness

A second action potential is difficult to produce immediately following the occurrence of an action potential; the cell is said to be *refractory*. The experimental paradigm used to illustrate refractoriness is indicated in Figure 1.12. Two pulses are applied to a cell. The intensity of the first pulse is adjusted to elicit an action potential ($I_1 > I_{1th}$, where I_{1th} is the threshold value of I_1). The intensity of the second pulse is adjusted to the value I_{2th}, which just elic-

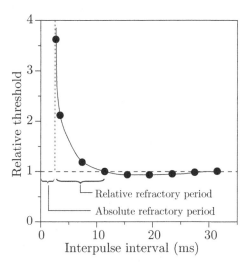

Figure 1.13 Measurement of the refractory period of a nerve. The threshold amplitude of a second pulse as a function of the interpulse interval between the second pulse and a suprathreshold first pulse was measured (adapted from Adrian and Lucas, 1912, Figure 18). These values are normalized to the threshold current for a single pulse of current. The preparation was a frog gastrocnemius muscle with its attached sciatic nerve at a temperature of 14.8°C. Threshold was determined by finding the current stimulus to the nerve that gave a just detectable muscle contraction. During the first 3 ms (the absolute refractory period) following the first pulse, no muscle contraction can be elicited no matter how large the amplitude of the second pulse. During the next 7 ms (the relative refractory period) a muscle contraction is elicited by the second pulse, but only with a higher threshold.

its an action potential (Figure 1.12). The experiment is repeated for different intervals of time between pulses, and I_{2th} is measured for each interpulse interval. The results indicate (Figure 1.13) that for a brief interval between pulses, called the *absolute refractory period,* a second action potential cannot be elicited for any intensity of the second pulse. For intervals greater than the absolute refractory period, there is a *relative refractory period* in which a second action potential can be elicited, but its threshold is elevated. These refractory properties indicate that the mechanism that generates the action potential requires some time to recover from the effects of producing an action potential. Absolute refractory periods are on the order of a few milliseconds in large invertebrate axons. Following the relative refractory period, the threshold for eliciting a second action potential depends in a complex manner on the history of previous stimulation and response of the axon. Effects lasting many minutes are known to occur.

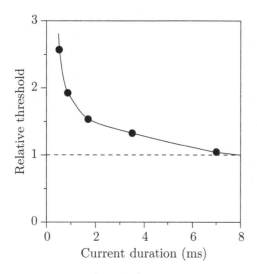

Figure 1.14 Threshold for eliciting excitation of a toad sciatic nerve (at 10°C) as a function of duration of the stimulating current (based on Lucas, 1906, Experiment 4). The preparation was a gastrocnemius muscle with its attached sciatic nerve. Shocks were applied to the nerve, and electrical excitation was signaled by a muscle contraction. Threshold was determined to be the lowest current level that just gave a muscle contraction. The threshold is normalized to the value of rheobase.

Relation of Strength to Duration

The relation between the threshold for eliciting an action potential and the duration of the current stimulus is called the *strength-duration relation,* and an example is shown in Figure 1.14. As the duration of a current stimulus is increased, the threshold for eliciting an action potential is decreased to an asymptotic value called the *rheobase,* which represents the threshold in response to a current of arbitrarily long duration.

Anode-Break Excitation

For brief pulses of suprathreshold current, an action potential occurs in response to the onset of the outward membrane current. However, if an intense inward membrane current of long duration is applied to the cell, an action potential can be initiated at the termination of the inward current, a phenomenon called *anode-break excitation.*

Accommodation

Action potentials can also be elicited by currents that increase linearly in time provided the slope is above some critical slope (Figure 1.15). If the slope is below this critical value, then no action potential is elicited no matter how large the depolarization of the membrane by the current. This property, called *accommodation,* demonstrates that the mechanism responsible for the action potential cannot be described as having a constant voltage threshold that, when exceeded, gives rise to an action potential.

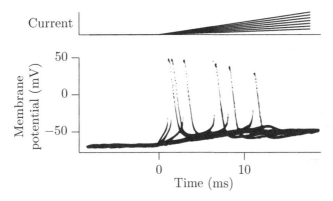

Figure 1.15   Responses of a myelinated fiber of the toad *Xenopus laevis* to linearly increasing currents of different slopes (adapted from Vallbo, 1964, Figure 1). The currents are shown schematically above. The membrane potential responses to 7 ramps of different slope are shown below. The first six elicit action potentials; the ramp with the lowest slope does not, even though the membrane potential exceeds the value of threshold for ramps with higher slopes.

Triggering of Action Potentials by Graded Potentials

As indicated in Figure 1.8, an extrinsic current stimulus can generate an action potential in an electrically excitable cell. However, since action potentials are elicited when the membrane potential change is sufficient (see Chapter 4), any source of a sufficient graded potential can elicit an action potential in an electrically excitable cell. So for example, an electrically excitable mechanoreceptor, such as the Pacinian corpuscle, that is subjected to a mechanical stimulus that generates a sufficient graded potential will generate an action potential. In contrast, electrically inexcitable mechanoreceptors, such as vertebrate hair cells in their in situ environments, do not produce action potentials no matter how large the receptor potentials elicited by displacement of their hair bundles.

Generation of Action Potentials by the Action of Specific Membrane Ionic Channels

Electrically inexcitable cells produce graded potentials, but not action potentials, while electrically excitable cells produce both graded and action potentials. As we shall see in Chapter 6, the production of action potentials results from the presence in the membranes of electrically excitable cells of a sufficient number and composition of special types of ion channels, voltage-gated ion channels. If some fraction of these channels is blocked by specific pharmacological agents, these cells cannot produce action potentials, but can still produce graded potentials. The membranes of electrically inexcitable cells do not normally contain a sufficient number of these voltage-gated ion channels. However, if a sufficient number of such channels are incorporated into the membrane from endogenous or exogenous sources, then formerly electrically inexcitable cells can be made electrically excitable. Thus, the difference between electrically excitable and electrically inexcitable cells results from the presence of an appropriate composition of membrane-bound voltage-gated ion channels.

1.2.2   Intracellular Transmission of Electric Potential in Large Cells

In small, uniform cells immersed in a homogeneous medium, the membrane potential has the same value over the whole surface of a cell. Thus, either graded or action potentials occur simultaneously across the entire membrane of a small, uniform cell. However, many cells in the body do not fit this description. For example, certain neurons and muscle fibers are extremely large cells. Neurons innervating the extremities of the body have their processes in the extremities and their cell bodies in the spinal cord, at a distance that can be several feet in a large animal. In such large cells, there can be large gradients in the membrane potential along the cell surface. We call such cells *electrically large cells*. There are two mechanisms by which a membrane potential change in a localized region of a cell can spread to neighboring regions: *decremental conduction* of graded potentials and *decrement-free conduction* of action potentials.

1.2.2.1   Decremental Conduction of Graded Potentials

A localized change in membrane potential produces a localized membrane current that is coupled to a neighboring patch of membrane via the cytoplasm of the cell and the extracellular medium. In this way a local current flows and produces a change in potential in a neighboring membrane patch. If the membrane patches can be represented as leaky insulators, which measurements show is a reasonable assumption, the pattern of potential change spreads out along the cell as a diffusion process. That is, the membrane potential as a function of space and time satisfies a diffusion equation (Chapter 3). Typically, the potential attenuates significantly over distances that are at most a few millimeters for the largest cells (see Figure 1.16). This mechanism, called decremental or electrotonic conduction, accounts for the local spread of potential in electrically inexcitable cells. This mechanism also operates in electrically excitable cells in which the membrane mechanism that generates action po-

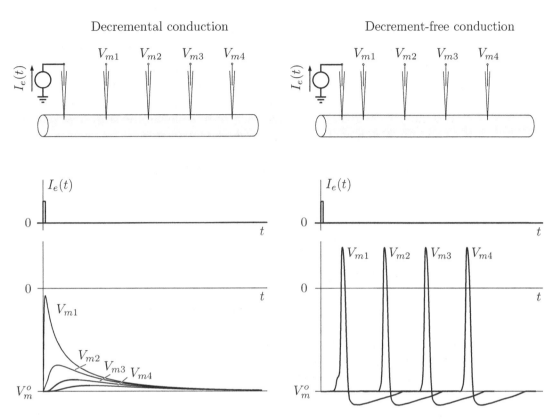

Figure 1.16 Schematic diagram showing decremental conduction along an electrically inexcitable cell by means of graded potentials (left panel) and decrement-free conduction along an electrically excitable cell by means of action potentials (right panel). The action potential recorded by the micropipette that is closest to the current source $V_{m1}(t)$ exhibits a graded potential (see inflection on rising edge of the action potential). This graded potential propagates decrementally and is not discernible on the electrodes located farther from the current source.

tentials has been blocked (e.g., pharmacologically) or in which the membrane potential changes are below the threshold for generating action potentials.

1.2.2.2   Decrement-Free Conduction of Action Potentials

In electrically excitable cells, an action potential generated along a patch of membrane can propagate along the cell in an undistorted manner over great distances (several feet). Because the membrane potential propagates without change in shape, the potential satisfies the wave equation (Chapter 4). The regenerative mechanism for producing action potentials is contained in the membrane throughout the cell's surface. Hence, an action potential in one region of the cell can produce a supra-threshold change in the membrane potential of an adjacent region that is then capable of generating the action potential (see Figure 1.16). Action potentials propagate along a cell with a velocity that is as high as 120 meters/sec (or mm/ms) in the largest vertebrate nerve fibers, which have the largest conduction velocities.

1.2.3 Intercellular Transmission of Electric Potential

Membrane potential changes in one cell can result in membrane potential changes in adjacent cells. Three distinct mechanisms of intercellular transmission can be distinguished.

1.2.3.1 *Ephaptic Transmission*

Ephaptic transmission refers to intercellular electrical interactions that are not mediated by a morphologically and physiologically specialized contact region between the cells. A change in the membrane potential of a cell is accompanied by a current flowing through the cell membrane and therefore extracellularly. Some of the extracellular current is shunted through the intercellular spaces surrounding the cell, but a portion of this current passes through the membranes of nearby cells. Although appreciable effects of such ephaptic transmission have been demonstrated in vitro, it is not known whether such transmission is normally physiologically significant in situ. It is generally felt that in the absence of specialized junctional regions, cells are effectively electrically insulated from each other.

1.2.3.2 *Electrical Transmission*

In certain tissues, including epithelia, smooth and cardiac muscle, and between some neurons, there exist morphologically specialized membrane junctions at which the membranes of adjacent cells are closely apposed. One such region of close apposition, called the *gap junction* because the outer membrane surfaces appear to be separated by a 2 nm gap, is the site of a low-resistance pathway that connects the cytoplasms of adjacent cells. Hence, if a current is injected into a cell in such a tissue, a significant fraction of the current is coupled via gap junctions to adjacent cells (see Figure 1.17). These

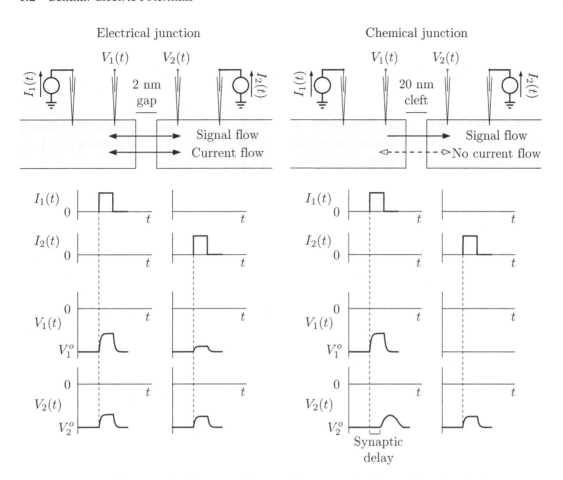

Figure 1.17 Schematic diagram illustrating the flow of signals and electric current at an electrical junction (left panel) and at a chemical junction (right panel).

electrical junctions (or synapses) conduct electric potential changes rapidly between cells. They occur widely in tissue where synchronization of electrical activity of many cells (such as in cardiac muscle) occurs. Such electrical connections may also be important for stabilization of resting potential in epithelia where the resting potential is important for secretory functions. Since gap junctions allow a variety of molecules to pass between cells, they may play a role in sharing metabolites and/or in the passage of chemical signals between cells.

1.2.3.3   Chemical Transmission

Another kind of specialized junction is found between two neurons and between a neuron and a receptor or effector (muscle, gland) cell. Each of these junctions has a relatively large extracellular space, called the *synaptic cleft,* between cells (ca. 20 nm) and has an asymmetric morphology. The mechanism of intercellular transmission of potential at these *chemical junctions* differs fundamentally from that at electrical junctions. In fact, the adjacent cells at a chemical junction are not well coupled electrically (Figure 1.17). However, a change in membrane potential in the cell on the prejunctional (or presynaptic) side results in the release of a chemical substance, called a *neurotransmitter,* that diffuses across the synaptic cleft between the cells. The neurotransmitter interacts with specialized receptor sites on the postsynaptic membrane of the adjacent cell, which opens ionic channels to allow the flow of current through the postsynaptic membrane. This current flow causes a change in membrane potential, called the *postsynaptic potential.* In this manner, a membrane potential change in one cell, the presynaptic cell, can result in a change in potential in an adjacent cell, the postsynaptic cell, via an intermediate chemical variable. The junction is physiologically polarized, so a change in potential in the postsynaptic cell causes virtually no change in potential in the presynaptic cell.

The relation between the presynaptic potential and the postsynaptic potential at a chemical junction can be quite involved and differs significantly at different types of chemical junctions. The combination of the characteristics of the neurotransmitter and those of the postsynaptic receptor determine the postsynaptic effect. So, for example, certain combinations produce short-acting postsynaptic effects, whereas other combinations produce long-acting effects. Synapses can be characterized as *excitatory synapses* or *inhibitory synapses.* Activation of excitatory synapses facilitates the production of action potentials in the postsynaptic cell, whereas activation of inhibitory synapses inhibits the production of action potentials in the postsynaptic cell. Inherent in chemical transmission is the secretion of a neurotransmitter by one cell and the chemoreception of that neurotransmitter by the neighboring cell. Thus, chemical transmission incorporates two very fundamental properties of cells: secretion and chemoreception.

1.2.4   Effects Produced by Electric Potential Changes

Membrane potential changes can have a wide variety of effects on cells. The most direct effect is to modulate the transport of charged particles across the membrane. In secretory cells, changes in the membrane potential initi-

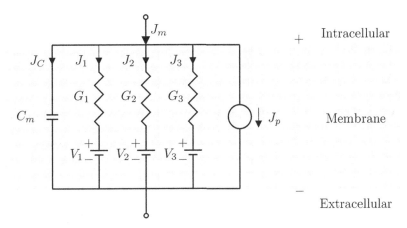

Figure 1.18 Macroscopic model of a unit area of membrane. C_m represents the capacitance of a unit area of membrane and represents the insulating properties of the lipid bilayer. G_n represents the conductance of a unit area of membrane for the nth ion. V_n is the Nernst equilibrium potential of the nth ion. The series combination of G_n and V_n represents conduction through a population of ion channels permeable to ion n. J_p represents the net current density due to other mechanisms, primarily the current carried by active transport mechanisms.

ate the secretion of a chemical substance. For example, at chemical junctions the neurotransmitter is released by the presynaptic cell following a change in its membrane potential. Changes in membrane potential can also affect the mechanical properties of cells. Membrane potential changes result in contractions in muscle fibers and affect the beating of cilia in ciliated cells. In neurons, synaptically induced changes in membrane potential at the inputs of neurons produce sequences of action potentials that propagate along the neuron outputs.

1.3 Mechanisms of Generation of Membrane Potentials

1.3.1 Macroscopic Mechanisms

The resting potential of a cell results from the difference in ionic composition across the semipermeable cell membrane as well as the presence of ionic pumps that maintain these concentrations (Weiss, 1996, Chapter 7). These characteristics are represented by an equivalent network of a patch of membrane shown in Figure 1.18. Rapid changes in the membrane potential generally result from changes in the conductance of the membrane to one or

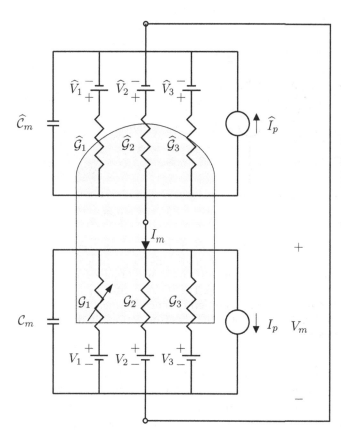

Figure 1.19 Schematic diagram illustrating the macroscopic mechanisms for the generation of receptor potentials. This cell is assumed to comprise two patches of membrane, each of which has an equivalent network of the form shown in Figure 1.18. The two patches differ quantitatively: one patch contains a conductance that depends upon the sensory stimulus (indicated with an arrow); the other patch does not. The conductances and capacitances are related to the specific conductances and capacitances as follows: $C_m = AC_m$, and $G_n = AG_n$, where A is the surface area of the membrane patch.

more ions rather than to changes in these pumps or in ion concentrations. For example, in many types of receptor cells the conductance of the membrane depends upon the sensory stimulus. In mechanoreceptors, the ion conductance depends upon the mechanical stimulus to the receptor surface. In chemoreceptors, the conductance depends upon the concentrations of specific chemical substances at the receptor sites.

In all of these cases, if the cell is small, the generation of a receptor potential can be represented macroscopically by an equivalent network representation of membrane patches of the form shown in Figure 1.19. The portion of the cell that contains the sensory transducer is coupled electrically to the portion of the cell that does not contain the transducer. The transducer portion has the special property that one (as indicated in Figure 1.19) or more ionic conductances depend upon the sensory stimulus to the cell. For example, in a sensory receptor G_1 might represent the ionic conductance to ion 1 in re-

Extracellular

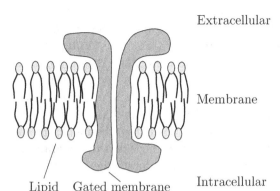

Membrane

Lipid   Gated membrane   Intracellular
bilayer   macromolecule

Figure 1.20   Schematic diagram of membrane structure illustrating the relation between the lipid bilayer and the gatable, integral membrane macromolecules that form channels.

sponse to a sensory input to the cell. The other ionic conductances, $\mathcal{G}_2$ and $\mathcal{G}_3$, do not depend on the sensory stimulus. Changes in the sensory stimulus result in changes in one or more conductances in the network and current flows in the network, and potentials are generated across the membrane of the cell. A similar mechanism occurs in the generation of a synaptic potential in the postsynaptic cell of a chemical synapse. The arrival of the neurotransmitter at the postsynaptic membrane causes a change in the ionic conductance of the membrane to one or more ions. This change in conductance produces a change in the membrane potential of the postsynaptic cell.

The same electrical network can be used to describe the properties of electrically excitable cells, except that the ionic conductance is a function of the membrane potential itself rather than of some extrinsic stimulus to the cell. This property allows regenerative action to take place across the membrane of such a cell, since a change in potential causes a change in conductance, which causes a change in potential. As we shall see in Chapter 4, this capability is responsible for the discontinuous, thresholdlike behavior of an electrically excitable membrane. The same membrane models are used to describe potential changes in large cells, but in these it is also necessary to represent the electrical connections between neighboring patches of membrane.

1.3.2   Microscopic Mechanisms: Gated Membrane Channels

It is now clear that the macroscopic ionic conductances of membranes, discussed in Section 1.3.1, result from the summation of the conductances of single ion channels, each of which is either opened or closed—i.e., gated—by some physicochemical variable. Transmembrane transport through a single ion channel apparently occurs via an intrinsic membrane macromolecule that spans the lipid bilayer and whose conformations determine the transport through the ionic channel (Figure 1.20). In the simplest such scheme, each channel can be in one of a discrete number of conformations, one of which is open and all the others of which are closed. When the channel is in its open conformation, ions can flow through the channel. When the channel is in one of its closed conformations, no ions can flow through the channel. Under the influence of thermal energy, the channel moves randomly among its conformations.

The transition rates between conformations depend on one or more gating variables. For example, the sodium channel responsible for the generation of sodium-activated action potentials in electrically excitable cells is gated by a depolarization of the membrane (see Chapter 6). This is an example of a *voltage-gated channel.* Many such voltage-gated channels have been identified in the last decade. The channel responsible for producing postsynaptic potentials at a cholinergic synapse is gated by the binding of acetylcholine at a postsynaptic receptor site. This is an example of a *ligand-gated channel.* Numerous ligand-gated channels occur. In mechanoreceptors it is the mechanical stress to the membrane that gates the channel. This is a *stress-gated channel.* It is becoming clear that single channels underlie the electrical properties of membranes and that there are a wide variety of channels gated by various physicochemical variables. Each type of channel is embodied by a specific membrane macromolecule.

1.4 Role of Electric Potentials in Information Coding

As we have seen, electric potential changes are involved in many cellular processes. However, electric potential changes have an additional important function. For neurons, electric potential changes are used to code information in the nervous system. In the nervous system, information comes in through the sensory organs that transduce the incoming mechanical, electrical, optic, or chemical signals into electric potential changes that are then processed to produce outputs that are primarily motor, but also secretory, photic (e.g., in fireflies) and electrical (e.g., in electric fish). Electric potential changes function as a common currency for representation of information.

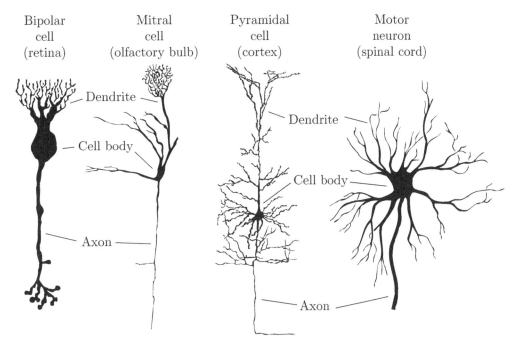

Figure 1.21 The variety of shapes of neurons (adapted from Nicholls et al., 1992, Figure 1-6). The scale for each neuron has been chosen to make the shape discernible.

1.4.1 The Neuron as an Information-Handling Element

The neuron is a basic element for coding, transmitting, transforming, and decoding information. The morphology of neurons varies greatly (see Figure 1.21). However, certain features of neurons are commonly found, and are illustrated in Figure 1.22. Because of the pattern of connectivity of neurons and the characteristics of synapses that polarize the flow of signals from one cell to another, signal flow through neurons normally has a preferred direction, and hence we can define the inputs and outputs of neurons.

Inputs to a neuron from the axons of other neurons occur at synaptic sites that are scattered over the surfaces of dendrites, the cell body, and the axonal tree. A neuron may contain one synaptic input or many thousands of synaptic inputs. Activity, coded as sequences of action potentials, in input axons produces local postsynaptic potentials in the neuron that then spread to the cell body. The spatiotemporal pattern of synaptic potentials produces a

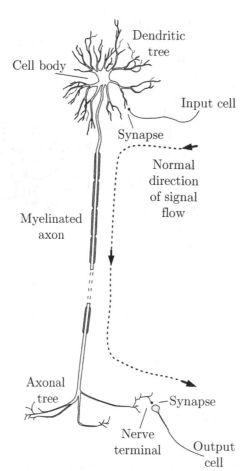

Cell body

Dendritic tree

Input cell

Synapse

Normal direction of signal flow

Myelinated axon

Axonal tree

Nerve terminal

Output cell

Synapse

Figure 1.22 Schematic diagram of a typical neuron indicating principal components.

membrane potential change that spreads either decrementally or decrement-free over the cell surface. If the depolarization is sufficiently large in the region of the cell where the threshold for eliciting an action potential is low, called the *trigger zone,* an action potential sequence may be produced. Such output action potentials propagate from the cell body via the axon into the axonal terminals. There are two major categories of axons: *unmyelinated axons* and *myelinated axons.* Myelinated axons (described in Chapter 5) occur predominantly in the vertebrates, whereas unmyelinated axons occur in both vertebrates and invertebrates. The two types of axons differ not only in their structures, but also in their modes of conduction of action potentials. The action potentials that reach the axon terminals via either type of axon form the input signals to other neurons. It is the elaboration of this basic scheme in a large variety of ways that contributes to the complexity of behavior of even the simplest nervous systems.

1.4.2 Organization of the Nervous System

The nervous system is a collection of about $10^{11\pm1}$ neurons arranged in a highly structured manner. The neurons in the nervous system can be categorized in a variety of ways. The nervous system can be divided into two portions: the *central nervous system,* which consists of the brain and the spinal cord, and the *peripheral nervous system,* which consists of all the neurons in the nervous system that are not in the central nervous system. In vertebrates, the cell bodies of neurons tend to be clustered in certain regions, which are called *nuclei* in the central nervous system and *ganglia* in the peripheral nervous system. Nuclei and ganglia are connected by bundles of axons (or nerve fibers) of cell bodies. Bundles of nerve fibers that connect nuclei that lie within the central nervous system are called *nerve tracts.* Bundles of fibers that connect nuclei to ganglia or that connect two ganglia are called *nerves.* Both ganglia and nuclei may contain heterogeneous populations of nerve cell bodies that have different structures and functions. Similarly, nerves and tracts may contain nerve fibers of different structures and functions.

The nervous system can also be categorized functionally into the *somatic nervous system* and the *autonomic nervous system.* The somatic nervous system has peripheral connections with the surface of the body, e.g., the voluntary skeletal musculature, skin, joints, special senses, etc. The autonomic nervous system has peripheral connections with the inside of the body, e.g., the involuntary muscles, internal organs, vascular system, glands, etc. Another important categorization of the nervous system is in terms of the direction of information flow. The *afferent nervous system* brings information from the peripheral nervous system toward the central nervous system. The *efferent nervous system* brings information from the central nervous system toward the peripheral nervous system.

1.4.3 Coding and Processing of Information

Information is transmitted along axons as a temporal sequence of identical action potentials or *spikes.* This property is most easily demonstrated in a peripheral sensory or motor neuron, where the sequence of action potentials can be related to some physical variable, either a sensory input or a mechanical output.

1.4.3.1 Motor Receptors

Muscles contain specialized mechanoreceptors that sense the mechanical muscle variables. Stretch receptors (muscle spindles or intrafusal muscle fibers) are located in the body of the muscle and are in parallel with the (extrafusal) muscle fibers that are responsible for muscle contractions. Stretch receptors are exceedingly sensitive to muscle stretch. Golgi tendon organs are located in the muscle tendons and are in series with the muscle. Golgi tendon organs sense muscle tension. As shown in Figure 1.23, a primary nerve fiber innervating a stretch receptor responds to a moderate passive stretch of the muscle with an increase in discharge rate of action potentials, whereas a fiber innervating a tendon organ, which is less sensitive, does not. Fibers from both types of receptors respond to a muscle twitch induced by stimulating the nerve to the muscle electrically. Since the muscle is contracted during the twitch, the muscle is being unstretched; hence, the stretch receptor decreases its rate of action potentials during the twitch. During the twitch, the tension the muscle applies to the tendon increases; hence, the Golgi tendon organ discharges vigorously during the twitch. Thus, these two muscle receptors signal the length and the tension of the muscle by temporal sequences of action potentials. This information propagates along the sensory fibers to the spinal cord, where it is distributed to a variety of different cells in the spinal cord.

1.4.3.2 The Auditory System

In Figure 1.24, the response of a cochlear nerve fiber is shown to a variety of sound stimuli ranging from simple sounds such as tones to complex sounds such as speech. This example shows that in the absence of sound stimulation, the fiber discharges spontaneously in an irregular fashion. When a sound stimulus is presented, the spike discharge pattern changes to reflect the temporal (and the spectral) content of the stimulus. The presentation of identical stimuli does not result in identical discharge patterns in this fiber; the relation between the stimulus and the response is intrinsically statistical. Nevertheless, the statistical properties of the action potential sequences encode the information about the sound stimulus carried by one fiber. The discharges of different cochlear nerve fibers in the auditory nerve differ. However, the statistical properties of the discharges of the population of fibers in the nerve contain all the information available to the organism about the sound stimulus.

The discharge patterns of these primary neurons are transmitted to secondary neurons located in the brain stem (in the cochlear nucleus). Here the sensory message is processed in a variety of ways and routed to different parts

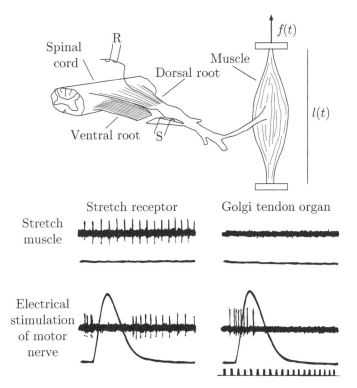

Figure 1.23 Recording of action potential trains from primary fibers innervating muscle receptors. The upper panel shows the stimulating and recording arrangement. A skeletal muscle is isolated and set in a mechanical device so that its length $l(t)$ can be controlled and so that the force on the muscle $f(t)$ can be recorded. The nerve innervating the muscle has a sensory component that enters the spinal cord in the dorsal root and a motor component that exits the spinal cord in the ventral root. The ventral root has been cut, and stimulating electrodes (S) have been placed on the peripheral portion of the ventral root to stimulate the motonerve fibers innervating the muscle. A small strand of the dorsal root has been isolated for recording electrical responses (R) of single motor receptor neurons. The lower panel shows the sequences of action potential recorded from two types of muscle receptors, a stretch receptor (left) and a Golgi tendon organ (right), during a stretch of the muscle and during a muscle twitch in response to electrical stimulation to the motonerve fibers. Each record includes the action potential sequences superimposed on the force on the muscle, $f(t)$. Both receptors signal the mechanical stimulus by a change in discharge rate (adapted from Patton, 1965, Figure 13).

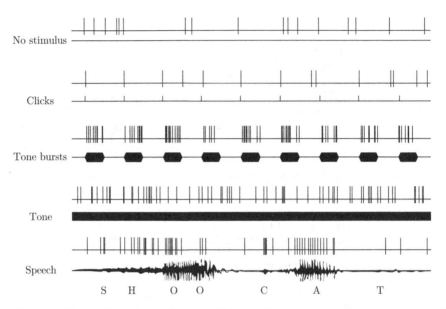

Figure 1.24 Coding of acoustic stimuli into action potential sequences in a cochlear nerve fiber. Each pair of traces shows the sound stimulus (below) and the action potential sequences (above) for one second in time. The frequency of the tone burst and tone stimuli is so high that the individual tone cycles cannot be discerned on this time scale, and the traces indicate the envelopes of these two stimuli. Action potential sequences are shown for the following conditions: no stimulus, clicks, tone bursts, a tone, and a speech utterance (SHOO CAT). The auditory neuron signals the presence of the different sound stimuli by characteristic changes in the sequence of action potentials (adapted from Kiang, 1975, Figure 3).

of the brain. This property is illustrated in Figure 1.25. Each primary cochlear neuron innervates several neurons of differing morphology in the cochlear nucleus. The response of the primary fiber to a tone burst results in a characteristically different response in each type of cochlear nucleus cell. These differences reflect differences in innervation, synaptic transmission, and integration in the cochlear nucleus neurons. The recoded messages are dispatched to other neurons in the central auditory system.

1.4.3.3 The Visual System

In the vertebrate retina, the photoreceptors (rods and cones) transduce light stimuli into graded receptor potentials. These graded receptor potentials act through a network of intermediate cells (horizontal, bipolar, and amacrine

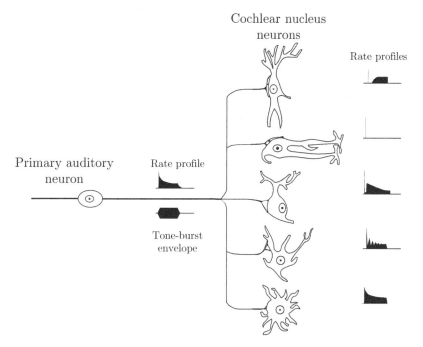

Figure 1.25 Processing of a sensory message of an incoming primary cochlear neuron by secondary neurons in the auditory system (adapted from Kiang, 1975, Figure 9). The rate profile of the primary auditory neuron is shown above the tone-burst stimulus. The rate profile is a plot of the average discharge rate of the neuron in response to the tone burst. The rate profiles of different types of cochlear nucleus neurons in response to the same tone-burst stimulus are shown on the right. The rate profiles show characteristic differences. For example, one cell type shows a rate increase in response to the onset of the tone burst only; others show rate increases throughout the duration of the tone burst.

cells) to produce action potentials in the retinal ganglion cells whose axons compose the optic nerve that carries the information about the light stimuli to the brain. Each ganglion cell responds to light in a specific part of the visual field called the *receptive field*. The receptive field can be determined by recording the response of a ganglion cell to illumination in different parts of the visual field. Each retinal ganglion cell has a receptive field that consists of two concentric circular regions, the center and the annular periphery. There are two types of ganglion cells, on-center cells and off-center cells. The responses of on-center cells are shown schematically in Figure 1.26.

In the absence of illumination, on-center retinal ganglion cells show a spontaneous discharge. If a small spot of light illuminates the center of the

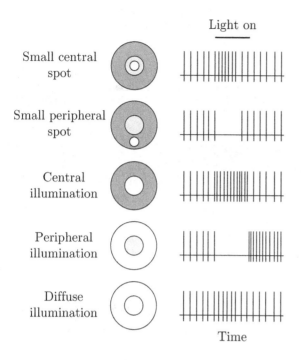

Figure 1.26 Schematic diagram showing responses of an on-center retinal ganglion cell to illumination in different parts of the receptive field (adapted from Nicholls et al., 1992, Figure 16-18). The left-hand panel shows a schematic diagram of the receptive field that consists of a central region (light gray) and a peripheral region (dark gray). The illumination pattern (shown in white) is superimposed on the receptive field. The right-hand panel shows the action potentials as a function of time of the retinal ganglion cell in response to the illumination pattern shown on the left. The illumination is on during the time indicated (Light on).

receptive field, the cell responds with an increased discharge rate of action potentials. A spot of light that just fills the central region of the receptive field is maximally effective in eliciting a response. If a spot of light illuminates the periphery, the spontaneous discharge rate is reduced. Maximal reduction is obtained with an annulus of light that just covers the periphery. A circular spot of light that overlaps both the center and periphery of the receptive field causes a submaximal response. Diffuse illumination of the receptive field causes only a slight increase in discharge rate. Off-center cells respond maximally to an annulus of light that just covers the periphery and are maximally inhibited by a spot of light that fills the center of the receptive field. Thus, each retinal ganglion cell sends information about the illumination of its receptive field to central visual nuclei in the form of a sequence of action potentials. For the simple receptive fields that we have described, the spatial pattern of illumination on the retina is coded by which groups of cells respond, and the temporal pattern of discharge codes the temporal pattern of illumination.

At higher levels of the central nervous system, receptive fields can again be measured and are more complex than those found in the retina. In the striate cortex, cells may respond best to bars of light of a particular orientation, to vertical edges of light of a particular orientation, or to moving patterns of light of particular orientations and locations. These more complex receptive fields can be synthesized conceptually by combining discharges of cells that have simple receptive fields. The cortical neurons send the information elsewhere in the nervous system as sequences of action potentials.

1.4.3.4 Summary

We have seen that encoding of messages from muscle receptors, auditory receptors, and visual receptors has some commonality. In all these systems, the information about the nature of the sensory stimulus is encoded as a sequence of action potentials in a population of sensory neurons. The information arrives in higher centers in the central nervous system and is recoded into other sequences of action potentials that are then distributed elsewhere in the nervous system.

1.4.4 Relation to Behavior: Signals in a Simple Reflex Arc

The relation between signals in the nervous system and organismic behavior is still a puzzle for all but the simplest behaviors. In recent years, there has been considerable progress toward understanding the neurophysiological bases of behavior in simple invertebrates where the behavior may be determined by a small number of neurons and where neurons that have a particular structure and function can be located unambiguously in different animals of the same species. In mammals, analysis of the simple knee-jerk (or patella) reflex—an example of a stretch reflex—reveals how the transmission and processing of neural signals determine an organism's behavior. For mammals, this behavioral response is virtually unique in that much of the anatomy and physiology subserving the behavior are comparatively well known.

The stretch reflex involves a minimum of four cells: a stretch receptor, a sensory neuron, a motoneuron, and an extrafusal muscle fiber (Figure 1.27). Tapping the patellar tendon causes the attached muscle to stretch. This stretch is transduced by the stretch receptor into a receptor potential. The receptor potential generates an action potential sequence in the sensory nerve fiber. The action potentials propagate to the endings of the sensory nerve fibers on a motoneuron (in the spinal cord), generating postsynaptic potentials in the motoneuron that summate to produce a new sequence of action potentials. These action potentials are conducted out of the spinal cord along the motonerve fiber to the muscle fibers in the muscle containing the stretch receptor. The motonerve action potentials generate synaptic potentials in

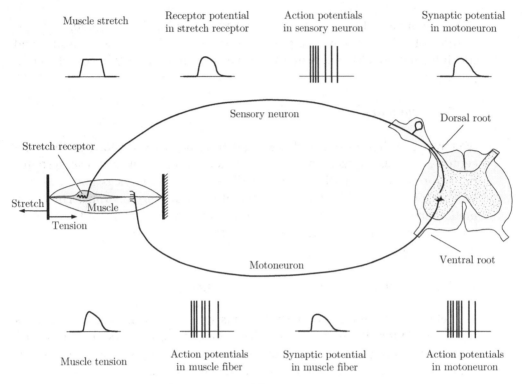

Figure 1.27 Structures and functions involved in the patella reflex (adapted from Nicholls et al., 1992, Figure 1-5).

muscle fibers that lead to a muscle action potential. The muscle action potential results in a contraction of the muscle. Thus, the stretch in the muscle has resulted in a muscle contraction. This reflex is used to maintain posture.

The examination of this reflex illustrates the interplay of graded potentials and action potentials in the successive transformation starting from a sensory stimulus and leading to a motor response. Needless to say, there are a large number of other factors that influence the efficacy of the reflex. The central nervous system (CNS) provides inputs to the stretch receptor that can modify the message transmitted along the sensory nerve fiber, and inputs to the motoneuron that can modify the message to the muscle fibers. Once again, these CNS messages are in the form of a sequence of action potentials along nerve fibers making synaptic contacts with the stretch receptor and the motoneuron, respectively. Hence, to understand the processes of information handling by the nervous system, it is essential to understand the processes by which electric potentials are generated in cells.

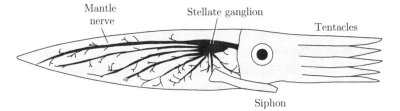

Figure 1.28 Gross anatomy of the squid showing the stellate ganglion and mantle nerves.

1.5   The Marvelous Giant Axon of the Squid

A good strategy for studying some universal biological mechanism is to find a biological preparation that is technically favorable. A particularly successful case in point is that of the giant axon of the squid. The squid, a cephalopod mollusc, is a marine invertebrate that pursues prey and avoids predators using jet propulsion to move through the water. The squid takes in seawater and ejects this water through a siphon by contracting its mantle muscle very rapidly, and this allows the animal to dart about in the water. The mantle muscle is controlled by neurons in the stellate ganglion from which axons radiate to innervate the mantle muscle (Figure 1.28). The axons are arranged so that the longest axon, the giant axon, which innervates the portion of the mantle muscle farthest from the stellate ganglion, has the largest diameter. The shortest axon, which innervates the portion of the mantle muscle nearest to the ganglion, has the smallest diameter. As we shall see in Chapter 2, larger-diameter axons conduct action potentials more rapidly than do smaller-diameter axons. Thus, the longest axons conduct faster than the shortest axons, and the fibers are arranged so that the nerve message arrives at all parts of the mantle muscle more or less synchronously so that the muscle contracts simultaneously along its entire length.

Invertebrates contain unmyelinated axons with large diameters. The giant axon of the squid is among the largest; the diameter approaches 1 mm in some species, and the axon can be several centimeters long. Hence, the giant axon of the squid is among the largest animal cells known. The technical advantage of studying excitation in the giant axon over that in vertebrate nerve fibers is made dramatically obvious by comparing the diameter of the squid giant axon with the diameter of an entire nerve from a vertebrate (Figure 1.29). The largest vertebrate nerve fibers are myelinated and have diameters of about 20 μm. The giant axon was observed and was shown to be a nerve cell in 1936 (Young, 1936a). The significance of this finding was appreciated very soon thereafter.

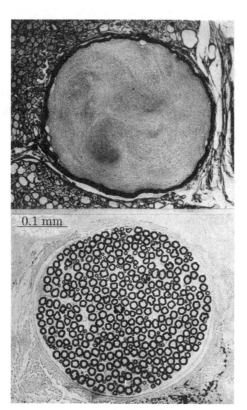

Figure 1.29 Comparison of a cross section of the giant axon of the squid (upper panel) with a cross section of the sciatic nerve of a rabbit (lower panel) showing numerous myelinated nerve fibers (Young, 1951). The unmyelinated fibers in the sciatic nerve are much smaller than the myelinated fibers and are not apparent using the histological method used to prepare the tissue. The giant axon is surrounded by smaller unmyelinated sensory nerve fibers.

The giant axon of the squid was the first animal cell for which the potential across the membrane was measured both at rest and during an action potential (Hodgkin and Huxley, 1939; Curtis and Cole, 1940). It was also found that the axon could be excised and still conduct action potentials (Figure 1.30). In the excised preparation it was easier to control a variety of experimental variables. Because of the axon's large diameter, a glass capillary tube of 100 μm diameter was inserted axially down the axon and used as an electrode to record the membrane potential (Weiss, 1996, Figure 7.2). Later (Marmont, 1949; Hodgkin et al., 1952) multiple electrodes were inserted axially down the axon to allow for the control of either membrane current, in which case the method is called a *current clamp*, or the membrane potential, in which case it is called a *voltage clamp*. This development proved to be a major technical advance, since it has turned out that the axon membrane is a voltage-controlled device, so measurement of membrane current under voltage control reveals the mechanisms more incisively than does measurement of the membrane potential under current control.

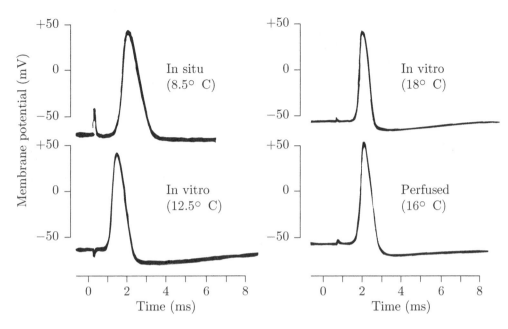

Figure 1.30 Comparison of action potentials recorded intracellularly from squid giant axons using different techniques. In each case, the action potential is elicited by a pulse of stimulating current. The left panel shows a comparison between an action potential recorded with a micropipette in situ with one recorded in vitro after the axon was dissected, tied off, and bathed in seawater (adapted from Hodgkin, 1958, Figure 3). The right panels show a comparison between an action potential recorded in vitro from a dissected axon using an axial electrode but filled with axoplasm with one perfused with an isotonic solution of potassium sulfate (adapted from Baker et al., 1961, Figure 3).

Figure 1.31 Method of extrusion of axoplasm from the giant axon of the squid (Baker et al., 1962).

The squid giant axon is also an abundant source of cytoplasm, also called *axoplasm,* which can be squeezed out of the axon with a miniature roller (Figure 1.31) or extruded with hydraulic pressure applied to one cut end of the axon. The squid giant axon was the source of early direct measurements of cytoplasmic composition. Studies of the composition of axoplasm continue to this day (Table 1.1).

Table 1.1 Composition of the cytoplasm (axoplasm) of the giant axon of the squid, squid blood, and seawater (Gainer et al., 1984). The water and protein compositions are expressed in gm/kg; the other components in mmol/kg.

Substance	Axoplasm	Blood	Seawater
Water	865	870	966
Protein	20	0.15	—
Major ions			
K^+	323–400	20–22	8.3–10
Na^+	44–65	440–456	423–460
Cl^-	40–151	560–578	506–580
Ca^{++}	0.4–7	10–11	9.3–10
Mg^{++}	6.4–20	54–55	48–53
Isethionate$^-$	164–250	1.6	—
SO_4^{++}	7.5	8.1	28.2
PO_4^{---}	2.5–17.8	—	—
Major amino acids			
Aspartate	33–100	—	—
Glutamate	6.2–8.4	—	—
Arginine	0.36–2.2	—	—
Alanine	1.7–16	—	—
Glycine	4.5–18.4	—	—
Taurine	81–106.7	3.5	—
Homarine	20.4	3.4	—
Betaine	73.7	4.4	—
Major carbohydrates			
Glycerol	4.35	—	—
Glucose	0.24	—	—
Mannose	0.92	—	—
Fructose	0.24	—	—
Succinate and fumarate	17	—	—
High energy phosphates			
ATP	0.7–1.7	—	—
Arginine phosphate	1.8–5.7	—	—

It is also possible to inject radioactively labeled substances into the axoplasm. Measurement of the passage of these substances through the membrane has supplemented the electrical measurements and has allowed investigations of the link between transport of ions through the membrane and the metabolism of the axon. In addition, diffusion and migration of these labeled ions in axoplasm have revealed the physical state of these ions in axoplasm.

In the course of experiments in which axoplasm was removed, it was found that the axons could be cannulated and reinflated with artificial solutions and that they conducted normal action potentials (Figure 1.30). Thus, the mechanisms responsible for generating action potentials could be investigated in perfused axons, which allowed for control of not only the potential difference across the membrane, but also the compositions of the solutions on both sides of the membrane.

The corpus of these experiments has led to the explanation of the electrical properties of the squid axon in terms of the conductances of the membrane to sodium and potassium ions. It has also been found that many of the principles of electrical excitation of the squid giant axon are widely applicable in cells throughout the animal kingdom. Hence, the squid giant axon has yielded information about universal mechanisms found in electrically excitable cells. Furthermore, attention has become focused on the molecular bases of the permeability changes first described in this preparation.

1.6 Preview

Chapters 2 through 4 deal with a succession of models of the electrical properties of cells and develop the principles of decremental conduction of graded potentials as well as decrement-free conduction of action potentials (Figure 1.32). In Chapter 2 we shall develop network topologies of both electrically small and electrically large cells. For large cells, we shall derive the core conductor model of a cylindrical cell. The core conductor model represents the intracellular and extracellular media as electrical conductors, but makes rather few assumptions about the voltage-current characteristics of the membrane. The core conductor model leads to a series of relations between voltage and current variables associated with a cell, relations that are relatively independent of membrane characteristics. Results include discernment of the pattern of current flow about an axon during a propagated action potential,

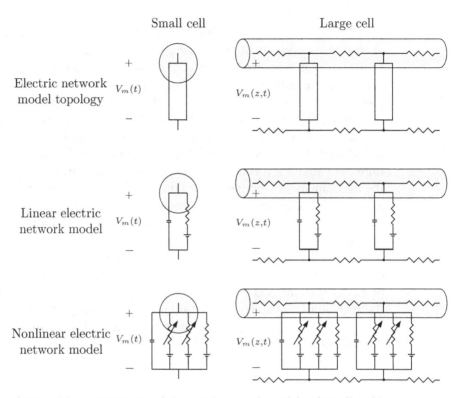

Figure 1.32 A succession of electrical network models of small and large cells.

the dependence of conduction velocity on physical constants of axons, and the relation of extracellular and intracellular potentials of cells. For small perturbations of the membrane potential from its resting value, the voltage-current characteristics of the membrane can be represented by a parallel combination of a membrane resistance representing the conducting channels in the membrane and a membrane capacitance representing the insulating lipid bilayer. The resulting resistance and capacitance model of a small cell and the cable model of a large cylindrical cell (Chapter 3) are developed. This development allows for the definition of electrically small and large cells and explains decremental conduction of graded potentials.

In Chapter 4 we shall consider the voltage-current characteristics of the membrane for large excursions of the membrane potential from its resting value. We shall investigate the measurements of Hodgkin and Huxley, which show that the electrical properties of the membrane of the squid giant axon can be described by a pair of parallel conduction pathways—one for potassium and the other for sodium—each of which contains voltage-dependent conductances with time-varying kinetics. The kinetics of these channels, when incorporated into the core conductor model, give rise to the Hodgkin-Huxley model of an axon, which explains the excitation and conduction of the action potential in the squid giant axon. In Chapter 5 we shall consider the mode of conduction in myelinated axons. In Chapter 6 we shall consider the molecular basis of the properties of the channels responsible for electrical excitability. We shall examine both gating currents and currents through single voltage-gated ion channels.

Exercises

1.1 What is the essential distinction between graded and action potentials?

1.2 When $I_e(t) > 0$ in Figure 1.1, which way does the current flow through the membrane of the cell? Make a sketch of the cell, and indicate the current flow through the membrane superimposed on the cell. If the membrane can be represented as a battery with a voltage V_m^o in series with a resistance, what is the polarity of the change in potential across the membrane? Indicate the polarity of the potential change on the sketch.

1.3 Determine whether each of the following assertions is true or false, and give a reason for your answer.

   a. Graded potentials do not occur in electrically excitable cells.

   b. Graded potentials are conducted decrementally.

   c. Action potentials occur in both electrically excitable and electrically inexcitable cells.

   d. The mechanism of generation of action potentials is the same in all electrically excitable cells.

   e. Graded potentials are not involved in signal generation in neurons.

1.4 What is the approximate peak-to-peak amplitude of the action potential?

1.5 Define the all-or-none property of the action potential.

1.6 Define *refractoriness*.

1.7 What is the distinction between electrical and chemical transmission between cells?

1.8 Compare the cross-sectional area of the largest myelinated fibers in vertebrates with the cross-sectional area of the giant axon of the squid. Discuss the significance of this for determining the chemical composition of the cytoplasm of nerve fibers.

References

Books and Reviews

Adelman, W. J. and Gilbert, D. L. (1990). Electrophysiology and biophysics of the squid giant axon. In Gilber, D. L., Adelman, W. J., and Arnold, J. M., eds., *Squid as Experimental Animals*, 93–132. Plenum, New York.

Aidley, D. J. (1989). *The Physiology of Excitable Cells*. Cambridge University Press, Cambridge, England.

Cajal, S. (1894). *Les Nouvelles idées sur la structure du système nerveux chez l'homme et chez les vertebrés*. C. Reinwald & Cie., Paris. Translated from Spanish by L. Azoulay and reviewed and augmented by the author.

Cajal, S. (1990). *New Ideas on the Structure of the Nervous System in Man and Vertebrates*. MIT Press, Cambridge, MA. Translation of Cajal, 1894 from French to English by N. Swanson and L. W. Swanson.

Cole, K. S. (1968). *Membranes, Ions, and Impulses*. University of California Press, Berkeley, CA.

Eccles, J. C. (1973). *The Understanding of the Brain*. McGraw-Hill, New York.

Gainer, H., Gallant, P. E., Gould, R., and Pant, H. C. (1984). Biochemistry and metabolism of the squid giant axon. In Baker, P. F., ed., *Current Topics in Membranes and Transport,* vol. 22, *The Squid Axon*, 57–90. Academic Press, New York.

Hille, B. (1977). Ionic basis of resting and action potentials. In Brookhart, J. M. and Mountcastle, V. B., eds., *Handbook of Physiology,* sec. 1, *The Nervous System,* vol. 1, *Cellular Biology of Neurons,* pt. 1, 99–136. American Physiological Society, Bethesda, MD.

Hille, B. (1992). *Ionic Channels of Excitable Membranes*. Sinauer, Sunderland, MA.

Hodgkin, A. L. (1958). Ionic movements and electrical activity in giant nerve fibres. *Proc. R. Soc. London, Ser. B*, 148:1–37.

Hodgkin, A. L. (1964). *The Conduction of the Nervous Impulse*. Charles C. Thomas, Springfield, MA.

Junge, D. (1992). *Nerve and Muscle Excitation*. Sinauer, Sunderland, MA.

Kandel, E. R., Schwartz, J. H., and Jessell, T. M. (1991). *Principles of Neural Science*. Elsevier, New York.

Katz, B. (1966). *Nerve, Muscle, and Synapse*. McGraw-Hill, New York.

Keynes, R. D. and Aidley, D. J. (1991). *Nerve and Muscle*. Cambridge University Press, Cambridge, England.

Loewenstein, W. R. (1960). Biological transducers. *Sci. Am.*, 203:98-108.

Meves, H. (1984). Hodgkin-Huxley: Thirty years after. In Baker, P. F., ed., *Current Topics in Membranes and Transport,* vol. 22, *The Squid Axon,* 279-329. Academic Press, New York.

Nicholls, J. G., Martin, A. R., and Wallace, B. G. (1992). *From Neuron to Brain: A Cellular and Molecular Approach to the Function of the Nervous System*. Sinauer, Sunderland, MA.

Patton, H. D. (1965). Reflex regulation of movement and posture. In Ruch, T. C., Patton, H. D., Woodbury, J. W., and Towe, A. L., eds., *Neurophysiology,* 181-206. W. B. Saunders, New York.

Patton, H. D., Fuchs, A. F., Hille, B., Scher, A. M., and Steiner, R. (1989). *Textbook of Physiology,* vol. 1, *Excitable Cells and Neurophysiology*. W. B. Saunders, Philadelphia.

Plonsey, R. and Barr, R. C. (1988). *Bioelectricity, A Quantitative Approach*. Plenum, New York.

Stanley, E. F. (1990). The preparation of the squid giant synapse for electrophysiological investigation. In Gilber, D. L., Adelman, W. J., and Arnold, J. M., eds., *Squid as Experimental Animals,* 171-192. Plenum, New York.

Weiss, T. F. (1996). *Cellular Biophysics,* vol. 1, *Transport*. MIT Press, Cambridge, MA.

Young, J. Z. (1951). *Doubt and Certainty in Science*. Oxford University Press, London.

Original Articles

Adrian, E. D. and Lucas, K. (1912). On the summation of propagated disturbances in nerve and muscle. *J. Physiol.*, 44:68-124.

Baker, P. F., Hodgkin, A. L., and Shaw, T. I. (1961). Replacement of the protoplasm of a giant nerve fibre with artificial solutions. *Nature*, 190:885-887.

Baker, P. F., Hodgkin, A. L., and Shaw, T. I. (1962). Replacement of the axoplasm of giant nerve fibres with artificial solutions. *J. Physiol.*, 164:330-354.

Baylor, D. A. and Fuortes, M. G. F. (1970). Electrical responses of single cones in the retina of the turtle. *J. Physiol.*, 207:77-92.

Baylor, D. A. and Nunn, B. J. (1986). Electrical properties of the light-sensitive conductance of rods of the salamander. *Ambystoma tigrinum. J. Physiol.*, 371:115-145.

Bernstein, J. (1902). Untersuchungen zur thermodynamik der bioelektrischen ströme. *Pflügers Arch. Ges. Physiol.*, 92:521-562.

Brock, L. G., Coombs, J. S., and Eccles, J. C. (1952). The recording of potentials from motoneurones with an intracellular electrode. *J. Physiol.*, 117:431-460.

Curtis, H. J. and Cole, K. S. (1940). Membrane action potentials from the squid giant axon. *J. Cell. Comp. Physiol.*, 15:147-157.

Dennis, M. J., Harris, A. J., and Kuffler, S. W. (1971). Synaptic transmission and its duplication by focally applied acetylcholine in parasympathetic neurons in the heart of the frog. *Proc. R. Soc. London, Ser. B*, 177:509-539.

Gaffey, C. T. and Mullins, L. J. (1958). Ion fluxes during the action potential in *Chara. J. Physiol.*, 144:505-524.

Gray, J. A. B. and Sato, M. (1953). Properties of the receptor potential in pacinian corpuscles. *J. Physiol.*, 122:610-636.

Hodgkin, A. L. and Horowicz, P. (1959). The influence of potassium and chloride on the membrane potential of single muscle fibres. *J. Physiol.*, 148:127–160.

Hodgkin, A. L. and Huxley, A. F. (1939). Action potentials recorded from inside a nerve fibre. *Nature*, 144:710–711.

Hodgkin, A. L., Huxley, A. F., and Katz, B. (1952). Measurement of current-voltage relations in the membrane of the giant axon of *Loligo. J. Physiol.*, 116:424–448.

Holton, T. and Weiss, T. F. (1983). Receptor potentials of lizard cochlear hair cells with free-standing stereocilia in response to tones. *J. Physiol.*, 345:205–240.

Hudspeth, A. J. and Corey, D. P. (1977). Sensitivity, polarity, and conductance change in the response of vertebrate hair cells to controlled mechanical stimuli. *Proc. Natl. Acad. Sci. U.S.A.*, 74:2407–2411.

Kiang, N. Y. S. (1975). Stimulus representation in the discharge patterns of auditory neurons. In Eagles, E. L. and Tower, D. B., eds., *The Nervous System*, vol. 3, *Human Communication and Its Disorders*, 81–96. Raven, New York.

Loewenstein, W. R. and Mendelson, M. (1965). Components of receptor adaptation in a pacinian corpuscle. *J. Physiol.*, 177:377–397.

Lucas, K. (1906). The analysis of complex excitable tissues by their response to electric currents of short duration. *J. Physiol.*, 35:310–331.

Marmont, G. (1949). Studies on the axon membrane: I. A new method. *J. Cell. Comp. Physiol.*, 34:351–382.

Mulroy, M. J., Altmann, D. W., Weiss, T. F., and Peake, W. T. (1974). Intracellular electric responses to sound in a vertebrate cochlea. *Nature*, 249:482–485.

Otis, T. S. and Gilly, W. F. (1990). Jet-propelled escape in the squid *Loligo opalescence*: Concerted control by giant and non-giant motor axon pathways. *Proc. Natl. Acad. Sci. U.S.A.*, 87:2911–2915.

Sato, T. and Beidler, L. M. (1983). Dependence of gustatory neural response on depolarizing and hyperpolarizing receptor potentials of taste cells in the rat. *Comp. Biochem. Physiol.*, 75A:131–137.

Vallbo, A. B. (1964). Accommodation related to inactivation of the sodium permeability in single myelinated nerve fibres. *Acta Physiol. Scand.*, 61:429–444.

Young, J. Z. (1936a). The giant nerve fibres and epistellar body of cephalopods. *Q. J. Microsc. Sci.*, 78:367–386.

Young, J. Z. (1936b). Structure of nerve fibers and synapses in some invertebrates. In *Cold Spring Harbor Symposium on Quantitative Biology*, vol. 4, 1–6. Long Island Biological Society, New York.

Young, J. Z. (1938). The functioning of the giant nerve fibres of the squid. *J. Exp. Biol.*, 15:170–185.

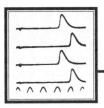

Lumped-Parameter and Distributed-Parameter Models of Cells

Nervous transmission is generally believed to depend upon an electrical process, but the precise way in which one section of a nerve fibre activates another is still uncertain. According to the membrane theory, restimulation is brought about by the local electric currents which spread in advance of the active region. In the conventional form of the theory the current is assumed to flow in one direction along the core of the fibre and to return through the conducting fluid outside. If this view is correct, the velocity of transmission should vary with the electrical resistance outside a nerve fibre.
—Hodgkin, 1939

2.1 Introduction

The notion of electrically small and electrically large cells was introduced in Chapter 1. In electrically small cells, the membrane potential is uniform along the surface of the cell. In electrically large cells, the membrane potential varies along the cell surface. In this chapter, we begin our detailed study of electrical properties of cells by examining the topologies of electrical network models of both small and large cells. We shall not be concerned in this chapter with the properties of the cell membrane.

2.2 Electrical Variables

2.2.1 Current, Current per Unit Length, and Current Density

We shall find it convenient to distinguish among three different current variables, which we shall illustrate with the thin annular conductor in which a radial current flows as shown in Figure 2.1. The potential across the annulus is V, and the current density J is the current per unit area in A/cm^2. J is

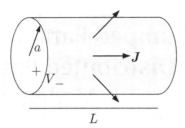

Figure 2.1   Thin annular conductor with a radial current.

radial and has magnitude J. If J is uniform along the surface of the annulus, then the total current flowing through the annulus of radius a and length L is $I = (2\pi a L)J$. The current per unit length is defined as $K = I/L = (2\pi a)J$.

2.2.2   Resistance, Resistivity, Resistance per Unit Length, and Resistance of a Unit Area

The constitutive law for a conductor is

$$\mathbf{J} = \sigma_e \mathcal{E}, \tag{2.1}$$

where $\mathcal{E}$ is the electric field intensity (V/m) and σ_e is the conductivity (S/m); $\sigma_e = 1/\rho$ where ρ is the resistivity ($\Omega \cdot$m). Equation 2.1 is Ohm's law expressed in terms of field quantities. This form of Ohm's law expresses the relation between flow and force variables at a point in space. Thus, the proportionality factor σ_e or ρ is a material property and is independent of the dimensions of the conductor. Equation 2.1 can be used to derive the voltage current relation for different conductor geometries. Three simple geometries, illustrated in Figure 2.2, will be explored because of their relevance to developments in this and in subsequent chapters.

2.2.2.1   *One-Dimensional Conductor*

Let the current density be uniform in a conductor of length l and surface area A (Figure 2.2, top panel). Then the total current flowing axially in the conductor is $I = JA$, so that $I = \mathcal{E}A/\rho$, where J and $\mathcal{E}$ are the magnitudes of the current density and electric field intensities. The electric field intensity is the gradient of the electric potential ψ in the conductor, i.e., $\mathcal{E} = -\nabla\psi$. Thus, we have $\mathcal{E} = -d\psi/dz = \rho I/A$, which we can integrate on z to yield

$$V = \psi(0) - \psi(l) = \frac{\rho l}{A}I,$$

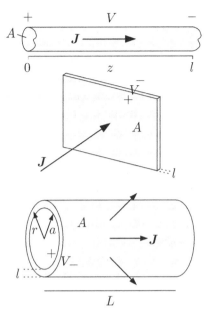

Figure 2.2   Geometries of three conductors: a uniform conductor with an axial current (top panel), a sheet of conducting material (middle panel), and an annular cylinder with a radial current (bottom panel). In the lower panel, r is the radius vector and a is the radius of the inner surface of the annulus.

where V is the voltage difference across the conductor. Therefore, we define the resistance of the conductor as

$$\mathcal{R} = \frac{V}{I} = \frac{\rho l}{A}. \tag{2.2}$$

The resistance is proportional to the resistivity and the length, but inversely proportional to the area; the larger the area of current flow, the smaller the resistance.

The resistance per unit length is defined as

$$r = \frac{\mathcal{R}}{l} = \frac{\rho}{A} \tag{2.3}$$

and has units of Ω/m. Thus, r is a specific resistance for a one-dimensional conductor. For example, consider copper wire, which can be thought of as a one-dimensional conductor, at least at low frequencies. The wire's gauge and composition determine the resistance per unit length. Thus, to obtain a given resistance with wire of a given gauge and composition, we need only specify the length of wire required.

2.2.2.2 Two-Dimensional Conductor

Now consider a sheet of resistive material of thickness l and surface area A with a current density that is normal to the surface of the sheet (Figure 2.2, middle panel). Equation 2.2 defines the resistance in terms of the dimensions. If the thickness is uniform, then we can define the resistance of a unit area as

$$R = \mathcal{R} \times A = \rho l, \tag{2.4}$$

which has units of $\Omega \cdot m^2$. Thus, R is a specific resistance for a two-dimensional conductor. For example, consider a sheet of copper of uniform thickness and composition. To obtain the resistance of the sheet made of material whose specific resistance is R, divide the specific resistance by the area of the sheet.

A two-dimensional conductor geometry that is relevant to cell physiology is a thin cylindrical annular conductor with a radial current (Figure 2.2, bottom panel). For example, the annulus might represent the membrane of a cylindrical cell. Suppose the total current flowing radially through the annulus at any radius r is I. Thus, the current density at r in the radial direction is $J = I/(2\pi r L)$. Since the current density is not constant for different values of r, we cannot simply apply Equation 2.2, but must proceed from more fundamental considerations. From Ohm's law in cylindrical coordinates, we obtain

$$\frac{I}{2\pi r L} = -\sigma_e \frac{d\psi}{dr},$$

which we can separate and integrate as follows:

$$\int_a^{a+l} \frac{I}{2\pi r L}\, dr = -\sigma_e \int_{\psi(a)}^{\psi(a+l)} d\psi'(r).$$

Evaluation of the integral yields

$$\frac{I}{2\pi L} \ln(\frac{a+l}{a}) = -\sigma_e\left(\psi(a+l) - \psi(a)\right),$$

which, after defining $V = \psi(a) - \psi(a+l)$, yields

$$\mathcal{R} = \frac{V}{I} = \frac{\rho}{2\pi L} \ln\left(1 + \frac{l}{a}\right).$$

Suppose the annular conductor is very thin, so that $l \ll a$. Then $\ln(1 + (l/a)) \approx l/a$. The resistance is $\mathcal{R} \approx \rho l/(2\pi a L)$, which is the same result we obtain from Equation 2.2, since the the area of the thin annulus is $A \approx 2\pi a L$. Thus, Equation 2.2 is a good approximation for a thin annulus. If the annulus is made of a conductor with specific resistance $R \ \Omega \cdot m^2$ and if the conductor is very thin $l \ll a$, then the resistance of a length L of this conductor is $R/(2\pi aL)$. A unit length of this annulus has a resistance of $r = R/(2\pi a) \ \Omega/m$.

2.2.2.3 Summary

The resistivity ρ is a material property. Material properties, such as the resistivity or mass density, are independent of dimensions. The resistance $\mathcal{R}$ is a proportionality factor between the current through and the voltage difference across a conductor. The quantities r and R are specific resistances; r is resistance per unit length of a one-dimensional conductor, and R is the resistance of a unit area of a two-dimensional conductor.

2.3 Electrically Small Cells

Figure 2.3 shows responses recorded with two micropipettes from *Paramecium caudatum* in response to a current injected through a third intracellular micropipette. The changes in membrane potential measured with the two intracellular micropipettes in response to the current are essentially identical. This result is consistent with the idea that the cytoplasm of this cell is an equipotential and, if the extracellular solution is also equipotential, the potential across the membrane is independent of position along the cell surface. The paramecium is an example of a cell that is an electrically small cell.

Any uniform, isolated small cell can be represented by the network topology shown in Figure 2.4. Since in an electrically small cell the difference of potential across the membrane is independent of position, we can treat the membrane as a lumped element with some electrical characteristic as yet not specified. Thus, the appropriate description of such a cell is that of a lumped-parameter electrical network.

2.4 Electrically Large Cells: The Core Conductor Model

In a uniform, isolated large cell, the difference of potential across the membrane depends upon position along the cell surface as shown in Figure 2.5. In this example, the nerve innervating a muscle fiber is stimulated to produce, via a chemical synaptic mechanism (see Chapter 1) located at a region of the

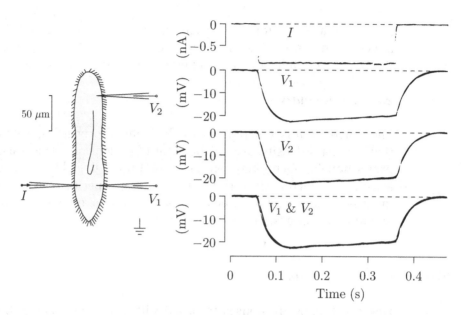

Figure 2.3 Responses recorded with micropipettes from *Paramecium caudatum* at two intracellular locations in response to a current passed through the membrane with a third micropipette (adapted from Eckert and Naitoh, 1970, Figure 1). The left panel shows a schematic diagram of the stimulating and recording arrangement. The right panel shows the current stimulus I (above) and the changes in potential from the resting potential V_1 and V_2 both individually and superimposed (lowest trace). The two electrodes used to measure the intracellular potential were 125 μm apart. The potentials are referred to a common extracellular reference electrode.

muscle called the *end plate*, a change in potential across the membrane of the muscle cell called the *end-plate potential* or *EPP*. The EPP is largest at the end plate and decays in amplitude as distance from the end plate increases. Thus, the membrane potential across the muscle cell is not uniform, but changes with position along the fiber. Thus, the muscle cell is by definition an electrically large cell.

The difference in membrane potential along a cell raises new questions about the spatial coupling of currents and potentials through neighboring portions of the cell. The electrical coupling of local regions of membrane is of interest with respect to graded potentials, such as the EPP shown in Figure 2.5, as well as to action potentials generated by large electrically excitable cells. In these cells, the passage of an action potential is accompanied by current flow through the cell membrane and through the intracellular and extracellular media. In this way, there is spatial coupling of current flow through adjacent

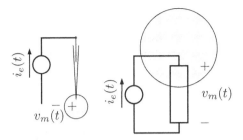

Figure 2.4 The left panel shows the arrangement for recording the membrane potential in response to a current through the membrane of a small cell. The right panel shows an equivalent network superimposed on the schematic diagram of the cell. The membrane characteristic is represented by a two-terminal device (schematized as a rectangle in the figure). The series resistance between the electrodes and the membrane, on both the intracellular and extracellular faces, is usually small and can be neglected for many purposes. The series resistance has not been included in this network.

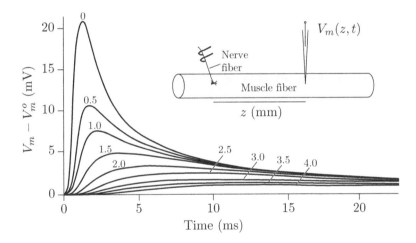

Figure 2.5 Membrane potential changes recorded at various locations along a sartorius muscle fiber of a frog (*Rana temporaria*) in response to electrical stimulation of the nerve innervating the muscle cell (adapted from Fatt and Katz, 1951, Figure 8). The parameter with each response is the distance from the end plate (×0.97) in mm. The recording arrangement is shown in the inset.

regions of the membrane as the action potential propagates along the cell. Several properties of the transmission of the action potential are a consequence of the geometry of the cell and the electrical characteristics of the intracellular and extracellular media, and these properties are essentially independent of the electrical characteristics of the membrane. In this section we shall explore these properties. Our analysis of large cells is confined to cells with cylindrical geometries (unmyelinated axons, muscle fibers, etc.), but the concepts are applicable more generally.

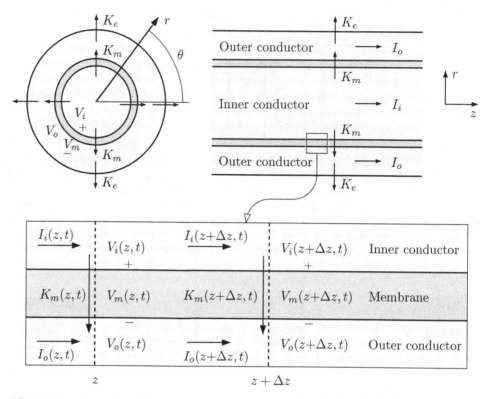

Figure 2.6 Geometry of the core conductor model of a cylindrical cell.

2.4.1 Assumptions of the Core Conductor Model

We shall make the following assumptions about the electrical characteristics of a cylindrical cell (Figure 2.6):

- The cell membrane is a cylindrical boundary that separates two conductors of electric current, the intracellular and extracellular solutions, which are assumed to be homogeneous and isotropic and to obey Ohm's law.

- All the electrical variables have cylindrical symmetry, i.e., all the electrical variables are independent of θ.

- A circuit theory description of currents and voltages is adequate. That is, the quasi-static terms of Maxwell's equations are sufficient, and electromagnetic radiation effects are negligible.

- Currents in the inner and outer conductors flow in the longitudinal direction only. Current flows through the membrane in the radial direction only.

- At a given longitudinal position along the cell, the inner and outer conductors are equipotentials, so the only variation in potential in the radial direction, r, occurs across the membrane.

The variables used to describe the electrical properties of a cylindrical cell are defined as follows:

- $I_o(z, t)$ is the total longitudinal current flowing in the positive z-direction in the outer conductor (A).
- $I_i(z, t)$ is the total longitudinal current flowing in the positive z-direction in the inner conductor (A).
- $J_m(z, t)$ is the membrane current density flowing from the inner conductor to the outer conductor (A/m^2).
- $K_m(z, t)$ is the membrane current per unit length flowing from the inner conductor to the outer conductor (A/m).
- $K_e(z, t)$ is the current per unit length due to external sources applied in a cylindrically symmetric manner (A/m). Inclusion of this current allows us to represent the current applied through external electrodes to the cell surface. A similar term could be added to represent the current supplied by an internal electrode (see Problem 2.6).
- $V_m(z, t)$ is the membrane potential, which is a positive quantity when the inner conductor has a positive potential with respect to the outer conductor (V).
- $V_i(z, t)$ is the potential in the inner conductor (V).
- $V_o(z, t)$ is the potential in the outer conductor (V).
- r_o is the resistance per unit length of the outer conductor (Ω/m).
- r_i is the resistance per unit length of the inner conductor (Ω/m).
- a is the radius of the cylindrical cell.

2.4.2 Derivation of the Core Conductor Equations

The relations among the electrical variables are derived most readily from an equivalent electrical network model of Figure 2.6, which is shown in Figure 2.7. The electrical variables must obey Kirchhoff's laws as well as the constitutive relations for the elements. Application of Kirchhoff's current law to node a yields

$$I_i(z, t) = I_i(z + \Delta z, t) + K_m(z, t)\Delta z. \tag{2.5}$$

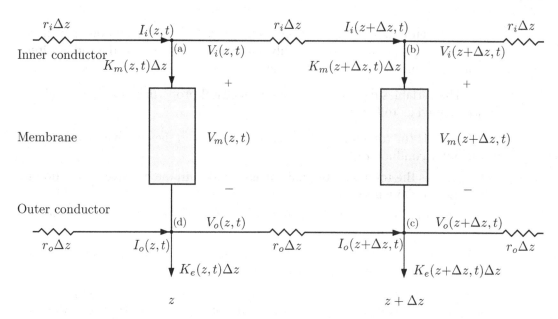

Figure 2.7 Electrical network model of an incremental portion of the core conductor model of a cylindrical cell.

To clarify the relation between the electrical network model (Figure 2.7) and the geometry (as indicated in Figure 2.6), consider Figure 2.8, which shows the relations among the current and voltage variables for an incremental portion of the inner conductor corresponding to node a in the network model. The longitudinal currents enter ($I_i(z, t)$) and leave ($I_i(z + \Delta z)$) the volume element through the circular ends, and the membrane current density ($J_m(z, t)$) leaves radially through the cylindrical volume element. The total current through the membrane can be expressed as

$$2\pi a J_m(z, t)\Delta z = K_m(z, t)\Delta z. \tag{2.6}$$

Equation 2.5 embodies the notion that charge does not accumulate in the cylindrical volume element, and hence that the sum of the currents entering the volume element must be zero.

Kirchhoff's current law applied to node d yields

$$I_o(z, t) + K_m(z, t)\Delta z = I_o(z + \Delta z, t) + K_e(z, t)\Delta z, \tag{2.7}$$

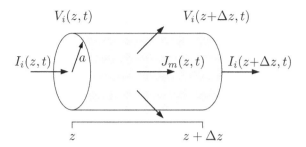

Figure 2.8 Relation of current and voltage variables in the inner conductor.

which expresses Kirchhoff's current law for the annular volume element that represents the outer conductor.

From the voltage-current characteristics of the resistances per unit length (r_i and r_o), we obtain

$$V_i(z, t) - V_i(z + \Delta z, t) = r_i \Delta z I_i(z + \Delta z, t). \tag{2.8}$$

Equation 2.8 can be interpreted by examining the relation between voltage and current in a cylindrical volume representing the inner conductor (Figure 2.8). The voltage difference across the volume element is equal to the product of the current through the element and the resistance of the element ($\Delta \mathcal{R}_i$):

$$V_i(z, t) - V_i(z + \Delta z, t) = \Delta \mathcal{R}_i I_i(z + \Delta z, t). \tag{2.9}$$

But the resistance is

$$\Delta \mathcal{R}_i = \frac{\rho_i \Delta z}{\pi a^2},$$

where ρ_i is the resistivity of the inner conductor (cytoplasm). Thus, for the resistance per unit length we obtain the expression

$$r_i = \frac{\Delta \mathcal{R}_i}{\Delta z} = \frac{\rho_i}{\pi a^2}. \tag{2.10}$$

In a similar manner, we obtain

$$V_o(z, t) - V_o(z + \Delta z, t) = r_o \Delta z I_o(z + \Delta z, t), \tag{2.11}$$

where r_o is the resistance per unit length of the annular external conductor.

Kirchhoff's voltage law yields

$$V_m(z, t) = V_i(z, t) - V_o(z, t). \tag{2.12}$$

Rearranging terms in Equations 2.5, 2.7, 2.8, and 2.11 and dividing by Δz yields the equations

$$\frac{I_i(z + \Delta z, t) - I_i(z, t)}{\Delta z} = -K_m(z, t), \tag{2.13}$$

$$\frac{I_o(z + \Delta z, t) - I_o(z, t)}{\Delta z} = K_m(z, t) - K_e(z, t), \tag{2.14}$$

$$\frac{V_i(z + \Delta z, t) - V_i(z, t)}{\Delta z} = -r_i I_i(z + \Delta z, t), \tag{2.15}$$

$$\frac{V_o(z + \Delta z, t) - V_o(z, t)}{\Delta z} = -r_o I_o(z + \Delta z, t). \tag{2.16}$$

$$\tag{2.17}$$

By taking the $\lim_{\Delta z \to 0}$ of the above equations, the following equations, known as the *core conductor equations,* are obtained

$$\frac{\partial I_i(z, t)}{\partial z} = -K_m(z, t), \tag{2.18}$$

$$\frac{\partial I_o(z, t)}{\partial z} = K_m(z, t) - K_e(z, t), \tag{2.19}$$

$$\frac{\partial V_i(z, t)}{\partial z} = -r_i I_i(z, t), \tag{2.20}$$

$$\frac{\partial V_o(z, t)}{\partial z} = -r_o I_o(z, t), \tag{2.21}$$

$$V_m(z, t) = V_i(z, t) - V_o(z, t). \tag{2.22}$$

These equations constitute a distributed-parameter model of the relation of voltages and currents. However, these equations have quite different general validities. Equations 2.18 and 2.19 follow directly from Maxwell's equations, i.e., $\nabla \cdot \mathbf{J} = 0$, without any further assumptions. They state that an increase in internal or external longitudinal current with distance along the cell, must be due to the membrane current per unit length. These two equations are due to Kirchhoff's current law, from which both were derived. Equations 2.20 and 2.21 are approximations of the exact solutions obtained by solving a three-dimensional field problem to which the core conductor model is a one-dimensional approximation (Clark and Plonsey, 1966; Rosenfalck, 1969). It can be shown that for typical parameters of cylindrical cells, Equation 2.20 is an excellent approximation of the relation of intracellular current to intracellular potential. However, Equation 2.21 is a poor approximation of the relation between extracellular current and extracellular potential unless the spatial extent of the extracellular space is limited.

Taking the partial derivative of V_m with respect to z in Equation 2.22 and substituting Equations 2.20 and 2.21 yields

$$\frac{\partial V_m(z,t)}{\partial z} = -r_i I_i(z,t) + r_o I_o(z,t). \tag{2.23}$$

Differentiating Equation 2.23 yields

$$\frac{\partial^2 V_m(z,t)}{\partial z^2} = r_o \frac{\partial I_o(z,t)}{\partial z} - r_i \frac{\partial I_i(z,t)}{\partial z}. \tag{2.24}$$

Substituting Equations 2.18 and 2.19 into Equation 2.24, we obtain

$$\frac{\partial^2 V_m(z,t)}{\partial z^2} = (r_o + r_i) K_m(z,t) - r_o K_e(z,t). \tag{2.25}$$

Equation 2.25 is sometimes referred to as the *core conductor equation*. It is an equation of the constraints placed upon the membrane variables, V_m and K_m, by the inner and outer conductors. The equation is valid independent of the properties of the membrane, which may be nonlinear and time-varying. In order to find the membrane current and voltage, we require another equation relating these variables, and this equation must result from the electrical properties of the membrane. We shall explore these equations for different membrane properties in Chapters 3 and 4. In the remainder of this chapter, we shall explore properties of the core conductor equations that are relatively independent of the electrical properties of the membrane.

2.4.3   Consequences of the Core Conductor Model

In this section, we shall explore consequences of the core conductor model. Since we have made rather weak assumptions about the electrical characteristics of the membrane (e.g., that there is no longitudinal current in the membrane), the consequences we shall derive are valid quite generally.

2.4.3.1   *Current Flow during a Propagated Action Potential:*
The Local Circuit Theory

During the passage of an action potential along a cell, there is a flow of current surrounding the cell. Using only the core conductor equations plus the assumption that the action potential travels at constant velocity, this flow of current can be discerned. The argument requires several distinct steps.

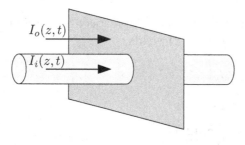

Figure 2.9 Relation of longitudinal currents.

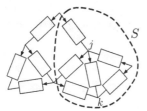

Figure 2.10 Network showing an arbitrary closed surface S.

Relation of Currents to Spatial Derivatives of the Membrane Potential

Suppose that an action potential is propagating along an axon and that there are no external electrodes along the axon, i.e., $K_e = 0$. Under these conditions, Equation 2.25 becomes

$$K_m(z,t) = \frac{1}{r_o + r_i} \frac{\partial^2 V_m(z,t)}{\partial z^2}. \tag{2.26}$$

Consider a plane perpendicular to the axon (Figure 2.9). The sum of the longitudinal currents entering that plane from either side must equal zero. This general property follows from network theory (Bose and Stevens, 1965) and can be readily verified for the generic network shown in Figure 2.10. The network consists of a set of branch elements, each having two terminals so that the same current flows into the element as flows out of the element. For any such network, the net current flowing into an arbitrary closed surface S is zero. The physical explanation of this result rests on the observation that no net charge accumulates in a network, i.e., currents always flow in closed loops. A more formal proof is readily given. Let the current flowing through the element from node j to node k be I_{jk}. The current entering node k must satisfy Kirchhoff's current law so that $\sum_{j=1}^{N} I_{jk} = 0$. Now consider the sum of all the Kirchhoff's current law equations for the nodes in S. This sum must also be zero. Since each branch current connects two nodes, each branch current in S appears in two node equations, once with a positive sign (e.g.,

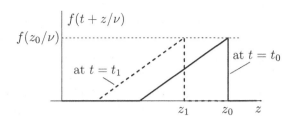

Figure 2.11 Solutions of the wave equation at two instants in time.

in the equation for the node that this current leaves) and once with a negative sign (in the equation for the node that this current enters). Thus, each such internal branch current will cancel out of the sum of the Kirchhoff's current law equations. Thus, the grand sum of the Kirchhoff's current law equations, which equals zero, will contain only those currents from nodes outside of S and these must sum to zero. Therefore,

$$I_i(z, t) + I_o(z, t) = 0. \tag{2.27}$$

From this condition together with Equation 2.23, we obtain

$$I_o(z, t) = \frac{1}{r_o + r_i} \frac{\partial V_m(z, t)}{\partial z}. \tag{2.28}$$

Implications of Propagation at Constant Velocity

If we knew the waveform of the action potential as a function of z, the currents could be sketched using Equations 2.26 and 2.28. However, suppose we knew the waveform of the action potential as a function of t. Could we then sketch the currents? If we assume that the action potential propagates at constant velocity, then, as we shall show, the spatial and temporal waveforms of the action potential are linked. Therefore, the currents can be sketched from either the temporal or the spatial waveform of the action potential.

Suppose an action potential propagates at constant velocity v in the $\mp z$-direction. Then

$$V_m(z, t) = f\left(t \pm \frac{z}{v}\right). \tag{2.29}$$

To see why a spatiotemporal dependence of $V_m(z, t)$ of the form $f(t \pm z/v)$ implies that $V_m(z, t)$ is a wave that travels at constant velocity v, consider the function shown in Figure 2.11. The solid line shows the dependence of $f(t + z/v)$ on z at $t = t_0$. The peak of this function occurs at the position z_0 at $t = t_0$ and has a value $f(t_0 + z_0/v)$. Now consider what the function should look like at some later time t_1. Let us define the new position of the peak as z_1 and find the relation between z_1 and z_0. The peak of the function f still occurs when its argument is $t_0 + z_0/v$. Hence, $t_0 + z_0/v = t_1 + z_1/v$, from which we see that $z_1 = z_0 - v(t_1 - t_0)$. Thus, the peak has moved in the negative z-direction by a distance proportional to the elapsed time, i.e., the peak has moved in the negative z-direction with velocity v. A similar argument applied to other points on the waveform leads to the dashed wave in Figure 2.11 and demonstrates that a function of the form $f(t + z/v)$ represents a wave traveling at constant velocity v in the negative z-direction. Similarly, a function of the form $f(t - z/v)$ represents a wave traveling in the positive z-direction.

Given that $V_m(z, t)$ has the form shown in Equation 2.29, it follows that

$$\frac{\partial V_m(z, t)}{\partial z} = \pm \frac{1}{v} \dot{f} \left(t \pm \frac{z}{v} \right) \quad \text{and} \quad \frac{\partial V_m(z, t)}{\partial t} = \dot{f} \left(t \pm \frac{z}{v} \right),$$

where $\dot{f}(\alpha) = df/d\alpha$. Hence,

$$\pm v \frac{\partial V_m(z, t)}{\partial z} = \frac{\partial V_m(z, t)}{\partial t}. \tag{2.30}$$

Also,

$$\frac{\partial^2 V_m(z, t)}{\partial z^2} = \frac{1}{v^2} \ddot{f} \left(t \pm \frac{z}{v} \right) \quad \text{and} \quad \frac{\partial^2 V_m(z, t)}{\partial t^2} = \ddot{f} \left(t \pm \frac{z}{v} \right).$$

Therefore, $V_m(z, t)$ satisfies the wave equation

$$\frac{\partial^2 V_m(z, t)}{\partial z^2} = \frac{1}{v^2} \frac{\partial^2 V_m(z, t)}{\partial t^2}. \tag{2.31}$$

Substitution of Equations 2.30 into 2.28 and 2.31 into 2.26 yields

$$I_o(z, t) = \pm \frac{1}{(r_o + r_i)v} \frac{\partial V_m(z, t)}{\partial t} \tag{2.32}$$

and

$$K_m(z, t) = \frac{1}{(r_o + r_i) v^2} \frac{\partial^2 V_m(z, t)}{\partial t^2}. \tag{2.33}$$

These equations have been derived from the core conductor model plus the assumption that the action potential propagates at constant velocity and therefore satisfies the wave equation.

Local Current Flow
Given the waveform of $V_m(z, t)$ as a function of time, then the currents $I_o(z, t)$, $I_i(z, t)$, and $K_m(z, t)$ can be computed from Equations 2.32, 2.27, and 2.33. Therefore, the current flow surrounding a propagated action poten-

tial can be visualized. For the purpose of sketching the currents, we assume that an action potential is traveling in the negative z-direction. Note that for this direction $V_m(z,t) = f(t + z/v)$ and the waveform of $V_m(z,t)$ is the same when plotted versus t (at a particular point, z) or versus z/v (at a particular time, t) except for a translation along the t or z/v axis. We shall construct a picture of the electrical variables from the shape of $V_m(z,t)$ (Figure 2.12). This picture can be regarded as a "snapshot" of the electrical variables along the axon at one point in time or as the time dependence of these variables at one point in space.

From Equations 2.32 and 2.33, we note that at the peak of the action potential the longitudinal currents I_o and I_i ($I_i = -I_o$) are zero and the membrane current is negative, $K_m < 0$; that is, it is inward. In the region to the left of the point of inflection on the rising phase of the action potential, the membrane current is positive, hence outward through the membrane. Between these two regions, the longitudinal current flows to the right in the outer conductor and to the left in the inner conductor. Hence, we can draw a closed current path (shown in the lower part of the figure) associated with the action potential. Current flows inward through the membrane in the region near the peak of the action potential. This current flows down the center conductor (the cytoplasm of the axon, or axoplasm) and outward through the membrane in front of the action potential. This outward current flow makes the membrane potential, in advance of the action potential peak, more positive inside, i.e., it depolarizes the membrane. As we shall see later, this depolarization triggers off the action potential at this spot. Hence, the action potential travels to the left. The action potential does not travel to the right, despite the fact that there is a region of outward current there, because, as we shall see, the action potential leaves the membrane in a refractory state in its wake. The ionic basis of this sequence of events will be discussed in Chapter 4.

2.4.3.2 Spatial Extent of the Action Potential

Because the membrane potential of an action potential that propagates at constant velocity must have a dependence on z and t of the form $f(t \pm z/v)$, if we know the waveform of the action potential in time, we can sketch the potential in space or vice versa. This property is illustrated schematically in Figure 2.13, which shows the membrane potential as a function of both z and t for an action potential propagating at velocity $v = 25$ m/s $= 2.5$ cm/ms in the positive z-direction. Therefore, $V_m(z,t) = f(t - z/2.5)$, where time is measured in ms and distance in cm. At $z = 5$ cm, the onset of the action

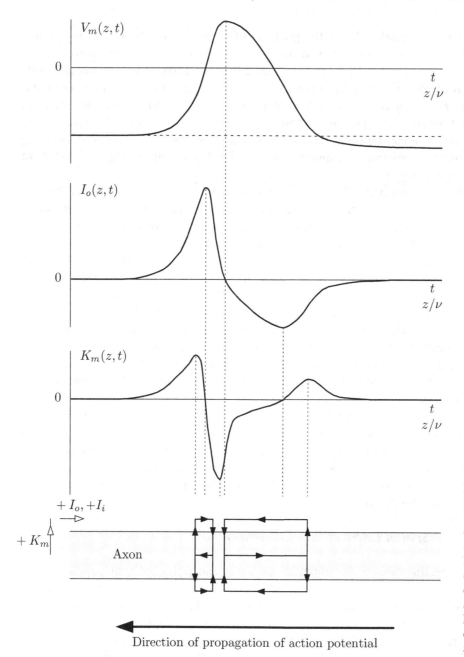

Figure 2.12 Local flow of current around a cylindrical cell during an action potential. Positive reference directions for current variables are shown in the lower left-hand corner with open arrows.

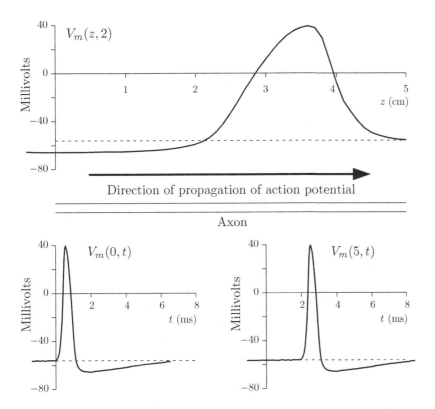

Figure 2.13 Relation between the spatial and temporal extents of the propagated action potential. The upper panel shows the spatial dependence of the action potential for a 5 cm length of axon at time $t = 2$ ms. The lower two panels show the temporal dependence of the action potential at the two ends of the axon. Time $t = 0$ corresponds to the time of onset of the action potential at $z = 0$.

potential is delayed by a time interval that equals (5 cm)/(2.5 cm/ms) = 2 ms. At $t = 2$ ms, the action potential just arrives at $z = 5$ cm, as can be seen in both the temporal and the spatial domains.

Suppose we define the interval over which the action potential is positive as the extent of the action potential. Measured in the time domain, the *temporal extent* is seen to be less than 0.5 ms for the schematic diagram shown. Measured in the space domain, the *spatial extent* is seen to be greater than 1 cm. Whatever definition we choose for the temporal extent, the commensurate spatial extent is simply the temporal extent multiplied by the propagation velocity.

To see what this means anthropomorphically, consider a sensory neuron that ends in a toe and has a cell body in the dorsal root of the spinal cord, about 1 m away. Suppose the conduction velocity is about 50 m/s and the duration of the action potential is 2 ms. Then the spatial extent of the action potential is 0.1 m or about 1/10 of the length of the neuron. Thus, the message that leaves the toe, say in response to a toe pinch, is spread out along an appreciable fraction of the neuron at any one time.

2.4.3.3 *Conduction Velocity of an Unmyelinated Axon*

The core conductor model also has implications for the dependence of the conduction velocity of an action potential on axon dimensions. We start with the relation

$$K_m(z,t) = \frac{1}{(r_o + r_i)v^2} \frac{\partial^2 V_m(z,t)}{\partial t^2}.$$

The membrane current density $J_m(z,t) = K_m(z,t)/(2\pi a)$, where a is the radius of the axon. Hence,

$$\frac{\partial^2 V_m(z,t)/\partial t^2}{J_m(z,t)} = 2\pi a(r_o + r_i)v^2. \tag{2.34}$$

Equation 2.34 is parsed into two terms: the left-hand side can be regarded as a constitutive relation for the membrane material, since it relates the (second derivative of the) membrane potential to the membrane current density. Therefore, the left-hand ratio in the equation is determined entirely by the electrical characteristics of a square centimeter of membrane. Hence, for a given axon this ratio has the same value for every patch of membrane independent of the surface area of the patch. Since the right-hand term in Equation 2.34 is constant (independent of z and t), the left-hand term must also be a constant (independent of z and t), designated as $\mathcal{K}_m$. The right-hand term in the equation contains a number of factors that are independent of the membrane properties plus the conduction velocity.

Now let us imagine that we have a sheet of membrane that we roll into axons of different diameters and fill and surround with electrolytes that give different resistances per unit length. In this case, $(\partial^2 V_m/\partial t^2)/J_m$ is the same for different axons, and we can determine the dependence of conduction velocity on r_o, r_i, and a for axons whose membranes have identical electrical characteristics. Therefore,

$$2\pi a(r_o + r_i)v^2 = \mathcal{K}_m, \tag{2.35}$$

where $\mathcal{K}_m$ is a specific property of the membrane that depends on the membrane material and its thickness, but not on the surface area. Equation 2.35 shows that the conduction velocity depends on the radius of the axon as well as on the resistance of the ionic media bathing the membrane.

Axon in:

seawater

oil

seawater

oil

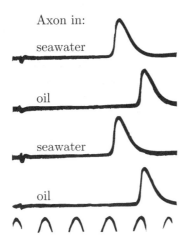

Figure 2.14 The effect of increasing the resistance of the extracellular solution on the conduction velocity of the action potential of an unmyelinated axon (adapted from Hodgkin, 1939, Figure 2). A crab (*Carcinus maenas*) axon of 30 μm diameter was immersed alternately in seawater and in oil and the action potential recorded 13 mm from the site of stimulation. The time scale is given by a simultaneously recorded sine wave of period 1 ms (lowest panel).

Equation 2.35 implies that increasing the external resistance, r_o, will reduce the conduction velocity. Measurements to test this notion are shown in Figure 2.14 for a crab axon immersed alternately in a large volume of seawater and then in oil. These data show that when an axon is immersed in oil, the time it takes an action potential to travel a fixed distance is increased, and therefore the conduction velocity is decreased. Since the external resistance, r_o, is much larger in oil than in seawater, these results show that an increase in external resistance per unit length leads to a decrease in conduction velocity. It has also been shown (Hodgkin, 1939) that if an external metallic conductor is used to decrease r_o, the conduction velocity is increased.

Alternatively, r_i can be manipulated experimentally. It has been shown that if r_i is increased by extruding some of the axoplasm, the conduction velocity also decreases (see Problem 2.11). One can appreciably reduce r_i by inserting a wire axially into the axon. If the axon is simultaneously immersed in a large bath of seawater, both r_o and r_i will be much reduced from their normal values and the conduction velocity can be made so large that the action potential effectively occurs simultaneously along the entire length of the axon (see Figure 2.15). Thus, insertion of the wire into the axon changes the cell from an electrically large cell to an electrically small cell. This result demonstrates that the determination of whether a cell is electrically small or large requires more information than just the cell geometry. In addition to demonstrating the relation between conduction velocity and resistance, this experiment illustrates the effectiveness of an important experimental technique called the *space clamp,* which has enabled experimenters to analyze

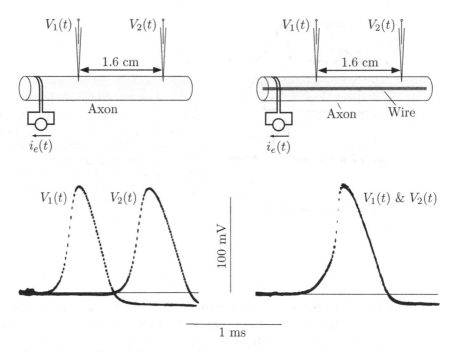

Figure 2.15 Effect of decreasing the internal longitudinal resistance on the conduction velocity of a squid giant axon (adapted from Del Castillo and Moore, 1959, Figure 1). The upper panels show schematic diagrams of the recording arrangements used to obtain the results shown in the lower panels. In both the left and right panels, two micropipettes impale the axon a distance 1.6 cm apart, and the voltage in response to a shock is measured. In the right panels, the experiment is performed with a low-resistance axial wire inserted into the axon. With the axial wire in place, the action potentials are virtually coincident. The amplifiers for recording the two potentials are set to have slightly different gains so that the two traces can be just discerned (note the thickened trace at the peak of the action potential).

the membrane properties that underlie the action potential. We shall describe these experiments further in Chapter 4.

From Equation 2.35 we see that the conduction velocity, v, depends inversely upon r_o and r_i. How does v depend on the diameter of the axon? Equation 2.35 shows that the conduction velocity depends upon a, but r_i also depends upon a. To proceed further, we shall make the assumption that $r_i \gg r_o$, which is an excellent assumption for giant axons of invertebrates, and an even better assumption for smaller unmyelinated fibers (see Problem 2.1). Hence, we have

$$2\pi a r_i v^2 = \mathcal{K}_m. \tag{2.36}$$

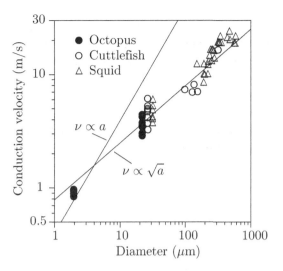

Figure 2.16 Dependence of the conduction velocity of the action potential on axon diameter for unmyelinated axons of three species of cephalopod molluscs measured at 20°C (adapted from Burrows et al., 1965). Two lines are drawn for comparison, one for proportionality between conduction velocity and radius and the other for proportionality with the square-root of the radius.

The internal resistance/length, r_i, can be expressed in terms of the resistivity of axoplasm, ρ_i, as

$$r_i = \frac{\rho_i}{\pi a^2}. \tag{2.37}$$

Thus, we have

$$2\pi a \frac{\rho_i}{\pi a^2} v^2 = \mathcal{K}_m.$$

Hence, we can write

$$v = \sqrt{\frac{\mathcal{K}_m a}{2\rho_i}}. \tag{2.38}$$

Thus, the theory predicts that for unmyelinated axons of identical specific properties (made from the same membrane material and having the same axoplasm), the conduction velocity is proportional to the square root of the radius of the axon. The data in Figure 2.16 from axons of three related species of cephalopod molluscs (octopus, cuttlefish, and squid) show that Equation 2.38 fits the data approximately for a range of fiber diameters of over two orders of magnitude. As is also shown in Figure 2.16, the conduction velocity deviates significantly from proportionality with fiber diameter. As we shall see in Chapter 5, myelinated nerve fibers, which are not uniform cylindrical structures as are unmyelinated fibers, do have a conduction velocity proportional to fiber diameter.

2.4.3.4 *Application of the Core Conductor Equations to the Recording of Extracellular Potentials*

The potentials recorded extracellularly from an axon can be calculated from the core conductor equations. Of course, the results obtained from this derivation are valid only if the assumptions on which the core conductor equations are based are valid. A critical assumption is that current in the outer conductor flows in the longitudinal direction only and that there is no radial component of current in the outer conductor. Thus, for any longitudinal position, z, the outer conductor is assumed to be equipotential in the radial direction. This condition is most certainly not met when the outer conductor is of large extent. In this case, the relation between potentials recorded in this outer conductor and membrane variables is more complex. However, if the outer conductor is of limited extent, such as can be achieved when an axon is bathed in a limited amount of normal extracellular fluid or when an axon is placed in oil,[1] the following analysis is valid.

The potentials in the inner and outer conductors, V_i and V_o, are

$$\Delta V_i(z,t) = V_i(z,t) - V_i(-\infty, t) = -\int_{-\infty}^{z} r_i I_i(\zeta, t)\, d\zeta \tag{2.39}$$

and

$$\Delta V_o(z,t) = V_o(z,t) - V_o(-\infty, t) = -\int_{-\infty}^{z} r_o I_o(\zeta, t)\, d\zeta, \tag{2.40}$$

where $V_i(-\infty, t)$ and $V_o(-\infty, t)$ are the potentials at some remote point along the axon. Now consider an external current source of total current $I_e(t)$ applied to the axon through an electrode situated at $z = 0$. The electrode is assumed to be cylindrical and coaxial with the axon. The other terminal of

---

1. When an axon is immersed in oil, a thin film of extracellular fluid adheres to the surface of the axon.

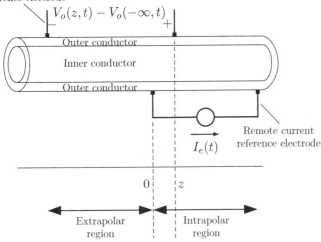

Remote voltage
reference electrode

$V_o(z,t) - V_o(-\infty, t)$

Outer conductor

Inner conductor

Outer conductor

$I_e(t)$

Remote current
reference electrode

0   z

Extrapolar
region

Intrapolar
region

Figure 2.17 Arrangement of electrodes for recording extracellular potentials from a cylindrical cell.

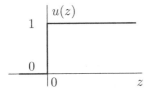

$u(z)$

1

0

0   z

Figure 2.18 The unit step function.

the current source is located at a remote position along the axon, as shown in Figure 2.17.

As we did in connection with Figure 2.9, we shall consider a plane perpendicular to the axon. The sum of the longitudinal currents entering that plane from either side must equal zero. However, we must consider the external current source as well as the longitudinal currents. Therefore,

$$I_i(z,t) + I_o(z,t) = 0 \quad \text{for} \quad z < 0,$$

$$I_i(z,t) + I_o(z,t) + I_e(t) = 0 \quad \text{for} \quad z > 0.$$

This result can be written more compactly by using the unit step function (Figure 2.18), defined as

$$u(z) = \begin{cases} 1 & \text{if } z > 0, \\ 0 & \text{if } z < 0. \end{cases}$$

Then we can write

$$I_i(z, t) + I_0(z, t) + I_e(t)u(z) = 0. \tag{2.41}$$

The membrane potential can be expressed as

$$\Delta V_m(z, t) = \Delta V_i(z, t) - \Delta V_0(z, t) = -\int_{-\infty}^{z} (r_i I_i(\zeta, t) - r_0 I_0(\zeta, t)) \, d\zeta. \tag{2.42}$$

Substitution of Equation 2.41 into 2.42 yields

$$\Delta V_m(z, t) = -\int_{-\infty}^{z} (-(r_i + r_0)I_0(\zeta, t) - r_i I_e u(\zeta)) \, d\zeta,$$

and substitution of Equation 2.40 yields

$$\Delta V_m(z, t) = -\frac{r_i + r_0}{r_0} \Delta V_0(z, t) + r_i I_e z u(z).$$

Therefore,

$$\Delta V_0(z, t) = -\frac{r_0}{r_i + r_0} \Delta V_m(z, t) + \frac{r_i r_0}{r_i + r_0} I_e z u(z). \tag{2.43}$$

Thus, for this case, the potential difference recorded between two electrodes in the outer conductor is simply related to the membrane potential. Assuming that at $z = -\infty$, $V_m(-\infty, t)$ is a constant (not a function of time), then $V_m(z, t) - V_m(-\infty)$ is the deviation of the membrane potential from its resting value. In the region that is not between the stimulus electrodes (commonly referred to as the *extrapolar region*), the potential in the outer conductor is the negative of the deviation of the membrane potential from its resting value weighted by a voltage-divider ratio, $r_0/(r_0 + r_i)$. Within the inter-electrode (or interpolar) regions, the potential in the outer conductor contains an additional term due directly to the externally applied current. This term is easily interpretable as the potential difference caused by the external current $I_e(t)$ flowing through a resistance per unit length of $r_i r_0/(r_i + r_0)$ over a length z cm long. In a typical situation of physiological interest, the two pairs of recording and stimulating electrodes are relatively remote from each other and the axon is placed in mineral oil. Under these conditions, an extracellular electrode can be used to measure the extracellular potential as an action potential travels past the electrode. Then

$$V_0(z, t) - V_0(-\infty) = -\frac{r_0}{r_0 + r_i}(V_m(z, t) - V_m(-\infty)). \tag{2.44}$$

This result gives insight into the recording of a monophasic action potential. The reference electrode is placed on an inactive end of the axon (either de-

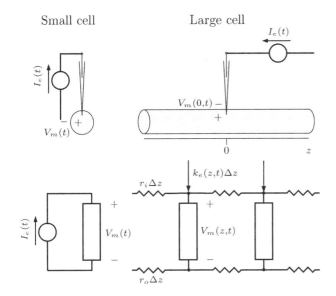

Figure 2.19 Comparison between recording arrangements (upper panel) and electrical network models (lower panel) for small and large cells. The small spherical cell is impaled with a micropipette to record the membrane potential $V_m(t)$. The large cylindrical cell is impaled at $z = 0$, and the membrane potential $V_m(0, t)$ is recorded at that location. The electrical network model for the small cell is a lumped-parameter network, whereas that for the large cell is a distributed-parameter network.

polarized with isotonic KCl or crushed). Thus, $V_m(-\infty) \approx 0$, and $V_o(z, t) - V_o(-\infty) = -(r_o/(r_o + r_i))V_m(z, t)$. If the action potential traverses both electrodes, the resultant recorded waveform is a superposition of two action potentials (a diphasic action potential), one positive and one negative, with the temporal separation determined by the propagation velocity of the action potential and the distance between recording electrodes (see Problem 2.10).

The importance of Equation 2.44 is that it suggests that it is possible to record potentials across the membrane of a cell without the necessity of an intracellular electrode. Because impaling a small cell with even the finest micropipettes can injure the cell and change its electrical properties, use of extracellular recording techniques to measure membrane potentials of small cells has been very valuable. The method depends upon making the extracellular resistance per unit length (r_o) as large as possible so that the voltage divider ratio ($r_o/(r_o + r_i)$) is close to unity. This result is achieved by placing the cell in media of low conductivity; air, oil, or deionized sucrose solutions have been used successfully.

2.5   Summary: A Comparison of Small and Large Cells

The principal difference in the treatment of electrically small and large cells is summarized in Figure 2.19. Electrically small cells can be represented by lumped-parameter electrical network models. The relation between membrane potential and current is dependent on time, but not on location along the cell. Electrically large cells can be represented by distributed-parameter electrical network models. The electrical variables depend upon space and time.

Exercises

2.1   The resistance of conductors is described by several different related variables: the resistivity, ρ; the resistance per unit length, r; the resistance of a unit area, R; and the total resistance, $\mathcal{R}$. Describe the meaning of each of these resistances, and give their units.

2.2   Define *electrically small cells* and *electrically large cells*.

2.3   Give a physical explanation of why the conduction velocity is larger in fibers of larger diameter if all other factors are the same.

2.4   Give a physical explanation of the meaning of Equation 2.18 without the use of equations.

2.5   I. M. Putz has obtained measurements of conduction velocity of unmyelinated nerve fibers from a variety of species, including both invertebrates and vertebrates. His purpose is to obtain fibers with a large range of fiber diameters, from less than 1 μm to 1 mm, and to test the notion that conduction velocity is proportional to the square root of fiber diameter. He finds that these measurements do not obey the square-root relation at all well. Assume that Putz's measurements are accurate. Discuss possible reasons why Putz's results might not be inconsistent with the theory developed in Section 2.4.3.

2.6   Consider the measurements of the action potential shown in Figure 2.14 in both oil and seawater.

    a. Find the conduction velocity of the action potential in oil.

    b. Find the conduction velocity of the action potential in seawater.

    c. Sketch the action potential as a function of distance along the axon in both oil and seawater.

2.7   For the measurements shown in Figure 2.15, determine the conduction velocity of the action potential in the absence of the platinum wire,

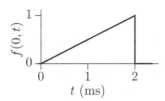

Figure 2.20 Solution of the wave equation as a function of time at one position ($z = 0$) (Exercises 2.10 and 2.11). The time function is nonzero for $0 < t < 2$ ms.

and determine a lower bound on the conduction velocity of the action potential with the platinum wire in place.

2.8 Suppose we have a nerve cell in a large bath of seawater. Does Equation 2.44 accurately describe the extracellular potential in the bath? Explain.

2.9 Equations 2.35 and 2.38 appear to imply different dependences of conduction velocity on axon radius. Equation 2.35 appears to imply that $v \propto 1/\sqrt{a}$, whereas Equation 2.38 appears to imply that $v \propto \sqrt{a}$. Resolve this dilemma, and determine which result is correct.

2.10 The function $f(z, t)$ is the solution to a wave equation. The solution is shown in Figure 2.20 as a function of time t at the position $z = 0$.

   a. Suppose that $f(z, t)$ is propagating in the positive z-direction at a propagation velocity of 100 mm/ms. Plot $f(z, t)$ versus z at time $t = 2$ ms.

   b. Suppose that $f(z, t)$ is propagating in the negative z-direction at a propagation velocity of 100 mm/ms. Plot $f(z, t)$ versus z at time $t = 2$ ms.

2.11 The function $f(z, t)$ is the solution to a wave equation. The solution is shown in Figure 2.20 as a function of time t at the position $z = 0$. Suppose that $f(z, t)$ is propagating in the positive z-direction at a propagation velocity of 10 mm/ms. What is the value of $f(z, t)$ at $t = 3$ ms and $z = 20$ mm?

2.12 An electrically large cylindrical cell has a radius of 100 μm. The internal longitudinal current is constant in time, and its spatial dependence is

$$I_i(z) = -e^{5z} \text{ for } z < 0,$$

where $I_i(z)$ has units of μA and z has units of cm. There are no external currents for $z < 0$.

a. Sketch $I_i(z)$ versus z for $z < 0$. Also make a schematic diagram of the cell, and sketch $I_i(z)$ on the cell using arrows. Let the direction of each arrow indicate the direction of the current and the length of each arrow indicate the magnitude of the current. Sketch the current every millimeter from $-0.6 < z < 0$.

b. Determine the external longitudinal current $I_o(z)$ for $z < 0$. Sketch $I_o(z)$ versus z for $z < 0$. On the same schematic diagram of the cell used in part a, sketch $I_o(z)$ on the cell using arrows to indicate the direction and magnitude of the current. Sketch the current every millimeter from $-0.6 < z < 0$.

c. Determine the current per unit length, $K_m(z)$ for $z < 0$. Sketch $K_m(z)$ versus z for $z < 0$. On the same schematic diagram of the cell used in part a, sketch $K_m(z)$ on the cell using arrows to indicate the direction and the magnitude of the current per unit length. Sketch the current per unit length every millimeter from $-0.6 < z < 0$.

2.13 An electrically large cylindrical cell of 100 μm radius has an internal longitudinal current that is constant in time with a spatial dependence of

$$I_i(z) = -e^{5z} \text{ for } z < 0,$$

where $I_i(z)$ has units of μA and z has units of cm. There are no external currents for $z < 0$.

a. Determine the longitudinal current density in the cytoplasm, $J_i(z)$.

b. Determine the current per unit length through the membrane, $K_m(z)$.

c. Determine the current density through the membrane, $J_m(z)$.

d. Determine the total current flowing through the membrane, I_m, in the segment $-1 < z < 0$.

2.14 An electrically large cylindrical cell has a radius of 100 μm. The internal longitudinal current is constant in time, and its spatial dependence is

$$I_i(z) = -e^{5z} \text{ for } z < 0,$$

where $I_i(z)$ has units of μA and z has units of cm. There are no external currents for $z < 0$. Consider a volume element of cytoplasm defined by $-1 < z < 0$.

a. Compute the longitudinal current flowing into the element at $z = -1$ and at $z = 0$.

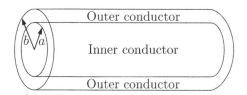

Figure 2.21 Relation between external and internal resistance per unit length (Problem 2.1).

b. Compute the total current flowing into this element through the membrane.

c. What is the relation between the results found in parts a and b?

Problems

2.1 A cylindrical cell of radius a is immersed in a cylindrical trough so that the external solution is cylindrical, with radius b as shown in Figure 2.21. The resistivities of the cytoplasm and the extracellular solutions are ρ_i and ρ_o, respectively.

a. Find the relation between b and a for the condition $r_o = r_i$.

b. Suppose that $r_o = r_i$ and $\rho_i = \rho_o$. Find the value of b for $a = 500$ and for $a = 1\ \mu m$.

c. What conclusions can you draw about the relation of r_i to r_o for unmyelinated fibers in situ? It may be helpful to examine the cross sections of nerves shown in Figure 1.29.

2.2 An action potential, $V_m(z, t)$, propagates at constant velocity along an unmyelinated axon and is recorded as a function of time at position $z = 3$ cm, as shown in Figure 2.22.

a. Assume that the action potential propagates in the positive z-direction with conduction velocity $v = 2$ cm/ms.

  i. Sketch $V_m(5, t)$, i.e., sketch the dependence of V_m on t at position $z = 5$ cm.

  ii. ii) Sketch $V_m(z, 5)$, i.e., sketch the dependence of V_m on z at time $t = 5$ ms.

b. Repeat part a for an action potential propagating in the negative z-direction with conduction velocity $v = 2$ cm/ms.

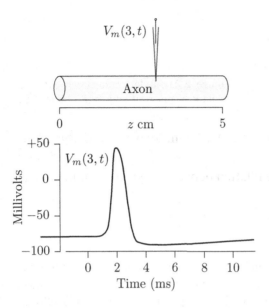

Figure 2.22 Action potential recorded from an axon (Problem 2.2).

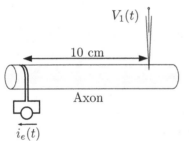

Figure 2.23 Arrangement for recording an action potential from an axon (Problem 2.3).

2.3 A squid axon (500 μm in diameter) is placed in seawater and stimulated electrically at $t = 0$ (Figure 2.23) to produce an action potential, $V_1(t)$, that is recorded at a site 10 cm from the point of stimulation, as shown in Figure 2.24. The resistivity of the axoplasm of this axon is (remarkably enough) 10π Ω-cm. The resistance of the external solution can be assumed to be negligibly small. A fine platinum wire with a resistance per unit length of 160 Ω/cm is inserted down the entire length of the axon (Figure 2.25). The wire takes up negligible space. The axon is stimulated electrically in an identical manner to produce an action potential, $V_2(t)$.

a. Find the conduction velocity, v_1, of the peak of the action potential under the conditions shown in Figure 2.23.

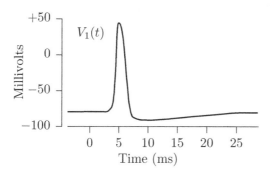

Figure 2.24 Action potential recorded as in Figure 2.23 (Problem 2.3).

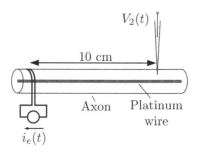

Figure 2.25 Arrangement for recording an action potential from a space-clamped axon (Problem 2.3).

    b. Find the conduction velocity, v_2, of the peak of the action potential under the conditions shown in Figure 2.25.

    c. Sketch $V_2(t)$ on the same time axis as $V_1(t)$.

    d. Write an expression for $V_2(t)$ in terms of $V_1(t)$.

2.4   Assume that an action potential is traveling at constant velocity, v, in the positive z-direction along an axon. Assume that the core conductor model is valid, so that

$$\frac{\partial^2 V_m(z,t)}{\partial z^2} = (r_i + r_o)K_m(z,t).$$

The waveform of the action potential at one point in space, $z = z_o$, is shown in Figure 2.26.

    a. Sketch $K_m(z_o, t)$ on the same time scale as $V_m(z_o, t)$.

    b. Prove that one cannot account for $K_m(z_o, t)$ by assuming that the membrane can be represented by the equivalent circuit for an incremental element of length Δz as shown in Figure 2.27 (g_m and c_m are constant conductances and capacitances per unit length). Consider the polarity of the current through the parallel combination of g_m and

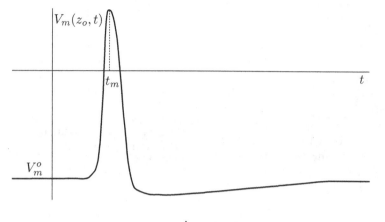

Figure 2.26 Waveform of a propagated action potential (Problem 2.4).

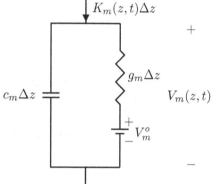

Figure 2.27 Equivalent network for the membrane of an axon (Problem 2.4).

c_m prior to the time of occurrence of the peak of the action potential, t_m.

2.5 I. M. Fumbler expertly dissects an invertebrate unmyelinated axon and places it in a recording chamber filled with oil. He mounts the stimulating electrodes on a micromanipulator so that he can stimulate the axon at different locations along its length. The location of the active stimulating electrode is indicated by a dark band in the schematic diagram shown in Figure 2.28. The reference electrode is located at a remote location on the axon that has been crushed to prevent the occurrence of an action potential there. Fumbler uses two bipolar recording electrodes, each of which consists of a pair of wires separated by a distance that is much smaller than the space constant of the axon. He measures the two extracellular voltage differences, $V_1(t)$ and $V_2(t)$, as shown in Figure 2.28. Fumbler stimulates the nerve with a brief superthreshold pulse

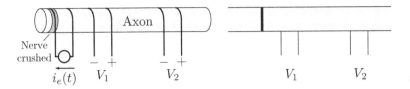

Figure 2.28 Schematic diagrams of recording arrangements (Problem 2.5). The left-hand panel shows the arrangement of stimulating and recording electrodes. The reference electrode for the current stimulus is placed on a crushed portion of the nerve; the active electrode is the cathode. The right-hand panel shows a schematic diagram of this arrangement indicating the positions of the active current electrode (dark band) and the two potential recording electrodes.

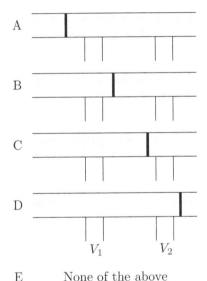

E   None of the above

Figure 2.29 Definitions of recording arrangements (Problem 2.5).

of current and measures $V_1(t)$ and $V_2(t)$ for various locations of the stimulating electrode. Unfortunately, after he obtains unlabeled printed copies of his electrode locations (Figure 2.29) and measurements (Figure 2.30), he clumsily drops the stack of papers on the floor. He picks up the papers and associates the voltage records into pairs, but is confused about which stimulating electrode locations correspond to which records. Your job is to help poor Fumbler assign the pairs of voltage records he has associated with the presumed locations of the stimulating electrodes. For each of the eight pairs of records shown in Figure 2.30, indicate which configuration of stimulating and recording electrodes (A through E in Figure 2.29) best fits each of the recorded potentials.

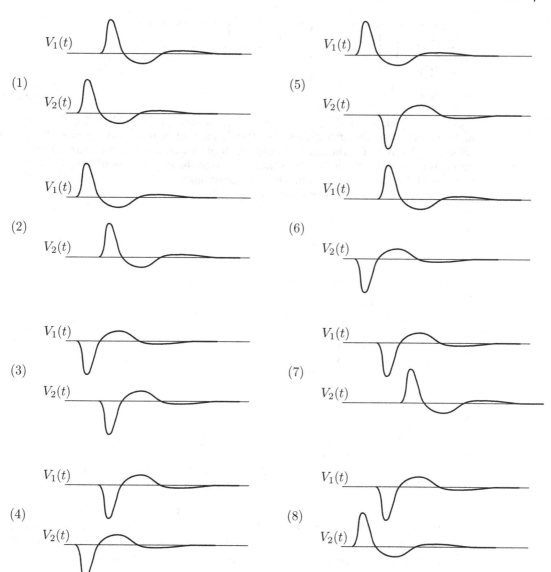

Figure 2.30   I. M. Fumbler's data (Problem 2.5).

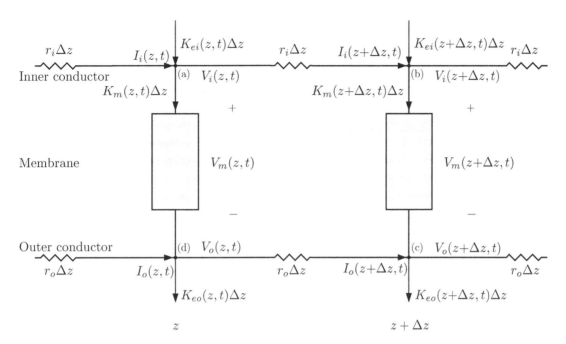

Figure 2.31 Equivalent network model of a core conductor with external currents applied externally and internally (Problem 2.6).

2.6 This problem concerns a generalization of the core conductor model, in which external currents can be applied both inside and outside of a cylindrical cell. Consider a core conductor model with external currents per unit length both inside, $K_{ei}(z, t)$, and outside, $K_{eo}(z, t)$, as shown in the incremental equivalent circuit shown in Figure 2.31, where

r_o = the resistance per unit length of the outer conductor,

r_i = the resistance per unit length of the inner conductor,

$I_o(z, t)$ = the external longitudinal current,

$I_i(z, t)$ = the internal longitudinal current,

$K_m(z, t)$ = the membrane current per unit length,

$K_{ei}(z, t)$ = is the external current per unit length applied to the inner conductor,

$K_{eo}(z, t)$ = is the external current per unit length applied to the outer conductor,

$V_m(z, t)$ = is the membrane potential.

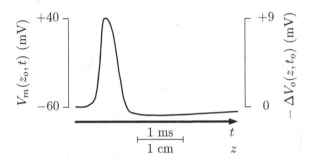

Figure 2.32 Propagated action potential (Problem 2.7).

a. Derive the core conductor equations, which are as follows:

$$\frac{\partial I_i(z, t)}{\partial z} = K_{ei}(z, t) - K_m(z, t),$$

$$\frac{\partial I_o(z, t)}{\partial z} = K_m(z, t) - K_{eo}(z, t),$$

$$\frac{\partial V_i(z, t)}{\partial z} = -r_i I_i(z, t),$$

$$\frac{\partial V_o(z, t)}{\partial z} = -r_o I_o(z, t),$$

$$V_m(z, t) = V_i(z, t) - V_o(z, t),$$

$$\frac{\partial V_m(z, t)}{\partial z} = -r_i I_i(z, t) + r_o I_o(z, t).$$

b. Combine the above equations to give the following relation between membrane potential and the current variables:

$$\frac{\partial^2 V_m(z, t)}{\partial z^2} = (r_o + r_i)K_m(z, t) - r_o K_{eo}(z, t) - r_i K_{ei}(z, t).$$

2.7 An unmyelinated axon is placed in a narrow trough of seawater and stimulated at one end to elicit a propagating action potential. Figure 2.32 shows two traces on two different scales chosen so that the two traces superimpose perfectly and only one curve can be distinguished. One trace is the membrane potential waveform plotted versus time, t, at one position, z_o, i.e. $V_m(z_o, t)$. The other trace is the external potential (note sign) at a particular time, t_o, plotted as a function of position, z, i.e., $-\Delta V_o(z, t_o)$, where $\Delta V_o(z, t_o) = V_o(z, t_o) - V_o(-\infty, t_o)$.

a. Is the action potential propagating in the positive or the negative z-direction? Explain your reasoning.

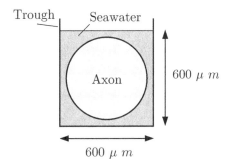

Trough Seawater

Axon

600 μ m

600 μ m

Figure 2.33 Cross section of an axon in a trough of seawater (Problem 2.9).

b. What is the propagation velocity of the action potential? Explain your reasoning.

c. What is the ratio of external to internal longitudinal resistance per unit length, i.e., r_o/r_i? Explain your reasoning.

2.8 The following two experiments are performed on a squid giant axon:

■ *Experiment 1*: The axon is placed in a large volume of seawater, and the size of the transmembrane action potential is measured by means of an intracellular micropipette and found to have a peak-to-peak value of 100 mV. The conduction velocity is 36 m/s.

■ *Experiment 2*: The axon is placed in oil, and the transmembrane potential is still found to be 100 mV peak to peak. The peak-to-peak size of the extracellular action potential is 75 mV.

Estimate the expected conduction velocity in Experiment 2. State your assumptions.

2.9 A squid axon of 500 μm diameter is placed in an insulated trough filled with seawater (a cross section through the trough is shown in Figure 2.33). Assume that the resistivity of seawater and axoplasm are both 25 Ω-cm. The peak-to-peak amplitude of the action potential measured across the membrane is 100 mV. What is the peak-to-peak amplitude of the action potential measured with an extracellular electrode (the reference electrode is placed on a remote inactive end of the axon). In this calculation, assume that the core conductor model is valid.

2.10 An extracellular action potential is recorded between two electrodes in response to an extracellular pulse of current as shown in Figure 2.34. The two voltage electrodes are separated by 5 mm. The longitudinal

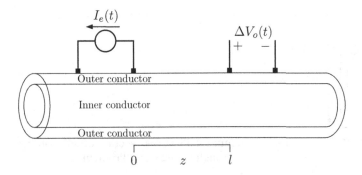

Figure 2.34 Recording of a diphasic action potential (Problem 2.10).

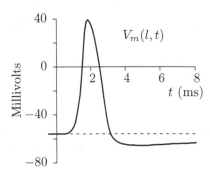

Figure 2.35 Potential across the membrane (Problem 2.10).

resistances are such that $r_i = 4r_o$. The membrane potential at location $z = l$ is shown in Figure 2.35.

a. For a conduction velocity of $v = 1$ m/s, sketch $\Delta V_o(t)$ as a function of time.

b. For a conduction velocity of $v = 10$ m/s, sketch $\Delta V_o(t)$ as a function of time.

2.11 Figure 2.36 shows the results of an experiment in which the extracellular potential is recorded from a squid giant axon in response to a suprathreshold current pulse that elicits an action potential under several different conditions. This response is recorded from an intact axon, then after the axon's cytoplasm has been extruded, and finally with the axon reinflated by perfusion. You may assume that the action potential recorded across the membrane had a peak-to-peak amplitude of 100 mV.

a. Based on the amplitude of the extracellular action potential, find r_o/r_i for the intact, extruded, and inflated axon.

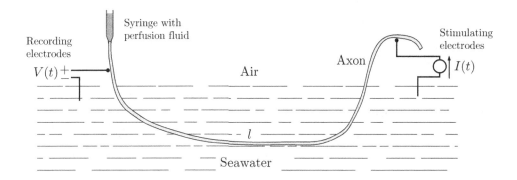

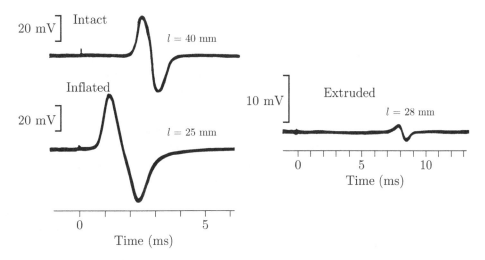

Figure 2.36 Extracellular action potential recorded from intact, extruded, and inflated squid axon (Problem 2.11). A schematic diagram of the stimulating and recording apparatus is shown above (adapted from Baker et al., 1962, Figure 3). The squid (*Loligo pealii*) giant axon is immersed in seawater for a length l and cannulated at one end to allow intracellular perfusion of the axon. Stimulating electrodes are attached at one end, and the extracellular potential is recorded at the other end of the axon. One voltage and one current electrode is immersed in the seawater bath. The measured responses are shown below for three conditions of the axon (adapted from Baker et al., 1962, Figure 6): intact after axoplasm had been extruded by the method illustrated in Figure 1.31, and after the extruded axon had been inflated by perfusion with isotonic K_2SO_4. The time of occurrence of the current stimulus is indicated by a small stimulus artifact at the beginning of each trace.

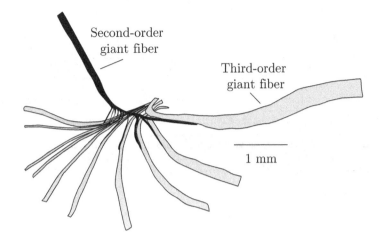

Second-order
giant fiber

Third-order
giant fiber

1 mm

Figure 2.37 Schematic diagram of the stellate ganglion of the squid (adapted from Young, 1939, Figure 11) (Problem 2.12). A second-order giant nerve fiber makes synaptic contact with 9 to 11 third-order giant fibers that innervate the mantle muscle.

b. Assume that extrusion and reinflation changes r_i only. Based on the results obtained in part a, what do you predict will be the change in conduction velocity on extrusion of axoplasm and on reinflation?

c. Based on the measurements, estimate the conduction velocity of the action potential in the intact, extruded, and inflated axon. Describe your estimation procedure. Compare the estimated values with the predicted values obtained in part b. Discuss possible explanations for any differences you obtain.

2.12 As indicated in Section 1.5, the squid giant axon is part of a system of giant fibers that control the mantle muscle of the squid, which is used for locomotion. As shown in Figure 2.37, a second-order giant fiber makes synaptic contact with nine to eleven third-order giant fibers in the stellate ganglion. These third-order giant fibers innervate the mantle muscle. The giant fibers have different lengths and diameters; the giant fibers innervating the most peripheral portions of the muscle are longer and have larger diameters. Measurements of the axon lengths and diameters are given in Table 2.1. Since larger-diameter fibers conduct action potentials more rapidly, it has been hypothesized that the arrangement of innervation of mantle muscles by giant fibers allows better synchronization of muscle contraction because conduction is more rapid in axons innervating more distant portions of the muscle. In this problem you will evaluate this suggestion quantitatively. For the purpose of this problem, assume that the giant fiber (the one with the largest diameter) conducts action potentials at 20 m/s and that all the giant fibers are identical except for their lengths and diameters.

Table 2.1 Measurements of the lengths and diameters of giant axons in the stellate ganglion of a squid (*Loligo pealii*) (adapted from Young, 1939, Table 1) (Problem 2.12).

Axon No.	Length mm	Diameter μm
1	23	131
2	22	90
3	22	86
4	24	148
5	43	227
6	48	291
7	76	291
8	91	317
9	142	447

    a. Compute the conduction time for each of the nine axons.

    b. Compute the conduction time for each of the nine axons assuming that they all have the same diameter as the smallest axon.

    c. Compute the conduction time for each of the nine axons assuming that they all have the same diameter as the giant axon.

    d. What do you conclude about the hypothesis?

2.13 An electrically large cylindrical cell is stimulated electrically by an extracellular electrode that imposes a potential $V_o(z, t)$ on the outside of the cell.

    a. Show that the relation between $V_m(z, t)$, $K_m(z, t)$, and $V_o(z, t)$ is given by

$$\frac{\partial^2 V_m(z, t)}{\partial z^2} = r_i K_m(z, t) - \frac{\partial^2 V_o(z, t)}{\partial z^2}.$$

    b. The extracellular potential acts on the membrane potential as an equivalent membrane current per unit length. Find this membrane current per unit length.

    c. It is proposed that this cell be stimulated by means of an extracellular potential that has a sinusoidal time function and that is constant in z, i.e., $V_o(z, t) = \sin 2\pi f_o t$. Determine the response of the cell to this stimulus.

d. Devise a stimulus $V_o(z, t)$ that is also a sinusoidal time function, but that is more effective in stimulating the cell than is the stimulus in part c.

References

Books and Reviews

Aidley, D. J. (1989). *The Physiology of Excitable Cells*. Cambridge University Press, Cambridge, England.

Bose, A. G. and Stevens, K. N. (1965). *Introductory Network Theory*. Harper & Row, New York.

Bullock, T. H. and Horridge, G. A. (1965). *Structure and Function in the Nervous Systems of Invertebrates*. W. H. Freeman, San Francisco.

Cole, K. S. (1968). *Membranes, Ions, and Impulses*. University of California Press, Berkeley, CA.

Hodgkin, A. L. (1964). *The Conduction of the Nervous Impulse*. Charles C. Thomas, Springfield, MA.

Original Articles

Baker, P. F., Hodgkin, A. L., and Shaw, T. I. (1962). Replacement of the protoplasm of giant nerve fibres with artificial solutions. *J. Physiol.*, 164:330–354.

Burrows, T. M. O., Campbell, I. A., Howe, E. J., and Young, J. Z. (1965). Conduction velocity and diameter of nerve fibres of cephalopods. *J. Physiol.*, 179:39–40.

Clark, J. and Plonsey, R. (1966). A mathematical evaluation of the core conductor model. *Biophys. J.*, 6:95–112.

Del Castillo, J. and Moore, J. W. (1959). On increasing the velocity of a nerve impulse. *J. Physiol.*, 148:665–670.

Eckert, R. and Naitoh, Y. (1970). Passive electrical properties of *Paramecium* and problems of ciliary coordination. *J. Gen. Physiol.*, 55:467–483.

Fatt, P. and Katz, B. (1951). An analysis of the end-plate potential recorded with an intracellular electrode. *J. Physiol.*, 115:320–370.

Goldman, L. (1964). The effects of stretch on cable and spike parameters of single nerve fibres: Some implications for the theory of impulse propagation. *J. Physiol.*, 175:425–444.

Hodgkin, A. L. (1939). The relation between conduction velocity and the electrical resistance outside a nerve fibre. *J. Physiol.*, 94:560–570.

Pumphrey, R. J. and Young, J. Z. (1938). The rates of conduction of nerve fibres of various diameters in cephalopods. *J. Exp. Biol.*, 15:453–466.

Rosenfalck, P. (1969). *Intra- and extracellular potential fields of active nerve and muscle fibres*. Akademisk Forlag, Copenhagen, Denmark.

Young, J. Z. (1939). Fused neurons and synaptic contacts in the giant nerve fibres of cephalopods. *Philos. Trans. R. Soc. London, Ser. B*, 229:465–503.

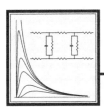

Linear Electrical Properties of Cells

An application of the theory of the transmission of electricity along a submarine telegraph-wire, shows how the question recently raised as to the practicability of sending distinct signals along such a length as the 2000 or 3000 miles of wire that would be required for America, may be answered. The general investigation will show exactly how much the sharpness of the signals will be worn down and will show what maximum strength of current through the apparatus, in America, would be produced by a specified battery action on the end in England, with wire of given dimensions, &c.
—Thomson, 1855

3.1 Introduction

As indicated in Chapter 1, electrically inexcitable cells produce graded electric potentials, and electrically excitable cells, such as neurons and muscle cells, produce both graded and action potentials. In this chapter, we shall examine graded electric potentials for both small and large cells. In large cells, we shall also examine the decremental propagation of graded potentials along the cell.

3.2 Electrical Properties of Cellular Membranes

3.2.1 Linearity of Voltage-Current Characteristics for Small Perturbations of Membrane Potential

Even in electrically excitable cells, whose membranes have highly nonlinear electrical properties for large membrane potential changes, the membrane voltage-current characteristic is linear for small perturbations of the membrane potential from its resting value. An example of measurements of mem-

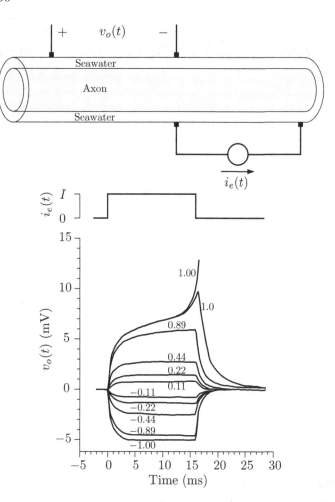

Figure 3.1 Measurements of extracellular potential of a crab (*Carcinus maenas*) nerve fiber in response to rectangular pulses of current of different amplitudes (adapted from Hodgkin and Rushton, 1946, Figure 5). The upper panel shows a schematic diagram of the recording arrangement. The axon is in oil (not shown), but a thin layer of seawater clings to the axon. The electrodes are in contact with this seawater. The middle panel shows the waveform of the current pulse, and the lower panel shows the potentials recorded for different amplitudes of the current pulse. The parameter is I/I_{th}, where I is the amplitude of the current pulse and I_{th} is the value of the current at the threshold for eliciting an action potential. For $I/I_{th} = 1.00$, an action potential is elicited but only its rising phase is shown.

brane electrical characteristics of a crab axon is shown in Figure 3.1. A rectangular pulse of current of amplitude I is delivered to an axon through thin surface electrodes, and the potential is recorded from the axon's surface. With the axon in oil, the recorded extracellular potential is proportional to the potential across the membrane (see Equation 2.44). Shown in Figure 3.1 is the external potential measured for various values of current amplitude expressed as a fraction of the current amplitude, I_{th}, required to just elicit an action potential from the axon. For $-1.00 \le I/I_{th} \le +0.44$, the amplitude of the external potential is proportional to the amplitude of the current applied. Doubling the current amplitude doubles the potential amplitude; reversing the current polarity reverses the polarity of the potential. The waveshape of the potential, except for multiplication by a scale factor, is the same independent of the stimulus. Hence, in this range the relation between the measured potential and the current is linear. It is this range of potential change with which we are concerned in this chapter. Experiments such as those shown in Figure 3.1 have been performed on numerous cell types. Linear behavior is invariably found for a range of membrane potentials, but this range varies from cell type to cell type.

3.2.2   Voltage-Current Characteristics of the Membrane for Small Perturbations: Membrane Conductance and Capacitance

The measurements shown in Figure 3.1 also show that the voltage change lags behind the change in current, i.e., a discontinuous current gives rise to a continuous membrane potential. Thus, it takes time for the membrane potential to charge to its final value. One simple interpretation of this measurement is that the cell membrane acts as a parallel capacitance that prevents a discontinuous membrane potential in response to a discontinuous membrane current.

3.2.2.1   *Historical Note*

At the beginning of the twentieth century, it was already known that cellular membranes had a high electrical impedance to the flow of alternating current and that this impedance decreased as the frequency of the current increased. These observations were based on rather indirect evidence obtained by suspending live cells, such as erythrocytes, in solutions of low conductivity, such as sucrose, and measuring the impedance of the suspension as a function of frequency (as shown schematically in the left panel of Figure 3.2). The electrical impedance was obtained from such measurements as follows. In the sinusoidal steady state, the voltage across and the current through the suspension can be expressed as $v_s(t) = \Re\{V_s \exp(j2\pi f t)\}$ and $i_s(t) = \Re\{I_s \exp(j2\pi f t)\}$, respectively, where $\Re\{a\}$ is the real part of the complex number a, f is the frequency, and V_s and I_s are complex amplitudes. The electrical impedance of the suspension is defined as

$$Z_s = \frac{V_s}{I_s}. \tag{3.1}$$

Results of the experiments of Höber (1910) (Cole, 1968) showed that for $f = 1$ kHz the magnitude of the impedance was high. At $f = 1$ MHz, the magnitude of the impedance was lower and equaled that obtained in similar ex-

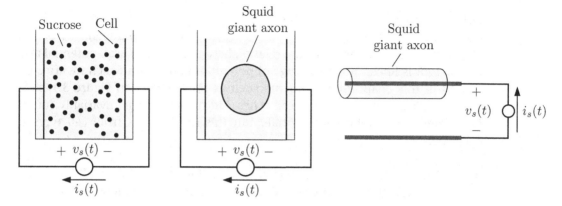

Figure 3.2 Schematic diagrams of methods of measuring membrane impedance of cells. The left panel shows a suspension of cells in a sucrose solution between two electrodes; the center panel shows a section through a trough containing a squid giant axon between extracellular electrodes; the right panel shows a squid giant axon with an intracellular electrode. In each case, a sinusoidal current is applied to the electrodes and the potential is measured, or vice versa. To minimize ambiguities caused by potentials across metal-solution interfaces in the presence of electric currents, it is usual to use two pairs of electrodes (rather than one pair as shown in these schematic diagrams)—one pair to deliver the current, and the other pair to measure the potential.

periments performed on erythrocytes whose membranes had been dissolved chemically. The latter value approached that obtained when the erythrocytes in the suspension were replaced by an equal quantity of physiological saline. The implications are that for normal living cells the membrane impedance is high at low frequencies and low at high frequencies. At high frequencies, the membrane impedance is so low that the cell's impedance is due just to its cytoplasm, which has a resistivity that is close to that of physiological saline. These results suggested that the electrical characteristics of the cell membrane for sinusoidal current could be represented by a parallel conductance and capacitance (see Figure 3.3).

By 1923, the capacitance of cellular membranes had been estimated for several cell types by Fricke (Cole, 1968) using measurements on cell suspensions together with a theory that related the measured impedance of the suspension, Z_s, to the membrane impedance of each cell, Z_m. It was found that the specific capacitance of the membranes (capacitance per cm^2 of membrane surface area) of a variety of cells was about $1\mu F/cm^2$. Fricke used these measurements to infer the thickness of the membrane. Fricke as-

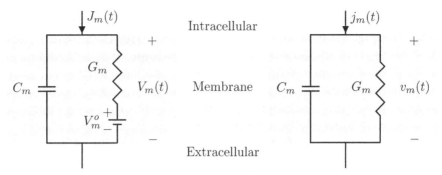

Figure 3.3 Network model of the relation of membrane current density to membrane potential of a unit area of cellular membrane. G_m and C_m are the conductance and capacitance of a square centimeter of membrane in units of S/cm^2 and μF/cm^2, respectively. The left panel shows an equivalent network in terms of the total variables $J_m(t)$ and $V_m(t)$. The right panel shows an equivalent network in terms of incremental variables $j_m(t)$ and $v_m(t)$. The total quantities are related to the incremental quantities as follows: $J_m(t) = J_m^o + j_m(t)$, and $V_m(t) = V_m^o + v_m(t)$, where J_m^o and V_m^o are the resting values of the membrane current density and the membrane potential, respectively.

sumed that the capacitance of a unit area of membrane could be represented as follows:

$$C_m = \frac{\kappa_m \epsilon_o}{d},\tag{3.2}$$

where κ_m is the dielectric constant of the membrane material, assumed to have a value of 3 (characteristic of oils, since it was known that membranes contained large amounts of lipid); ϵ_o is the permittivity of free space; and d is the membrane thickness. Solving for d, Fricke obtained

$$d = \frac{\kappa_m \epsilon_o}{C_m} = \frac{3 \times 10^{-13}}{10^{-6}} = 3 \times 10^{-7} \mathrm{cm} = 30\text{Å}.\tag{3.3}$$

This estimate was the first indication that cellular membranes were of molecular dimensions, and it was the first estimate based on physicochemical measurements. This thickness is within a factor of 2.5 of the value obtained from electron microscope images and from X-ray diffraction studies.

The early methods based on cell suspensions were not sensitive enough to measure the membrane conductance. Accurate measurements of the membrane conductance began to become available when membrane biophysicists focused on large invertebrate nerve fibers. The first accurate measurements of membrane conductance were obtained with extracellular electrodes in the squid giant axon (Cole and Hodgkin, 1939) by a method explored in Problem 3.6. Measurements of membrane impedance obtained on single cells (see middle panel of Figure 3.2) avoided the ambiguities of measurements based on cell populations. However, with extracellular recording the relation between the measured impedance and the membrane impedance is complex. With intracellular recording, this last difficulty was mitigated (see right panel of Figure 3.2). Subsequently, the membrane impedance of many cell types has been obtained by passing current through intracellular electrodes and measuring the membrane potential with methods similar to those used to record the resting membrane potential (Weiss, 1996, Figure 7.2 and Section 7.1).

3.2.2.2 *Survey of Membrane Impedance*

Subsequent to these early measurements, the membrane impedances of a variety of cell types had been measured more directly. To a first-order approximation, the linear electrical properties of many cellular membranes can be represented by a parallel conductance and capacitance network. The conductance represents ionic conduction through the membrane predominantly via gated ion channels (Chapter 6), and the capacitance represents the insulating properties of the lipid bilayer. We shall use the symbol G_m to represent the conductance of a square centimeter of membrane (S/cm^2) and C_m to represent the capacitance of a square centimeter of membrane (F/cm^2). G_m is typically in the range of 0.01–1 mS/cm^2, with a "typical" cell having a G_m of about 1 mS/cm^2. Similarly, C_m is about 1 μF/cm^2 for most cells, although muscle cells have larger membrane capacitances (see Table 3.1).

3.3 Electrically Small Cells

A small, uniform, isolated cell has a simple electrical equivalent network as shown in Figure 3.4. Since the cell has uniform electrical characteristics, the total capacitance of the cell is simply $C = C_m A$, and the total conductance is $G = G_m A$, where C_m and G_m are the specific capacitance and conductance and A is the surface area of the cell. The relation between current and membrane potential is derived by writing Kirchhoff's current law at the node inside the

Table 3.1 Electrical parameters of a variety of cell types (Florey, 1966; Katz, 1966; Rall, 1977). Here d is the diameter of the cell; C_m and G_m are the specific membrane capacitance and conductance; r_i is the internal resistance of the cell cytoplasm per unit length; λ_C is the cell space constant; and τ_M is the membrane time constant. The space constant is computed under the assumption that $r_i \gg r_o$.

Species	d (μm)	C_m (μF/cm^2)	G_m (mS/cm^2)	r_i MΩ/cm	λ_C (mm)	τ_M (ms)
Invertebrate giant axons						
Squid	500	1	1	0.015	6.5	1
Loligo pealii						
Lobster	75	1	0.5	1.4	2.5	2
Homarus vulgaris						
Crab	30	1	0.14	13	2.3	7
Carcinus maenas						
Lobster	100	—	0.13	1	5.1	—
Homarus americanus						
Earthworm	105	0.3	0.083	2.3	4	3.6
Lumbricus terrestris						
Marine worm	560	0.75	0.012	0.023	5.4	0.9
Myxicola infundibulum						
Nerve cell bodies						
Cat motoneuron	—	2	0.4	—	—	5
Muscle fibers						
Frog sartorius	75	2.5-6	0.25	4.5	2	10-24

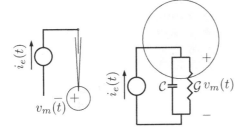

Figure 3.4 The left panel shows the arrangement for recording the membrane potential in response to a current through the membrane of a small cell. The right panel shows an equivalent network superimposed on the schematic diagram of the cell. The series resistance between the electrodes and the membrane has not been included.

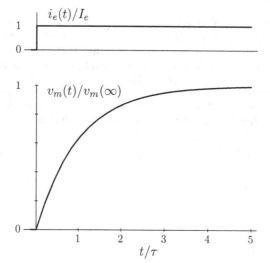

Figure 3.5 Step response of a small cell.

cell. The total current is the current through the capacitance and the current through the conductance

$$i_e(t) = \mathcal{C}\frac{dv_m(t)}{dt} + \mathcal{G}v_m(t).$$ (3.4)

Suppose the external current is $i_e(t) = I_e u(t)$. Since Equation 3.4 is a first-order ordinary differential equation with constant coefficients, the solution has the form of an exponential:

$$v_m(t) = v_m(\infty) + (v_m(0) - v_m(\infty))\, e^{-t/\tau_c} \text{ for } t \geq 0.$$

If the cell is initially at rest, then $v_m(0) = 0$ and we need only find $v_m(\infty)$ and τ_c. The final value of $v_m(t)$ is $v_m(\infty)$. Therefore, the final value of the derivative of $v_m(t)$ is zero. Substitution of a zero derivative into Equation 3.4 and the value of $i_e(t)$ yields $v_m(\infty) = I_e/\mathcal{G} = I_e\mathcal{R}$. The time constant can be obtained either by substituting the solution into the differential equation or by noting the units of the terms in the differential equation. Therefore, $\tau_c = \mathcal{C}/\mathcal{G} = (C_m A)/G_m A) = C_m/G_m = \tau_M$. Thus, the cell time constant is independent of the dimensions of the cell and only a function of the membrane electrical characteristics. Therefore, it is called the *membrane time constant*.

Inserting the constants into the general solution yields

$$v_m(t) = \mathcal{R}I_e \left(1 - e^{-t/\tau_c}\right) \text{ for } t \geq 0,$$ (3.5)

which is shown in Figure 3.5.

Thus, we see that for a small cell the membrane potential lags behind the change in current by first-order kinetics characterized by one parameter—the membrane time constant. This result is qualitatively consistent with the results seen in Figure 3.1, but it is not quantitatively consistent with those results because the crab axon, whose responses are shown in the figure, cannot be regarded as an electrically small cell. To understand the results in the figure, we will need to investigate the kinetics for electrically large cells.

3.4 Electrically Large Cells: The Cable Model

For large cells, the membrane potential varies with position along the cell. Thus, the starting point for our discussion of electrically large cells is the core conductor model, which was derived in Chapter 2 without making any explicit assumptions about the voltage-current characteristic of cellular membranes. The geometry of the cell and the properties of the media bathing the membrane led to the core conductor equations (Equations 2.18 to 2.22), which relate voltages and currents along the core conductor. We shall now combine the core conductor equations with the equations that characterize the membrane by a parallel conductance and capacitance. The resulting model is called the *cable model*.

3.4.1 Derivation of the Cable Equation

We can represent an incremental length of a cylindrical cell as indicated in Figure 3.6, where g_m is the membrane conductance per unit length (S/cm), c_m is the membrane capacitance per unit length (F/cm), $V_m(z, t) = V_m^o + v_m(z, t)$, and $K_m(z, t) = K_m^o + k_m(z, t)$. Since $G_m = 1/R_m$ is the conductance of a square centimeter of membrane (S/cm^2), $g_m = 2\pi a G_m$. Similarly, $c_m = 2\pi a C_m$, where a is the radius of the axon. $V_m(z, t)$ is the total membrane potential and is equal to its resting or quiescent value, V_m^o, plus an incremental change in membrane potential, $v_m(z, t)$. The notation for all variables will be as follows: the total quantities are indicated in upper-case letters, the resting or quiescent values are in upper-case letters with superscript o, and incremental quantities are in lower-case letters.

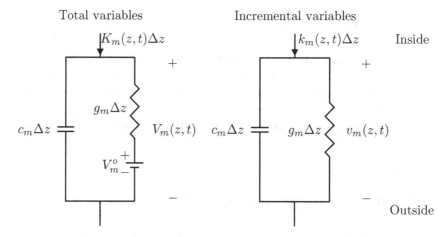

Figure 3.6 Model of the voltage-current characteristics of an incremental length of axonal membrane. The left-hand panel shows the voltage-current characteristics in terms of the total variables; the right-hand panel shows these characteristics in terms of incremental variables.

By applying Kirchhoff's current law to the incremental equivalent circuit shown in Figure 3.6, we obtain

$$k_m(z, t) = g_m v_m(z, t) + c_m \frac{\partial v_m(z, t)}{\partial t}, \tag{3.6}$$

which expresses the incremental membrane current as the sum of a capacitance current and a conductance current (carried by ions). The core conductor equation is

$$\frac{\partial^2 V_m(z, t)}{\partial z^2} = (r_i + r_o) K_m(z, t) - r_o K_e(z, t), \tag{3.7}$$

which can be expressed in terms of quiescent and incremental variables as

$$\frac{\partial^2 \left(V_m^o + v_m(z, t) \right)}{\partial z^2} = (r_i + r_o) \left(K_m^o + k_m(z, t) \right) - r_o \left(K_e^o + k_e(z, t) \right). \tag{3.8}$$

But for quiescent conditions we set the incremental variables in Equation 3.8 to zero, which yields

$$0 = (r_i + r_o) K_m^o - r_o K_e^o. \tag{3.9}$$

Subtracting Equation 3.9 from 3.8 yields

$$\frac{\partial^2 v_m(z, t)}{\partial z^2} = (r_i + r_o) k_m(z, t) - r_o k_e(z, t), \tag{3.10}$$

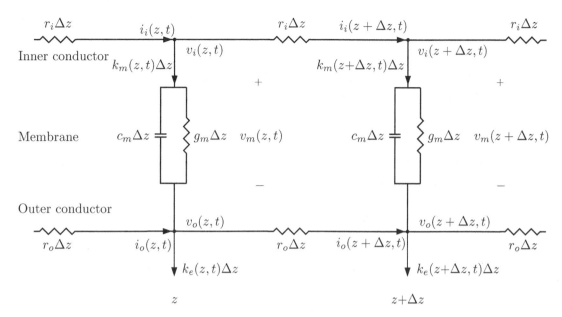

Figure 3.7 Cable model of an incremental length of a cylindrical cell.

which demonstrates that the incremental variables also obey the core conductor equation.

The so-called *cable equation* is derived by substituting Equation 3.6 into 3.10, thus eliminating the variable $k_m(z,t)$, to yield

$$\frac{\partial^2 v_m(z,t)}{\partial z^2} = (r_i + r_o)\, g_m v_m(z,t)$$

$$+ (r_i + r_o)\, c_m \frac{\partial v_m(z,t)}{\partial t} - r_o k_e(z,t). \tag{3.11}$$

This equation has been obtained from the core conductor model of incremental quantities of a membrane represented by the equivalent network of Figure 3.6. Alternatively, the cable equation could be obtained directly by combining the core conductor model of incremental quantities with a network model of the membrane, as shown in Figure 3.7, and writing the equilibrium equations for this network.

It is useful to define two constants: the time constant,

$$\tau_M = \frac{c_m}{g_m}, \tag{3.12}$$

and the space constant,

$$\lambda_C = \frac{1}{\sqrt{(r_i + r_o)g_m}}. \tag{3.13}$$

We can express the time constant in terms of specific membrane capacitance and conductance as follows: $\tau_M = (2\pi a C_m)/(2\pi a G_m) = C_m/G_m$. Thus, the time constant of a cylindrical cell is the same as the time constant of a unit area of membrane, and it has the dimensions of seconds. That is, the time constant is independent of the dimensions of the cell and is the same for small and large cells comprised of the same membrane material. Equation 3.13 shows that the space constant has units of length, since r_o and r_i have units of resistance per unit length and g_m has units of conductance per unit length. However, in contrast to the time constant, which is independent of the dimensions of the cell, the space constant does depend upon the cell dimensions. For simplicity, let us assume that $r_o \ll r_i$. Since $r_i = \rho_i/\pi a^2$ and $g_m = G_m 2\pi a$, we obtain

$$\lambda_C = \frac{1}{\sqrt{(\rho_i/(\pi a^2))(G_m 2\pi a)}} = \sqrt{\frac{a}{2\rho_i G_m}}, \tag{3.14}$$

which shows that λ_C decreases as a decreases. Whereas the time constant is a property of the membrane, the space constant is a property of the cell. Thus, we refer to them as the *membrane time constant* and the *cell space constant*. The time constant and the space constant are the characteristic time and the characteristic length of a linear cable. Therefore, it is useful to express the cable equation in terms of these two constants as follows:

$$\lambda_C^2 \frac{\partial^2 v_m(z,t)}{\partial z^2} = v_m(z,t) + \tau_M \frac{\partial v_m(z,t)}{\partial t} - \lambda_C^2 r_o k_e(z,t). \tag{3.15}$$

Equation 3.15 is called the cable equation because it was first solved by William Thomson, also known as Lord Kelvin, in connection with the laying of the Atlantic submarine cable to be used for intercontinental telegraphy (Thomson, 1855). This equation describes the potential along a leaky submarine cable (Figure 3.8), where r_i represents the resistance/length of the (inside) copper conductor, g_m and c_m represent the leaky insulation of the cable, and r_o is the resistance per unit length of the outside seawater. For a cylindrical cell, r_i represents the resistance/length of the cytoplasm, g_m represents conduction through membrane ionic channels, c_m represents the insulation of lipid bilayer, and r_o is the resistance per unit length of the external interstitial fluid.

In connection with submarine cables, it was of interest to know how much attenuation and delay would occur if a pulse of potential were applied at one

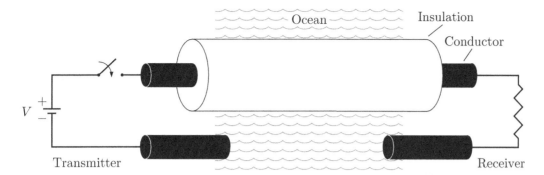

Figure 3.8 Model of a submarine cable for telegraphy. Depression of the switch in the transmitter charges the cable and causes currents and voltages in the receiver.

end of the cable and received at the other end (Figure 3.8). We are interested in essentially the same question with respect to transmission of small potential changes along an axon or muscle. Is a nerve fiber or a muscle fiber a good electrical cable? If a small potential change occurs in a nerve fiber innervating the big toe, how large a potential change would we expect to find at the cell body of this neuron (located in the spinal cord) if the fiber acted as a simple linear cable? We shall find that for short distances, comparable to the dimensions of short dendritic processes of some cells, the mode of transmission given by the cable model is adequate to produce appreciable electrophysiological effects. However, for long cells, such as those found in the peripheral nervous system, this mode of transmission is inadequate.

3.4.2 Time-Independent Solutions

We can gain some understanding of cable properties of cells by examining solutions of the cable equation for several simple cases, each involving a constant current applied to a point along a cable. In these cases, all variables in Equation 3.15 are time independent and the equation becomes an ordinary differential equation.

3.4.2.1 Solution for an Infinite Cable in Response to a Constant External Current

We first shall consider an infinite cable, which is a plausible model for large cells (or portions of cells) such as axons or muscle fibers. An external current source is assumed to deliver a constant current, I_e, to the cell, which is represented as a linear cable, at $z = 0$ via an infinitesimal electrode (Fig-

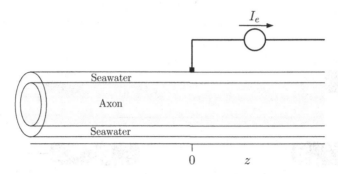

Figure 3.9 External constant current applied at a point along a cylindrical cell. One terminal of the current source is connected to the active electrode at $z = 0$; the other terminal is shown with a horizontal line to the right, denoting that this terminal is connected to an electrode located at a remote location on the axon, so remote that electrical responses at the remote location do not affect the responses near the active electrode.

ure 3.9). The other (reference) electrode is located remotely. After we solve the cable equation, we shall be able to specify more precisely what is meant by an infinitesimal electrode and by a remote reference electrode. We shall also be able to specify more precisely what is meant by an electrically small cell and what is meant by an electrically large cell.

We can represent the external current source as a spatial impulse (or Dirac delta function) of current per unit length, i.e.,

$$k_e(t) = I_e \delta(z), \quad \text{where } \delta(z) = \frac{d}{dz} u(z). \tag{3.16}$$

The derivative in Equation 3.16 is interpreted in the distributional sense (Lighthill, 1960). Since the variables are assumed to be time independent, we can rewrite Equation 3.15 as

$$\lambda_C^2 \frac{d^2 v_m(z)}{dz^2} - v_m(z) = -\lambda_C^2 r_0 I_e \delta(z). \tag{3.17}$$

To obtain a solution to Equation 3.17, we first solve the homogeneous equation,

$$\lambda_C^2 \frac{d^2 v_m(z)}{dz^2} - v_m(z) = 0, \tag{3.18}$$

which is valid in two semi-infinite regions, $z < 0$ and $z > 0$. Then we connect the solutions in the two regions by matching the boundary condition at $z = 0$. Since we know that linear, homogeneous, ordinary differential equations with constant coefficients have exponential solutions, we assume that the solution has the form

$$v_m(z) = A e^{pz}. \tag{3.19}$$

Substitution of Equation 3.19 into 3.18 yields the characteristic equation

$$\lambda_C^2 p^2 - 1 = 0, \tag{3.20}$$

which has two roots, called natural frequencies, characteristic frequencies, or eigenvalues,

$$p = \pm \frac{1}{\lambda_C}. \tag{3.21}$$

Since there are two regions for which the homogeneous equation applies, the homogeneous solution can be written in general as

$$v_m(z) = \begin{cases} A_1 e^{-z/\lambda_C} + A_3 e^{z/\lambda_C}, & z > 0, \\ A_2 e^{z/\lambda_C} + A_4 e^{-z/\lambda_C}, & z < 0. \end{cases} \tag{3.22}$$

On physical grounds, we expect a bounded solution, i.e., a solution for which $\lim_{|z| \to \infty} |v_m(z)|$ is bounded. Hence, $A_3 = A_4 = 0$ and

$$v_m(z) = \begin{cases} A_1 e^{-z/\lambda_C}, & z > 0, \\ A_2 e^{z/\lambda_C}, & z < 0. \end{cases} \tag{3.23}$$

From the differential equation (Equation 3.17), it can be shown that $v_m(z)$ is continuous for $z = 0$. To see this, assume the contrary, namely, that $v_m(z)$ is discontinuous at $z = 0$. If this were so, $dv_m(z)/dz$ would contain an impulse at $z = 0$ and $d^2v_m(z)/dz^2$ would contain a doublet (the derivative of an impulse) at $z = 0$. No such term appears on the right-hand side of Equation 3.17. Therefore, $v_m(z)$ cannot be discontinuous at $z = 0$, i.e., $v_m(z)$ is continuous at $z = 0$, which implies that $A_1 = A_2$ and hence that $v_m(z)$ can be expressed as

$$v_m(z) = A e^{-|z|/\lambda_C}. \tag{3.24}$$

The value of A is found most simply by matching impulses as illustrated in Figure 3.10. Substituting $d^2v_m(z)/dz$ and $v_m(z)$ as shown in Figure 3.10 into Equation 3.17 and matching impulse areas on the two sides of the equation shows that $A = (r_o\lambda_C/2)I_e$. Therefore,

$$v_m(z) = \frac{r_o\lambda_C}{2} I_e e^{-|z|/\lambda_C}. \tag{3.25}$$

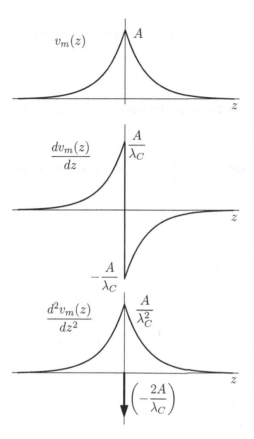

Figure 3.10 Matching boundary conditions at $z = 0$. Note that $v_m(z)$ is continuous at $z = 0$ but has a discontinuous derivative at $z = 0$. Therefore, $d^2v_m(z)/dz^2$ contains an impulse at $z = 0$.

Thus, in response to a constant current applied through infinitesimal electrodes, the membrane potential decays exponentially along the cable (Figure 3.11). The value of the potential at the location of the electrode is $v_m(0) = (r_o\lambda_C/2)I_e$, i.e., it is equivalent to the potential difference of a current I_e through a resistance of value $r_o\lambda_C/2$. Thus, the relation of the membrane potential to the external current is a transfer resistance that can be interpreted as the parallel combination of two resistances, each having a resistance per unit length r_o and a length λ_C. Each of these represents the transfer resistance seen looking into a semi-infinite cable in the z-direction.

The solution for $v_m(z)$ can be used to find all the currents and voltages for the cylindrical cell. The membrane current per unit length is proportional to the membrane potential,

$$k_m = g_m v_m(z) = \frac{r_o\lambda_C}{2} g_m I_e e^{-|z|/\lambda_C}.$$

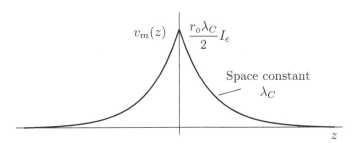

Figure 3.11 Spatial distribution of membrane potential that results from applying a steady current at one point along a cable.

The internal longitudinal current can be found from the core conductor relations for incremental quantities (Equation 2.18),

$$i_i(z) = -\int_{-\infty}^{z} k_m(z')\,dz' = -\frac{r_o\lambda_C}{2}g_m I_e \int_{-\infty}^{z} e^{-|z'|/\lambda_C}\,dz',$$

which can be written as

$$i_i(z) = \begin{cases} -\frac{r_o\lambda_C}{2}g_m I_e \int_{-\infty}^{z} e^{z'/\lambda_C}\,dz' & \text{for } z < 0, \\ -\frac{r_o\lambda_C}{2}g_m I_e \left(\int_{-\infty}^{0} e^{z'/\lambda_C}\,dz' + \int_{0}^{z} e^{-z'/\lambda_C}\,dz'\right) & \text{for } z \geq 0 \end{cases}$$

and integrated to yield

$$i_i(z) = \begin{cases} -\frac{r_o}{2(r_i+r_o)}I_e e^{z/\lambda_C} & \text{for } z < 0, \\ -\frac{r_o}{2(r_i+r_o)}I_e \left(2 - e^{-z/\lambda_C}\right) & \text{for } z \geq 0. \end{cases}$$

The equation for $i_i(z)$ can be written compactly as

$$i_i(z) = \frac{r_o}{2(r_i+r_o)}I_e \left(e^{-|z|/\lambda_C}\,\mathrm{sgn}(z) - 2u(z)\right),$$

where

$$\mathrm{sgn}(z) = \begin{cases} -1 & \text{for } z < 0, \\ +1 & \text{for } z > 0. \end{cases}$$

As shown in Section 2.4.3, for any z, $i_i(z) + i_o(z) + I_e u(z) = 0$. Therefore, after simplification we obtain

$$i_o(z) = -\frac{r_o}{2(r_i+r_o)}I_e \left(e^{-|z|/\lambda_C}\,\mathrm{sgn}(z) + 2\left(\frac{r_i}{r_o}\right)u(z)\right).$$

The intracellular potential is obtained from the core conductor equation (Equation 2.20) for incremental quantities,

$$v_i(z) - v_i(-\infty) = -r_i \int_{-\infty}^{z} i_i(z')\,dz',$$

which yields, after integration and simplification,

$$v_i(z) - v_i(-\infty) = \frac{r_i r_o \lambda_C}{2(r_i + r_o)} I_e \left(e^{-|z|/\lambda_C} - 2 \left(\frac{z}{\lambda_C} \right) u(z) \right).$$

The extracellular potential can be derived in a similar manner starting with the core conductor equation (Equation 2.21) for incremental quantities,

$$v_o(z) - v_o(-\infty) = -r_o \int_{-\infty}^{z} i_o(z') \, dz',$$

which yields

$$v_o(z) - v_o(-\infty) = -\frac{r_o^2 \lambda_C}{2(r_i + r_o)} I_e \left(e^{-|z|/\lambda_C} + 2 \left(\frac{r_i}{r_o} \right) \left(\frac{z}{\lambda_C} \right) u(z) \right).$$

As a consistency check on these results, note that if $v_i(-\infty) - v_o(-\infty) = 0$, then $v_i(z) - v_o(z) = v_m(z)$.

Two simple cases for these variables are when $r_o = 0$ and when $r_i = 0$. Substitution of these constraints into the equations yields the solutions shown in Table 3.2 and sketched schematically in Figure 3.12. When $r_o = 0$, all the longitudinal current flows through the outer conductor, but only in the interpolar region to the right of the active current electrode. Since there is no longitudinal current in the inner conductor, there is no current through the membrane. Hence, there is no potential difference across the membrane. This result guarantees that $v_i(z) = 0$. When $r_i = 0$, the potential in the inner conductor is zero, i.e., $v_i(z) = 0$. Therefore, $v_o(z) = -v_m(z)$.

In general, the voltages and currents are functions of four parameters: r_i, r_o, g_m, and I_e. However, because of the linearity of the cable equations, all the currents and voltages are proportional to I_e. Thus, it is necessary to determine the dependence of the solutions on only three parameters. The dimensional-

Table 3.2 Time-independent solutions for an infinite cable with $r_i = 0$ and $r_o = 0$. We assume that $v_i(\infty) = v_o(\infty) = 0$.

$r_o = 0$	$r_i = 0$
$v_m(z) = 0$	$v_m(z) = (r_o \lambda_C / 2) I_e e^{-|z|/\lambda_C}$
$k_m(z) = 0$	$k_m(z) = (r_o \lambda_C / 2) g_m I_e e^{-|z|/\lambda_C}$
$i_i(z) = 0$	$i_i(z) = (I_e/2) \left(e^{-|z|/\lambda_C} \operatorname{sgn}(z) - 2u(z) \right)$
$i_o(z) = -I_e u(z)$	$i_o(z) = -(I_e/2) e^{-|z|/\lambda_C} \operatorname{sgn}(z)$
$v_i(z) = 0$	$v_i(z) = 0$
$v_o(z) = 0$	$v_o(z) = -(r_o \lambda_C / 2) I_e e^{-|z|/\lambda_C}$

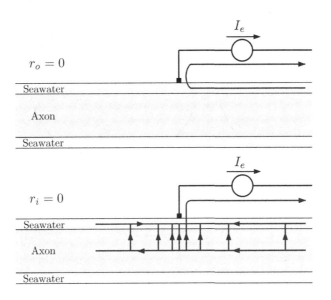

Figure 3.12 Schematic diagram of current flow around a cylindrical cable driven by a constant current for two conditions, $r_o = 0$ and $r_i = 0$. All the currents, including the current source, have cylindrical symmetry about the axis of the cylinder, but are drawn for the upper half of the cylinder only.

ity of the parameter space is further reduced by expressing the solutions in normalized form and noting that these normalized solutions depend upon only one parameter, $\alpha = r_i/r_o$. Let us define the normalized variables as follows: the normalized distance is $\lambda = z/\lambda_C$; the normalized membrane potential and membrane current are $\hat{v}_m = v_m/(r_o\lambda_C I_e/2)$ and $\hat{k}_m = k_m/(r_o\lambda_C g_m I_e/2)$, respectively; the normalized longitudinal currents are $\hat{i}_i = i_i/I_e$ and $\hat{i}_o = i_o/I_e$; the normalized intracellular and extracellular potentials are $\hat{v}_i = v_i/(r_o\lambda_C I_e/2)$ and $\hat{v}_o = v_o/(r_o\lambda_C I_e/2)$. Thus, the normalized variables are

$$\hat{v}_m(\lambda) = e^{-|\lambda|},$$

$$\hat{k}_m(\lambda) = e^{-|\lambda|},$$

$$\hat{i}_i(\lambda) = \frac{1}{2(\alpha+1)}\left(e^{-|\lambda|}\,\mathrm{sgn}(\lambda) - 2u(\lambda)\right),$$

$$\hat{i}_o(\lambda) = -\frac{1}{2(\alpha+1)}\left(e^{-|\lambda|}\,\mathrm{sgn}(\lambda) + 2\alpha u(\lambda)\right),$$

$$\hat{v}_i(\lambda) = \frac{\alpha}{\alpha+1}\left(e^{-|\lambda|} - 2\lambda u(\lambda)\right),$$

$$\hat{v}_o(\lambda) = -\frac{1}{\alpha+1}\left(e^{-|\lambda|} + 2\alpha\lambda u(\lambda)\right).$$

All the potentials and currents are plotted as a function of position in normalized coordinates in Figure 3.13 for three values of α. Note that the spatial dependence of the normalized membrane potential and current per

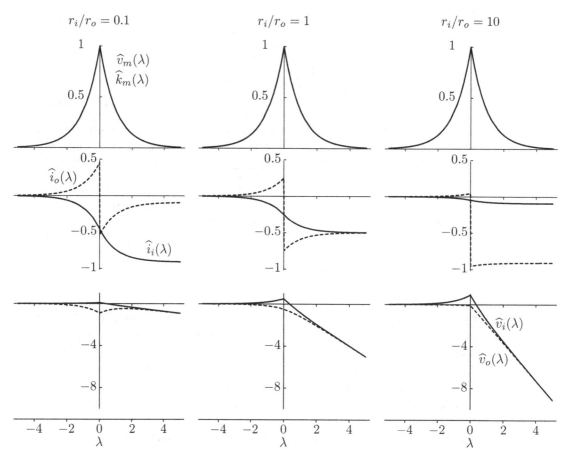

Figure 3.13 Time-independent solutions of the cable equation in normalized coordinates for different values of r_i/r_o.

unit length are independent of α. However, both the longitudinal currents and the external and internal potentials depend upon α. Note that the pattern of the longitudinal currents and inner and outer potentials are simpler in the extrapolar region ($\lambda < 0$). In this region, all these variables have magnitudes that decay exponentially from their values at $\lambda = 0$. In the interpolar region ($\lambda > 0$), there is also an exponentially decaying component, but in addition there is a current component due to the external current. This component manifests itself as a negative unit step in $\hat{i}_o(\lambda)$. The ratio of longitudinal currents in the inner and outer conductors at locations in the interpolar region that are far from the active electrode depends upon α. When $\alpha = r_i/r_o$ is large, a larger

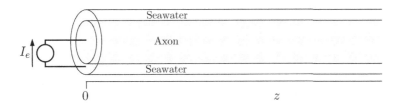

Figure 3.14 Constant current applied intracellularly at one end of a cylindrical cell.

fraction of the longitudinal current flows in the outer conductor. When α is small, a larger fraction of the longitudinal current flows in the inner conductor. The longitudinal currents flowing through the inner and outer resistances per unit length produce a component of the potential whose magnitude grows linearly with distance. However, since $v_m(z)$ decays exponentially with increasing distance from the active electrode, $v_i(\lambda)$ and $v_o(\lambda)$ approach the same potential.

3.4.2.2 Solution for a Semi-Infinite Cable in Response to a Constant Internal Current

We shall next consider the spread of membrane potential along a cylindrical cell when a current is applied at one end of the cell as shown in Figure 3.14. In the steady state, the cable equation is

$$\lambda_C^2 \frac{d^2 v_m(z)}{dz^2} - v_m(z) = 0 \text{ for } z > 0. \tag{3.26}$$

Hence, the solution for the membrane potential is

$$v_m(z) = Ae^{-z/\lambda_C}, \tag{3.27}$$

where A is determined from the boundary condition at $z = 0$. The membrane current per unit length is $k_m(z) = g_m v_m(z)$, and, since $di_i(z)/dz = -k_m(z)$, the longitudinal currents can be gotten by integrating the membrane current to obtain

$$i_i(z) = A g_m \lambda_C e^{-z/\lambda_C}.$$

Since $i_i(0) = I_e$, $A g_m \lambda_C = I_e$; therefore, the solution is

$$v_m(z) = \frac{1}{g_m \lambda_C} I_e e^{-z/\lambda_C}.$$

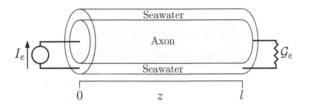

Figure 3.15 Schematic diagram of a cable of length l driven by a current at $z = 0$ and loaded by a conductance of $\mathcal{G}_e$ at $z = l$.

Note that $v_m(0) = I_e/(g_m\lambda_C)$, so that $I_e/v_m(0) = g_m\lambda_C$. Therefore, we define the Thévenin equivalent conductance looking into an infinite cable (also called the *characteristic conductance* of the cable) as $\mathcal{G}_\infty = g_m\lambda_C$, which can be written as

$$\mathcal{G}_\infty = g_m\lambda_C = \sqrt{\frac{g_m}{r_i + r_o}}.$$

If we assume that $r_i \gg r_o$, then the characteristic conductance takes on the simple form

$$\mathcal{G}_\infty \approx \sqrt{\frac{g_m}{r_i}}.$$

For a cylindrical cell of radius a, we have

$$\mathcal{G}_\infty \approx \sqrt{\frac{g_m}{r_i}} = \sqrt{2\pi a G_m\left(\frac{\pi a^2}{\rho_i}\right)} = \pi a^{3/2}\sqrt{\frac{2G_m}{\rho_i}}.$$

Hence, the characteristic conductance of the cell varies as $a^{3/2}$, so that larger-diameter cells have larger characteristic conductances.

3.4.2.3 Solution for a Finite Cable in Response to a Constant Internal Current

Axons and muscle fibers are cells whose dimensions can be many space constants long. Hence, an infinite-length cable may be an appropriate model for such a cell. However, dendrites of neurons can be much shorter, so it may be inappropriate to represent these structures with infinite cables. Therefore, we shall consider the spread of potential in a cable of length l as shown in Figure 3.15. The finite cable is driven by a current at $z = 0$ and is terminated by a load conductance $\mathcal{G}_e$, which represents either the conductance of the membrane that caps the end of the cylindrical cell or the conductance of a cell body attached to the end of the cylindrical portion of the cell.

The membrane potential satisfies the time-invariant cable equation, Equation 3.18, for the interval $0 < z < l$. The general solution of this equation can be expressed either as a sum of exponentials or, as can be ascertained by direct substitution, as a sum of hyperbolic functions. It is most convenient to express the general solution as

$$v_m(z) = A \cosh\left(\frac{l-z}{\lambda_C}\right) + B \sinh\left(\frac{l-z}{\lambda_C}\right) \text{ for } 0 \le z \le l,$$

where the constants A and B are determined by the boundary conditions. Since $k_m(z) = g_m v_m(z)$,

$$k_m(z) = A g_m \cosh\left(\frac{l-z}{\lambda_C}\right) + B g_m \sinh\left(\frac{l-z}{\lambda_C}\right),$$

and since $di_i(z)/dz = -k_m$, the internal longitudinal current is

$$i_i(z) = A g_m \lambda_C \sinh\left(\frac{l-z}{\lambda_C}\right) + B g_m \lambda_C \cosh\left(\frac{l-z}{\lambda_C}\right).$$

At $z = l$, the relation between the longitudinal current and the membrane potential is constrained by the load conductance, i.e., $i_i(l) = \mathcal{G}_e v_m(l)$. From the solution to the cable equation, we have $i_i(l) = B g_m \lambda_C$ and $v_m(l) = A$. We can combine these results to obtain $B = A \mathcal{G}_e/(g_m \lambda_C)$. At $z = 0$, the longitudinal current $i_i(0) = I_e$. Therefore,

$$I_e = A g_m \lambda_C \sinh\left(\frac{l}{\lambda_C}\right) + A \mathcal{G}_e \cosh\left(\frac{l}{\lambda_C}\right),$$

which can be solved for A. Substitution for A, B, and $\mathcal{G}_\infty = g_m \lambda_C$ yields the solution for $v_m(z)$ as follows:

$$v_m(z) = \frac{I_e}{\mathcal{G}_\infty} \left(\frac{\cosh\left(\frac{l-z}{\lambda_C}\right) + \left(\frac{\mathcal{G}_e}{\mathcal{G}_\infty}\right) \sinh\left(\frac{l-z}{\lambda_C}\right)}{\sinh\left(\frac{l}{\lambda_C}\right) + \left(\frac{\mathcal{G}_e}{\mathcal{G}_\infty}\right) \cosh\left(\frac{l}{\lambda_C}\right)} \right) \text{ for } 0 \le z \le l. \tag{3.28}$$

Therefore, the Thévenin equivalent conductance looking into the cable at $z = 0$ is

$$\mathcal{G}_T(l) = \frac{I_e}{v_m(0)} = \mathcal{G}_\infty \left(\frac{\sinh\left(\frac{l}{\lambda_C}\right) + \left(\frac{\mathcal{G}_e}{\mathcal{G}_\infty}\right) \cosh\left(\frac{l}{\lambda_C}\right)}{\cosh\left(\frac{l}{\lambda_C}\right) + \left(\frac{\mathcal{G}_e}{\mathcal{G}_\infty}\right) \sinh\left(\frac{l}{\lambda_C}\right)} \right).$$

It is simpler to examine $v_m(z)$ and $\mathcal{G}_T(l)$ in terms of normalized variables. Let $\lambda = z/\lambda_C$, $L = l/\lambda_C$, $\hat{\mathcal{G}}_e = \mathcal{G}_e/(g_m\lambda_C)$, $\hat{v}_m(\lambda) = v_m(\lambda\lambda_C)\mathcal{G}_\infty/I_e$, and $\hat{\mathcal{G}}_T(L) = \mathcal{G}_T(L\lambda_C)/G_\infty$. Then we can express the membrane potential in normalized coordinates as

$$\hat{v}_m(\lambda) = \frac{\cosh(L-\lambda) + \hat{\mathcal{G}}_e \sinh(L-\lambda)}{\sinh L + \hat{\mathcal{G}}_e \cosh L}$$

and the Thévenin conductance as

$$\hat{\mathcal{G}}_T(L) = \left(\frac{\sinh L + \hat{\mathcal{G}}_e \cosh L}{\cosh L + \hat{\mathcal{G}}_e \sinh L} \right).$$

Special Cases
We shall consider three special cases of this result.

Load Conductance Equals the Characteristic Conductance   If $\hat{\mathcal{G}}_e = 1$, then

$$\hat{v}_m(\lambda) = \frac{\cosh(L-\lambda) + \sinh(L-\lambda)}{\sinh L + \cosh L},$$

which can be simplified to yield

$$\hat{v}_m(\lambda) = e^{-\lambda} \text{ for } 0 \le \lambda \le 1. \tag{3.29}$$

Thus, if the finite cable is terminated by its characteristic conductance, the voltage distribution behaves as if the cable were infinite. Furthermore, the conductance of the cable as seen at $\lambda = 0$ is $\hat{\mathcal{G}}_T(L) = 1$, i.e., the input conductance of the finite cable equals the characteristic conductance.

Load Is an Open Circuit   Suppose we assume that the cable is capped at $z = l$ with a patch of membrane whose specific conductance is identical to that of the rest of the cylindrical cell. Since the surface area of the patch is small compared to that of the cylinder, the conductance of the cap will be small compared to the rest of the cell. If, as a consequence, little current flows through the cap, we can represent the cap as an open circuit and explore the limiting case when $\hat{\mathcal{G}}_e \to 0$. In this case,

$$\hat{v}_m(\lambda) = \frac{\cosh(L-\lambda)}{\sinh L} \text{ for } 0 \le \lambda \le 1,$$

and

$$\hat{\mathcal{G}}_T(L) = \tanh L. \tag{3.30}$$

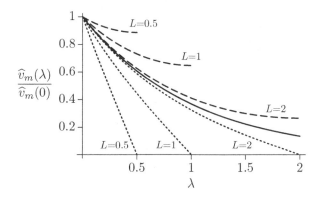

Figure 3.16 Dependence of membrane potential on position in a finite cable in normalized coordinates. Here $\hat{v}_m(\lambda)/\hat{v}_m(0)$ is plotted versus λ. The dashed lines are for a termination that is an open circuit ($\hat{\mathcal{G}}_e = 0$), the dotted lines are for a termination that is a short circuit ($\hat{\mathcal{G}}_e = \infty$), and the solid line is for a termination equal to the characteristic conductance of the cable ($\hat{\mathcal{G}}_e = 1$) which is also the solution for a semi-infinite cable.

Load Is a Short Circuit   Another extreme possibility is if the conductance of the cap is high, so that the cap is represented as a short circuit. Under these circumstances, we can assume that $\hat{\mathcal{G}}_e \to \infty$. In this case,

$$\hat{v}_m(\lambda) = \frac{\sinh(L - \lambda)}{\cosh L} \text{ for } 0 \leq z \leq l,$$

and

$$\hat{\mathcal{G}}_T(L) = \coth L.$$

Dependence of Solution on Length of Cable and on Load Conductance

The dependence of the membrane potential on position along the cable for cables of different lengths and for different cable terminations is shown in Figure 3.16. The results are shown in normalized coordinates, so the lengths of the cables are expressed in units of space constants and the position along the cable is also expressed in units of space constants. For a cable terminated in its characteristic conductance, the response attenuates exponentially along the cable. For open circuit terminations the response attenuates more slowly than exponentially, and for short circuit terminations the response attenuates more rapidly than exponentially and becomes zero at the end of the cable. The longer the cable and the larger the conductance of the termination, the larger the attenuation at the end of the cable. These trends are depicted more directly in Figure 3.17, which shows the ratio of the membrane potential at the end of the cable to that at the beginning as a function of the length of the cable for different values of the termination conductance. This figure shows that the attenuation increases as the cable length increases and as the termination conduction increases. It is also apparent that for $L < 0.2$ and for $\hat{\mathcal{G}}_e < 0.25$, the attenuation along the cable is small. In this range, the finite cylindrical cell

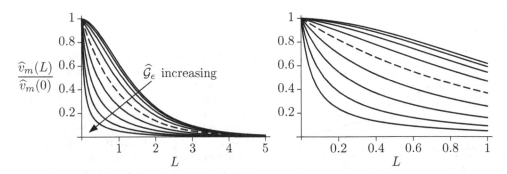

Figure 3.17 Attenuation of the membrane potential from one end to another of a finite cable, $\hat{v}_m(L)/\hat{v}_m(0)$, plotted versus the length of the cable, L, in normalized coordinates. The parameter is the normalized termination conductance of the cable, $\hat{\mathcal{G}}_e$, which ranges in multiples of two from 0.0625 to 16. The dashed curve corresponds to $\hat{\mathcal{G}}_e = 1$. The results are shown on two scales of L.

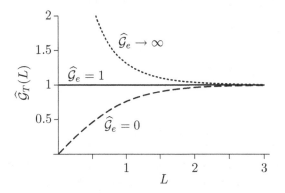

Figure 3.18 Dependence of Thévenin conductance ($\hat{\mathcal{G}}_T(L)$) on cable length (L) in normalized coordinates for different cable terminations.

shows little difference in membrane potential as a function of position along the cell. Hence, a cell with these dimensions approaches the behavior of an electrically small cell.

The normalized Thévenin input conductance of the cell as a function of the length of the cell is shown in Figure 3.18 for different termination conductances. $\hat{\mathcal{G}}_T(L) = 1$ when the $\hat{\mathcal{G}}_e = 1$; that is, when the cable is terminated in its characteristic conductance, its input conductance also equals the characteristic conductance independent of the length of the cable. When $\hat{\mathcal{G}}_e \neq 1$, then in general $\hat{\mathcal{G}}_T(L) \neq 1$. However, $\lim_{L \to \infty} \hat{\mathcal{G}}_T(L) = 1$; that is, as the length of the cable increases, the input conductance becomes independent of the terminating conductance and approaches the characteristic conductance of the cable. For a short cable terminated in a short circuit, the input conductance approaches that of a short circuit, i.e., the input conductance grows arbitrarily large. For a short cable terminated in an open circuit, the input conductance grows linearly with the length of the cable. This behavior can be seen from an examination of Equation 3.30. For small values of L, $\hat{G}_T(L) \approx L$. Unnormalizing these variables gives

$$\frac{G_T(l)}{g_m \lambda_C} \approx \frac{l}{\lambda_C},$$

which yields

$$G_T(l) \approx g_m l,$$

which is the conductance of an electrically small cylindrical cell of length l with a membrane conductance per unit length of g_m.

3.4.2.4   Implications of the Time-Independent Solutions

Small Cells, Infinitesimal Electrodes, Remote Electrodes

The space constant measures the distance along a cylindrical cell over which the potential difference across the membrane varies. Hence, if a cell's dimensions are small compared to its space constant, the membrane potential difference will not vary much with position along the cell. Therefore, a cell is electrically small when its dimensions are small compared to a space constant. Similarly, an electrode has infinitesimal width when its width is small compared to a space constant. An electrode is located at a location remote from another electrode if it is many space constants away. Analysis of a finite cable shows that, in addition to the relation between cell dimensions and the cell space constant, the electrical termination of a cylindrical cell also affects the extent to which the cell can be regarded as electrically small.

Linearity

Because the cable equation is a linear partial differential equation and because the transformation from external current to membrane potential (Equation 3.15) is linear, knowledge of the potential that results from a point source of current can be used to determine responses to more complex external currents. In particular, we can find the time-independent solution for an arbitrary arrangement of electrodes. For example, suppose we apply two electrodes to an axon, each with respect to remote reference electrodes (Figure 3.19). The resulting membrane potential is simply a sum of the responses obtained from each electrode applied alone. The results can be further generalized to obtain the potential due to an arbitrary electrode configuration. Since $k_e(z)$ can be expressed as a superposition of spatial impulses and because the relation be-

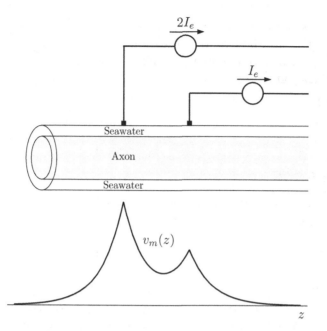

Figure 3.19 Superposition of constant currents applied at two points along a cylindrical cell. The current stimulating arrangement is shown above, and the resulting membrane potential is shown below. One current has amplitude $2I_e$, and the other amplitude I_e. The membrane potential is a sum of two terms, each having the form given in Equation 3.25, but appropriately scaled and shifted in space.

tween membrane potential and external current is linear and space invariant, the membrane potential can be expressed in terms of a superposition integral (Weiss, 1996, Section 3.5, Chapter 3). Thus, to obtain the membrane potential for an arbitrary electrode geometry, it is necessary only to convolve the external current distribution with the spatial impulse responses (Green's function) of the cable to obtain

$$v_m(z) = \int_{-\infty}^{\infty} k_e(z') \left(\frac{r_o \lambda_C}{2} e^{-|z-z'|/\lambda_C} \right) dz'. \tag{3.31}$$

From this integral, we can also note that if the spatial distribution of $k_e(z)$ is narrow compared to the exponential function, $v_m(z)$ will resemble the exponential function. In addition, Equation 3.31 can be used to analyze the effect of a finite electrode width on the membrane potential distribution.

3.4.2.5 *Comparison with Measurements*

The spatial dependence of the potential resulting from an electrode delivering a constant current has been studied in a number of cells. A schematic diagram of the apparatus for delivering stimuli and recording responses for one such study is shown in Figure 3.1 (top panel). The two stimulating electrodes are fixed in space; one voltage-recording electrode is also fixed in space. The other voltage-recording electrode, mounted on a micromanipulator, is used to record the extracellular potential at different locations along the cell. The responses to rectangular current pulses are shown in Figure 3.1. In this section, we shall consider current stimuli for which the potential response is in the linear range, and we shall consider the final value of the potential after the transient response has come to completion. Since the response was recorded extracellularly in the extrapolar region, the core conductor model implies that the potential response has the form

$$v_o(z) = -\frac{r_o}{r_o + r_i} \left(\frac{r_o \lambda_C}{2} I_e e^{-|z|/\lambda_C} \right) \tag{3.32}$$

and that

$$v_o(0) = -\frac{r_o^2}{r_o + r_i} \frac{\lambda_C I_e}{2}. \tag{3.33}$$

Therefore,

$$\frac{v_o(z)}{v_o(0)} = e^{-|z|/\lambda_C}. \tag{3.34}$$

Taking the logarithm of both sides, we obtain

$$\log_{10} \left(\frac{v_o(z)}{v_o(0)} \right) = -(|z|/\lambda_C) \log_{10} e. \tag{3.35}$$

Hence, the normalized potential should plot as a straight line on logarithmic coordinates. The measurements are shown in Figure 3.20 for a single axon. The fit to a straight line is adequate. Different axons have different space constants and different resistances. Therefore, to compare results obtained from a number of axons, it is convenient to compute the responses on normalized scales as shown in Figure 3.21 for a collection of axons. Here the ordinate scale is linear, and it can be seen that the measurements are well fit by an exponential as predicted by the cable model.

3.4.2.6   Significance of the Space Constant

Measurements for a number of different cell types (Table 3.1) show that λ_C is on the order of a few millimeters for the largest cells. However, λ_C decreases as the radius of the cell decreases (Equation 3.14). If we extrapolate the value of the space constant for large cells to that for small unmyelinated vertebrate nerve fibers ($a < 1\mu m$), we get a value that is less than 100 μm. It is clear

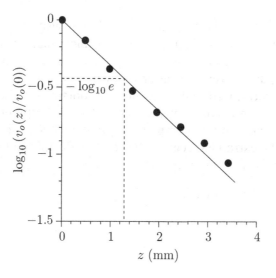

Figure 3.20 Spatial distribution of extracellular potential along a single unmyelinated axon in response to a steady current (adapted from Hodgkin and Rushton, 1946, Figure 7). The measurements (points) were obtained from an axon in the walking leg of a lobster, *Homarus vulgaris*. The line has been fit to these points. The space constant can be computed from the distance required for the potential to decrease by $1/e$ of its value at $z = 0$. For this axon, $\lambda_C \approx 1.3$ mm.

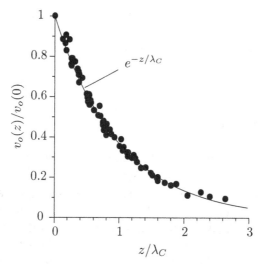

Figure 3.21 Spatial distribution of extracellular potential along unmyelinated lobster axons in response to a steady current (adapted from Hodgkin and Rushton, 1946, Figure 8). The distance from the current electrode is normalized by the space constant for each axon. The measurements (points) are compared to an exponential function.

that unmyelinated axons in particular and cells in general are poor cables for sending small potentials for long distances (many space constants) in the body. As we shall see, the spatial attenuation of small potentials along a fiber is even greater for transient potential changes than for the constant potentials that we have considered thus far.

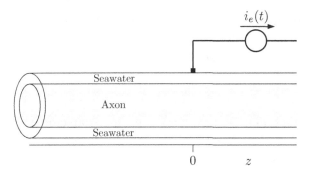

Figure 3.22 Arrangement of electrodes for delivering a time-varying current to a cylindrical cell at a single position.

3.4.3 Time-Dependent Solutions

So far we have considered the membrane potential produced by a constant current applied through external electrodes along a cable. We have seen that measurements on cells are described rather accurately by the cable model. However, the physiologically more interesting situation is that in which membrane potentials of cells vary with time. In this section, we shall investigate the time-dependent potentials in response to transient current stimuli.

3.4.3.1 *Impulse Response of an Infinite Cable*

First we shall consider the response to a brief pulse of current, i.e., an impulse or Dirac delta function. This solution itself has physiological relevance, as we shall see, and in addition this solution can be used to characterize the response of a cable to an arbitrary spatial and temporal distribution of external currents.

Derivation
The case we shall consider is shown schematically in Figure 3.22 with $i_e(t) = Q_e\delta(t)$, where $\delta(t)$ is the unit impulse function in time. This current stimulus is a brief current that delivers an amount of charge Q_e to the external surface of a cable at a single point. The reference electrode for the current source is effectively at $z = \infty$, i.e., many space constants away.

We start with the cable equation (Equation 3.15), where v_m and k_e are functions of z and t and assume that

$$\int_{-\infty}^{\infty} \int_{-\infty}^{\infty} k_e(z, t) \, dz \, dt = Q_e, \tag{3.36}$$

where $k_e(z, t)$ is an impulse in z and t, so that $k_e(z, t) = 0$ for $z \neq 0$, $t \neq 0$, and Q_e is the total charge delivered by this current per unit length. To solve this problem, we shall solve the homogeneous equation for $t > 0$,

$$\lambda_C^2 \frac{\partial^2 v_m}{\partial z^2} = v_m + \tau_M \frac{\partial v_m}{\partial t}, \tag{3.37}$$

and then match initial boundary conditions. It is convenient to define normalized variables $\lambda = z/\lambda_C$, $\tau = t/\tau_M$, and $v_m(z, t) = v_m(\lambda \lambda_C, \tau \tau_M) = \hat{v}_m(\lambda, \tau)$. Therefore, the homogeneous cable equation can be written as

$$\frac{\partial^2 \hat{v}_m}{\partial \lambda^2} = \hat{v}_m + \frac{\partial \hat{v}_m}{\partial \tau}. \tag{3.38}$$

This equation can be transformed into the ordinary diffusion equation by a simple change in variables:

$$w(\lambda, \tau) = \hat{v}_m(\lambda, \tau) e^\tau. \tag{3.39}$$

Note that

$$\frac{\partial^2 w}{\partial \lambda^2} = \frac{\partial^2 \hat{v}_m}{\partial \lambda^2} e^\tau,$$

and

$$\frac{\partial w}{\partial \tau} = \hat{v}_m e^\tau + \frac{\partial \hat{v}_m}{\partial \tau} e^\tau.$$

Hence, substitution into Equation 3.38 yields

$$\frac{\partial^2 w}{\partial \lambda^2} = \frac{\partial w}{\partial \tau}. \tag{3.40}$$

We have reduced the problem of solving the cable equation to the problem of solving the homogeneous diffusion equation (Equation 3.40). The solution to this equation is (Weiss, 1996, Chapter 3)

$$w(\lambda, \tau) = \frac{A}{\sqrt{4\pi\tau}} e^{-\lambda^2/4\tau} \text{ for } \tau > 0, \tag{3.41}$$

as can be seen by direct substitution of Equation 3.41 into Equation 3.40. This solution satisfies the homogeneous diffusion equation for any value of A. Using Equation 3.39, the general form of the solution for $\hat{v}_m$ is

$$\hat{v}_m(\lambda, \tau) = \frac{A}{\sqrt{4\pi\tau}} e^{-\lambda^2/4\tau} e^{-\tau} \text{ for } \tau > 0, \tag{3.42}$$

and

$$v_m(z, t) = \frac{A}{\sqrt{4\pi(t/\tau_M)}} e^{-(z/\lambda_C)^2/(4t/\tau_M)} e^{-t/\tau_M} \text{ for } t > 0. \tag{3.43}$$

The value of A can be obtained by substituting the solution (Equation 3.43) into the cable equation. Let $k_e(z, t) = Q_e\delta(z, t)$, where $\delta(z, t)$ is a two-dimensional impulse function whose integral on z and t is 1. It is most convenient to evaluate A from the inhomogeneous cable equation for $w(\lambda, \tau)$, for which we need to find $k_e(\lambda, \tau)$. Changing the independent variables in the delta function changes its area, as can be seen by evaluating $\int_{-\infty}^{\infty} \int_{-\infty}^{\infty} \delta(\lambda, \tau)\, d\lambda\, d\tau$, so that $k_e(\lambda, \tau) = (Q_e/\lambda_C\tau_M)\delta(\lambda, \tau)$. Therefore, the cable equation, including the term for the external current, is

$$\frac{\partial^2 w(\lambda, \tau)}{\partial \lambda^2} = \frac{\partial w(\lambda, \tau)}{\partial \tau} - \frac{r_o\lambda_C Q_e}{\tau_M}\delta(\lambda, \tau)e^{\tau},$$

but $\delta(\lambda, \tau)e^{\tau} = \delta(\lambda, \tau)$. The solution for all τ is

$$w(\lambda, \tau) = \frac{A}{\sqrt{4\pi\tau}} e^{-\lambda^2/4\tau} u(\tau).$$

To find the value of A, we integrate the cable equation on λ to obtain

$$\int_{-\infty}^{\infty} \frac{\partial^2 w(\lambda, \tau)}{\partial \lambda^2}\, d\lambda = \int_{-\infty}^{\infty} \frac{\partial w(\lambda, \tau)}{\partial \tau}\, d\lambda - \frac{r_o\lambda_C Q_e}{\tau_M} \int_{-\infty}^{\infty} \delta(\lambda, \tau)\, d\lambda.$$

The term on the left-hand side is zero. We change the order of integration and differentiation in the first term on the right-hand side and evaluate the second term to give

$$\frac{\partial}{\partial \tau} \int_{-\infty}^{\infty} w(\lambda, \tau)\, d\lambda = \frac{r_o\lambda_C Q_e}{\tau_M}\delta(\tau).$$

Since the integral on the left-hand side is $Au(\tau)$ and its derivative on τ is $A\delta(\tau)$, we have

$$A = \frac{r_o\lambda_C Q_e}{\tau_M}. \tag{3.44}$$

We combine Equations 3.43 and 3.44 to yield the total solution,

$$v_m(z, t) = \frac{r_o\lambda_C Q_e/\tau_M}{\sqrt{4\pi(t/\tau_M)}} e^{-(z/\lambda_C)^2/(4t/\tau_M)} e^{-t/\tau_M} u(t). \tag{3.45}$$

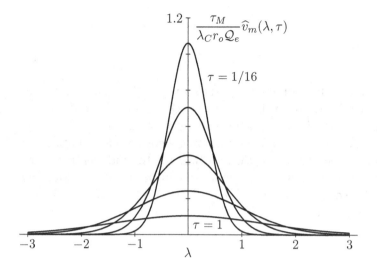

Figure 3.23 Spatial distribution of membrane potential, plotted at different instants in time, in response to a fixed charge delivered to a cable at one point in space and time. The times are $\tau = 1/16$, $1/8$, $1/4$, $1/2$, and 1.

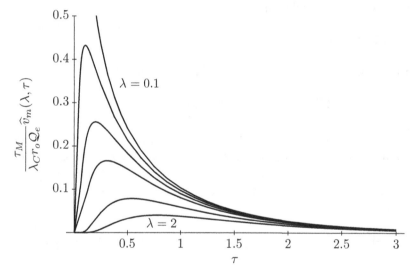

Figure 3.24 Temporal distribution of membrane potential, plotted at different locations along the cable, in response to a fixed charge delivered to a cable at one point in space and time. The locations are $\lambda = 0.1$, 0.5, 0.75, 1, 1.5, and 2.

The solution is plotted versus z/λ_C in Figure 3.23 and versus t/τ_M in Figure 3.24.

We can see that a few space constants away from the electrode, the potential decays to a small value after a few time constants. Note that v_m does *not* decay exponentially in space; instead, for any one instant in time, v_m decays much faster than exponentially, at a rate $v_m \propto e^{-\beta z^2}$. Furthermore, any temporal variations in the current stimulus that are fast compared with the time constant, τ_M, will be smoothed out and not transmitted down the cable. As indicated in Table 3.1, τ_M is on the order of a few milliseconds for many nerve cells and somewhat longer for muscle cells.

Comparison to the Time-Independent Solutions
The spatial dependences of the time-independent (Equation 3.25) and time-dependent (Equation 3.43) responses to a point source of current are compared directly in Figure 3.25. The time-independent solution is shown for a constant current, and the time-dependent solution is shown for a temporal impulse of current. The figure graphically demonstrates the much more rapid spatial attenuation of the time-dependent solution.

Superposition
The time-dependent solution exhibits superposition in space, just as does the time-independent solution. Thus, the response to two electrodes, each with an impulse of current, yields a spatial superposition of the responses from each electrode obtained separately, as is illustrated in Figure 3.19. However, in addition the time-dependent solution shows superposition in the time domain, as is illustrated in Figure 3.26. At each location along the cable, the response to two current impulses delivered through an infinitesimal electrode consists of the sum of responses to each impulse alone. If the duration between impulses is brief, then there is appreciable overlap between the individual responses. If the duration is increased, overlap decreases. However, at a given duration between impulses, the overlap of responses is greater for locations farther from the electrode location. This results because the response duration increases as distance from the electrode increases.

Effect of Membrane Conductance
It is of interest to examine the effect of the membrane conductance on cable theory. The left panel of Figure 3.27 summarizes major results of cable theory that have just been derived. The right panel shows the results that are obtained when the membrane conductance is zero. With zero membrane conductance, the partial differential equation changes from the cable equation, which is a modified diffusion equation, to the ordinary diffusion equation. The Green's function changes from a Gaussian function multiplied by an exponential to a Gaussian function. Since the potential across the membrane is proportional to the charge on the membrane, we can interpret the integral of

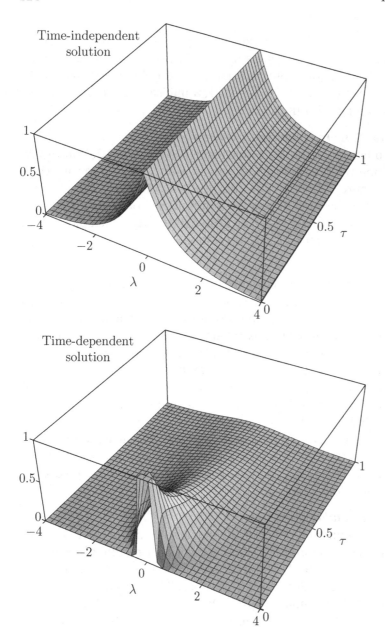

Time-independent
solution

Time-dependent
solution

Figure 3.25 Comparison
of waveforms of the time-
independent (upper panel)
and time-dependent (lower
panel) solutions of the cable
equation in response to point
current sources. The membrane
potential responses are shown
normalized. The upper panel
shows the function $e^{-|\lambda|}$, and the
lower panel shows the quantity
$(1/\sqrt{4\pi\tau})e^{-\lambda^2/4\tau}e^{-\tau}$, both plotted
versus λ and τ.

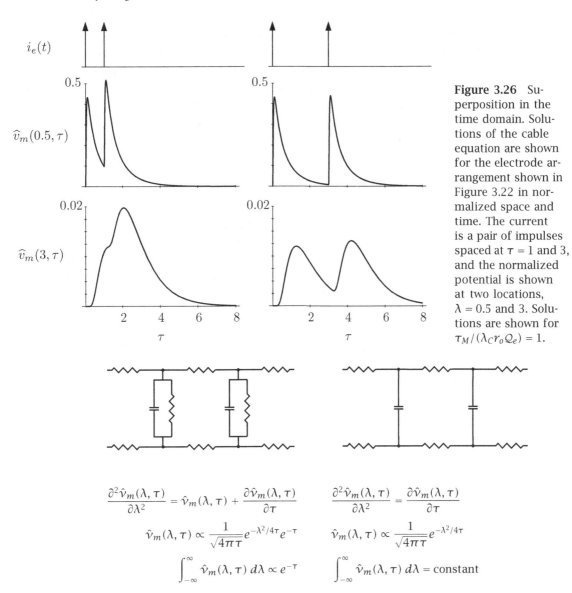

Figure 3.26 Superposition in the time domain. Solutions of the cable equation are shown for the electrode arrangement shown in Figure 3.22 in normalized space and time. The current is a pair of impulses spaced at $\tau = 1$ and 3, and the normalized potential is shown at two locations, $\lambda = 0.5$ and 3. Solutions are shown for $\tau_M / (\lambda_C r_o \mathcal{Q}_e) = 1$.

$$\frac{\partial^2 \hat{v}_m(\lambda, \tau)}{\partial \lambda^2} = \hat{v}_m(\lambda, \tau) + \frac{\partial \hat{v}_m(\lambda, \tau)}{\partial \tau} \qquad \frac{\partial^2 \hat{v}_m(\lambda, \tau)}{\partial \lambda^2} = \frac{\partial \hat{v}_m(\lambda, \tau)}{\partial \tau}$$

$$\hat{v}_m(\lambda, \tau) \propto \frac{1}{\sqrt{4\pi\tau}} e^{-\lambda^2/4\tau} e^{-\tau} \qquad \hat{v}_m(\lambda, \tau) \propto \frac{1}{\sqrt{4\pi\tau}} e^{-\lambda^2/4\tau}$$

$$\int_{-\infty}^{\infty} \hat{v}_m(\lambda, \tau)\, d\lambda \propto e^{-\tau} \qquad \int_{-\infty}^{\infty} \hat{v}_m(\lambda, \tau)\, d\lambda = \text{constant}$$

Figure 3.27 Effect of membrane conductance on cable theory. The left panels show key results for a cable with nonzero membrane conductance; the right panel shows the same for zero membrane conductance.

the potential on space as proportional to the total charge on the membrane. With a nonzero membrane conductance, the total charge on the membrane decays exponentially in time; with zero membrane conductance, the total charge on the membrane is constant in time. Thus, with nonzero membrane conductance, the charge dissipates, i.e., the total charge goes to zero. However, with a zero membrane conductance, the charge on the membrane diffuses along the membrane, but the total charge on the membrane remains constant, i.e., the charge is not dissipated.

3.4.3.2 Step Response of an Infinite Cable

To find the response to an external current, $i_e(t) = I_e u(t)$, where $u(t)$ is the unit step function, we make use of the linearity of the relation between external current and membrane potential. Since the step function is the integral of the impulse function, the response to the step will be the integral of the response to the impulse. Thus, we integrate and scale Equation 3.45 to obtain the response to the step of current as follows:

$$v_m(z,t) = \frac{r_o \lambda_C I_e}{\tau_M} \int_0^t \frac{1}{\sqrt{4\pi(t'/\tau_M)}} e^{-(z/\lambda_C)^2/(4t'/\tau_M)} e^{-t'/\tau_M} \, dt' \text{ for } t \geq 0. \qquad (3.46)$$

We express the potential in the normalized coordinates $\tau = t/\tau_M$, $\tau' = t'/\tau_M$, and $\lambda = z/\lambda_C$ as

$$\hat{v}_m(\lambda, \tau) = r_o \lambda_C I_e \int_0^\tau \frac{1}{\sqrt{4\pi\tau'}} e^{-(\lambda^2/4\tau' + \tau')} \, d\tau' \text{ for } \tau \geq 0. \qquad (3.47)$$

To evaluate this expression, we make the substitutions $\tau' = w^2$ and $a = |\lambda|$, and then after combining terms we obtain

$$\hat{v}_m(\lambda, \tau) = \frac{r_o \lambda_C I_e}{\sqrt{\pi}} \int_0^{\sqrt{\tau}} e^{-(a^2/4w^2 + w^2)} \, dw \text{ for } \tau \geq 0.$$

Now we complete the square in the exponent in two ways as follows:

$$\hat{v}_m(\lambda, \tau) = \frac{r_o \lambda_C I_e}{2\sqrt{\pi}} \int_0^{\sqrt{\tau}} \left(e^{-(a/2w + w)^2 + a} + e^{-(a/2w - w)^2 - a} \right) \, dw \text{ for } \tau \geq 0.$$

The two terms in the integrand are equal, so that

$$\hat{v}_m(\lambda, \tau) = \frac{r_o \lambda_C I_e}{2\sqrt{\pi}} \int_0^{\sqrt{\tau}} \Big((1 - a/2w^2) e^{-(a/2w + w)^2 + a} +$$

$$(1 + a/2w^2) e^{-(a/2w - w)^2 - a} \Big) \, dw \text{ for } \tau \geq 0, \qquad (3.48)$$

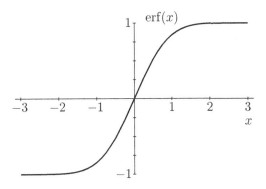

Figure 3.28 The error function.

which can be written as

$$\hat{v}_m(\lambda, \tau) = \frac{r_o \lambda_C I_e}{2\sqrt{\pi}} \left(e^a \int_0^{\sqrt{\tau}} e^{-(w+a/2w)^2} \, d(w + a/2w) + \right.$$

$$\left. e^{-a} \int_0^{\sqrt{\tau}} e^{-(w-a/2w)^2} \, d(w - a/2w) \right) \text{ for } \tau \ge 0. \tag{3.49}$$

Now we let the dummy variable of integration $y = w + a/2w$ in the first integral and $y = w - a/2w$ in the second integral to yield

$$\hat{v}_m(\lambda, \tau) = \frac{r_o \lambda_C I_e}{2\sqrt{\pi}} \left(e^a \int_\infty^{\sqrt{\tau}+a/2\sqrt{\tau}} e^{-y^2} \, dy + e^{-a} \int_{-\infty}^{\sqrt{\tau}-a/2\sqrt{\tau}} e^{-y^2} \, dy \right) \text{ for } \tau \ge 0.$$

This equation can be expressed in terms of the *error function,* defined as

$$\text{erf}(x) = \frac{2}{\sqrt{\pi}} \int_0^x e^{-y^2} \, dy \tag{3.50}$$

and shown in Figure 3.28. Figure 3.28 shows that $\text{erf}(-\infty) = -1$, $\text{erf}(0) = 0$, and $\text{erf}(\infty) = 1$ and that the error function is an odd function of its argument.[1] Therefore, if we make use of the error function, we can express the step response in normalized coordinates as

$$\hat{v}_m(\lambda, \tau) = \frac{r_o \lambda_C I_e}{4} \left(e^{|\lambda|} \left(\text{erf}(\sqrt{\tau} + \frac{|\lambda|}{2\sqrt{\tau}}) - 1 \right) + \right.$$

$$\left. e^{-|\lambda|} \left(\text{erf}(\sqrt{\tau} - \frac{|\lambda|}{2\sqrt{\tau}}) + 1 \right) \right) \text{ for } \tau \ge 0. \tag{3.51}$$

---

1. Some properties of the Gaussian function, on which the error function is based, are discussed elsewhere (Weiss, 1996, Appendix 3.2).

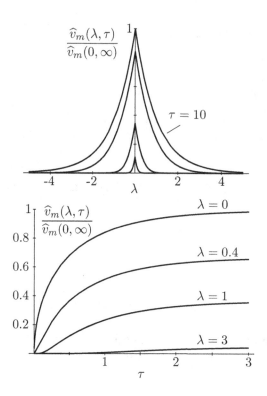

Figure 3.29 The step response of the cable equation in normalized coordinates, with $\hat{v}_m(\lambda, \tau)/\hat{v}_m(0, \infty)$ plotted versus λ for values of τ = 0.01, 0.1, 1, and 10 (upper panel) and versus τ for λ = 0, 0.4, 1, and 3 (lower panel).

The step response can be expressed somewhat more compactly in terms of the complementary error function, defined as erfc$(x) = 1 - erf(x)$, as follows:

$$\hat{v}_m(\lambda, \tau) = \frac{r_o \lambda_C I_e}{4} \left(e^{-|\lambda|} \text{erfc} \left(\frac{|\lambda|}{2\sqrt{\tau}} - \sqrt{\tau} \right) \right.$$

$$\left. - e^{|\lambda|} \text{erfc} \left(\frac{|\lambda|}{2\sqrt{\tau}} + \sqrt{\tau} \right) \right) \quad \text{for } \tau \geq 0.$$

(3.52)

The step response is plotted in normalized coordinates in Figure 3.29. At the onset of the current, the membrane potential is small and its spatial extent is also small. As time increases, the potential increases and the spatial extent approaches a limiting time-independent form that is easily found from Equation 3.51 by letting $\tau \to \infty$. Note that in this limit, the error functions in both parts of the equation approach 1. Therefore,

$$\lim_{\tau \to \infty} \hat{v}_m(\lambda, \tau) = \hat{v}_m(\lambda, \infty) = \frac{r_o \lambda_C}{2} I_e e^{-|\lambda|},$$

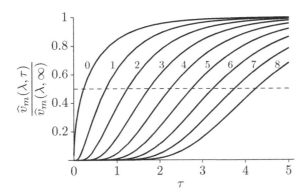

Figure 3.30 The normalized step response of the cable equation, with $\hat{v}_m(\lambda, \tau)/\hat{v}_m(\lambda, \infty)$ plotted versus τ for the indicated values of λ.

which agrees with the result we obtained by solving the time-independent cable equation (Equation 3.25).

How does the time it takes to charge the cable depend upon location along the cable? Since different locations along the cable charge to different steady-state voltages, we will normalize the step response of the cable to its steady-state value to obtain

$$\frac{\hat{v}_m(\lambda, \tau)}{\hat{v}_m(\lambda, \infty)} = \frac{1}{2}\left(\text{erfc}\left(\frac{|\lambda|}{2\sqrt{\tau}} - \sqrt{\tau}\right) - e^{2|\lambda|}\text{erfc}\left(\frac{|\lambda|}{2\sqrt{\tau}} + \sqrt{\tau}\right)\right) \text{ for } \tau \geq 0. \quad (3.53)$$

The normalized membrane potential of Equation 3.53 is shown plotted in Figure 3.30 as a function of τ for different values of λ. Figure 3.30 indicates that the normalized time, τ, it takes to charge the membrane potential to half its steady-state value appears to be related linearly to normalized distance, λ. Figure 3.31 shows that for $\lambda \geq 0$, $\lambda \approx 2\tau_{1/2} - 1/2$, where $\tau_{1/2}$ is the normalized time at which the membrane potential is half its steady-state value (Jack et al., 1975). Thus, we can define the "velocity" of the point at which the membrane potential is half its steady state as $d\lambda/d\tau_{1/2} = 2$. This velocity can be expressed in unnormalized coordinates as $v_{1/2} = dz/dt_{1/2} = 2\lambda_C/\tau_M$. Assuming that $r_i \gg r_o$, we obtain

$$v_{1/2} = \frac{2\lambda_C}{\tau_M} = 2\frac{1/\sqrt{g_m r_i}}{c_m/g_m} = \frac{2}{c_m}\sqrt{\frac{g_m}{r_i}}.$$

For a cylindrical cell of radius a, we obtain

$$v_{1/2} = \frac{2}{2\pi a C_m}\sqrt{\frac{2\pi a G_m}{\rho_i/(\pi a^2)}} = \sqrt{\frac{2aG_m}{\rho_i C_m^2}}.$$

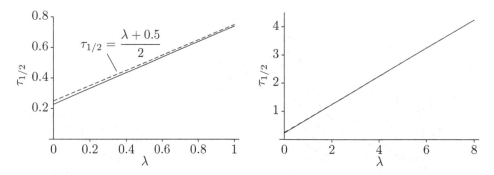

Figure 3.31 The time required for the step response of the cable equation to reach half its steady-state value, $\tau_{1/2}$, as a function of location along the cable, λ. The left panel shows the relation for small values of λ, and the right for larger values of λ. The dashed line shows the equation $\tau_{1/2} = (\lambda + 0.5)/2$; the solid line shows solutions for the equation $\hat{v}_m(\lambda, \tau_{1/2})/\hat{v}_m(\lambda, \infty) = 1/2$. The dashed line is not appreciably different from the exact solution for $\lambda > 1$.

The dependence of this velocity on the cell parameters can be understood through the dependence of the cell space constant and the membrane time constant on these parameters. For example, the velocity of this decrementally conducted step response is proportional to the square root of the cell radius, because the cell space constant is proportional to the square root of the radius and the membrane time constant is independent of the radius. Thus, all other factors being equal, larger cells have a step response that spreads more rapidly along the cell. The velocity is also inversely proportional to the specific membrane capacitance, because the cell space constant is independent of the capacitance and the membrane time constant is proportional to the capacitance. Thus, a larger membrane capacitance produces a slower response, acting over the same distance scale to produce a smaller velocity. The velocity is inversely proportional to the square root of the intracellular resistivity, because this resistivity affects the cell space constant. The effect of the specific membrane conductance is more complex. An increase in the specific membrane conductance decreases both the cell space constant and the membrane time constant, but the space constant decreases proportionally with $1/\sqrt{G_m}$, whereas the time constant decreases proportionally with $1/G_m$. Hence, the velocity increases as the square root of the specific membrane conductance.

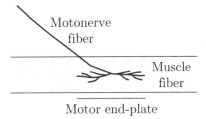

Figure 3.32 Schematic diagram of the motor end-plate region of a muscle fiber.

3.4.3.3 Comparison with Measurements

The solution of the cable equation for a brief current stimulus applied at a point along a cable has a number of physiologically important applications. We shall briefly describe one of these. At a nerve-muscle junction, an action potential in a motor nerve fiber triggers the release of a neurotransmitter substance (acetylcholine at the vertebrate neuromuscular junction) at a synaptic region of the muscle called the *end-plate* (Figure 3.32). The transmitter is released in multimolecular units called *quanta*. Each action potential releases many quanta of neurotransmitter. The neurotransmitter acts on the muscle membrane to open specific ionic channels, causing a brief flow of current through the muscle membrane in the region of the end-plate. This current produces a graded potential across the muscle membrane at the end-plate, called the *end-plate potential* (EPP). The EPP spreads along the muscle fiber and triggers an action potential that travels down the muscle fiber. The muscle action potential produces a muscle contraction. In addition to the action potential-evoked release of the neurotransmitter, there is also a spontaneous release of neurotransmitter in quantal units. The arrival of these spontaneously released neurotransmitter quanta causes a *miniature end-plate potential* (MEPP) across the muscle membrane. The waveforms of the MEPP and the EPP are virtually identical, but the amplitude of the EPP exceeds that of the MEPP by about two orders of magnitude. Here we shall explore one step in this sequence of events—the spread of the EPP and the MEPP along the muscle fiber.

In a toad sartorius muscle fiber, the end-plate region has a spatial extent on the order of 10^2 μm, and the ionic channels opened by the transmitter are open for about 1 to 2 ms. The space constant of this muscle is about 2 mm, and the membrane time constant is about 8 ms. Hence, for this muscle fiber the spatial extent of the end-plate region, and therefore the region over which current flows through the muscle membrane due to the

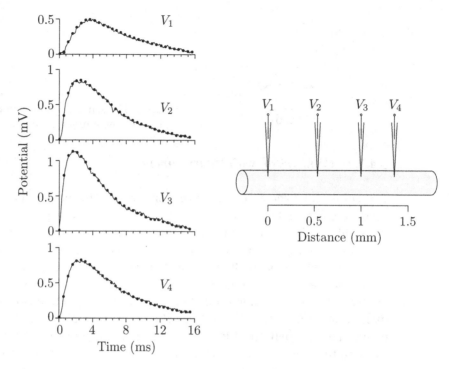

Figure 3.33 Miniature end-plate potentials (MEPPs) recorded simultaneously at four positions along a muscle fiber (adapted from Gage and McBurney, 1973, Figure 6). The measured potentials are indicated by the line, and calculations based on the impulse response of a cable are indicated by points. The distances of successive electrodes from the leftmost electrode were 0.55, 1, and 1.315 mm. The electrodes straddled the end-plate region, which appears to have been closest to electrode 3 judging from the magnitude of the MEPP.

transmitter, is much less than the space constant of the muscle cell. Also, the duration of this current is much shorter than the time constant of the muscle membrane. Therefore, the physical situation is approximated rather well by a brief pulse (an impulse) of current delivered at a point along the muscle.

Figure 3.33 shows measurements of MEPPs recorded simultaneously at four locations across the membrane of a frog sartorius muscle fiber, together with the predictions of the cable equation with an impulse of current applied at the location of the end-plate. The cable parameters of the muscle (r_i, τ_M, and λ_C), were measured independently by passing current through an intracellular micropipette and recording the voltage responses. The location of the end-plate was determined from the measured MEPPs by plotting the magnitude of the response versus distance, fitting the curves with straight lines on semilogarithmic coordinates, and interpolating. With this estimate of the location of the end-plate and the cable parameters, the impulse response was computed as a function of time at the known locations of the electrodes. The potential predicted by the cable model fits the measured MEPPs to within the variations in the measurements. The agreement between theory and experiment indicates that these MEPPs spread along the muscle according to the cable model.

3.4.4 Implications of Cable Properties

Cable properties have implications for the integration of graded potentials in neurons, a basic step neurons use to process incoming information. Recall from Chapter 1 that neurons transmit information over long distances in the form of a sequence of action potentials. These action potentials arrive at synaptic sites on a target cell and cause graded postsynaptic potentials. These postsynaptic potentials either trigger action potentials that propagate without decrement along the cell's surface membrane or propagate along the target cell according to the cable model. Both mechanisms are known to occur. In this section we shall consider only the latter possibility. A quantitative description of the propagation of graded potentials in a dendritic tree is beyond the scope of this chapter. This topic has been illuminated in the many publications of Rall and his colleagues. Their results are reviewed in a number of publications (Jack et al., 1975; Rall, 1977; Rall, 1989; Johnston and Wu, 1995), and Rall's papers have recently been reprinted (Segev et al., 1995).

Because graded potentials summate, both the timing of postsynaptic potentials and their spatial distribution on the target cell are critical in determining whether the trigger zone of the target cell produces an outgoing action potential. A number of factors are important in determining the efficacy of synaptic input to the target cell. Some of these can be understood in terms of the cable model.

3.4.4.1 *Electrotonic Distance*

As we have seen, solutions of the cable equation are the same for all cables if computed in normalized coordinates in which distance is divided by

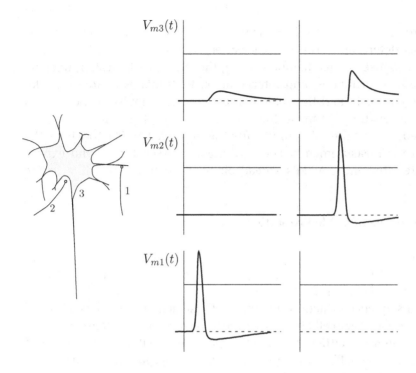

Figure 3.34 Schematic diagram illustrating the concept of electrotonic distance. Each column of waveforms shows electric potentials that occur when an action potential propagates on an axon that terminates on the target cell (shown shaded). $V_{m1}(t)$ and $V_{m2}(t)$ are the potentials across the membranes of two input cells at locations 1 and 2, respectively. $V_{m3}(t)$ is the potential across the membrane at the trigger zone of the target cell (location 3).

the space constant and time is divided by the time constant. Therefore, the cable model implies that if all other things are equal, the efficacy of individual inputs to a target cell is determined by the *electrotonic distance* of the synaptic input from the trigger zone of the cell, as shown schematically in Figure 3.34. The electrotonic distance is defined as the distance divided by the space constant. An ending near the trigger zone produces a large postsynaptic potential at the trigger zone, whereas an ending located on a distal site of a dendrite many space constants away from the trigger zone produces a much smaller response. Nevertheless, a sufficient number of distal synapses on the target cell, each producing small postsynaptic potentials, can produce a larger potential at the trigger zone than a single input near the trigger zone.

3.4.4.2 *Temporal Integration*

The timing of incoming action potentials along an input axon can determine whether the postsynaptic cell produces an action potential (Figure 3.35). If the time between two input action potentials is relatively long, there is no output

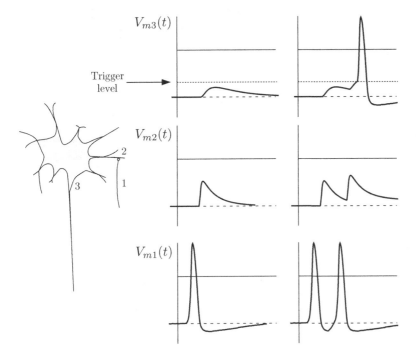

Figure 3.35 Schematic diagram illustrating temporal integration. The left column of waveforms shows the responses that occur when a single action potential occurs in the input cell. The right column shows the responses when two closely spaced action potentials occur in the input cell. Membrane potentials are shown in the input cell at location 1 (V_{m1}), as well as at two locations in the target cell—in the dendrite at location 2 (V_{m2}) and at the trigger zone at location 3 (V_{m3}).

action potential. If the time is short, the postsynaptic potentials summate, so that the resultant potential at the trigger zone of the cell exceeds threshold and an output action potential is generated. This type of summation of temporal signals is called *temporal integration*. The temporal metric is the time constant. Thus, inputs separated by a fraction of a time constant will produce a larger net effect than inputs separated by many time constants.

3.4.4.3 *Spatial Integration*

Graded postsynaptic potentials generated in different parts of a cell also summate to produce a graded potential at the trigger zone. Thus, input action potentials generated individually along two different inputs may cause a graded potential at the trigger zone that is insufficient to trigger an output action potential. However, if action potentials occur in both cells within a short time interval, then an action potential can be generated in the target cell (Figure 3.36). This type of integration is called *spatial integration*.

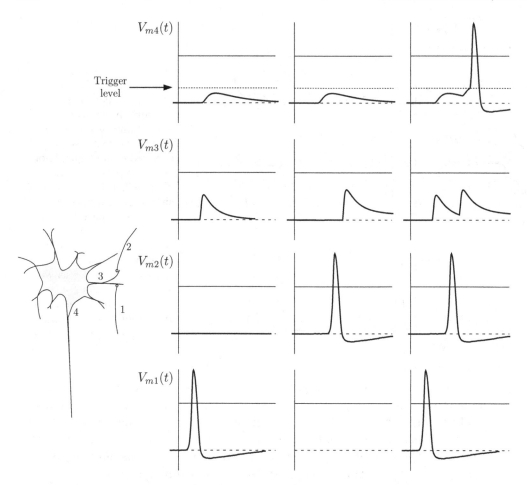

Figure 3.36 Schematic diagram illustrating spatial integration. The three columns of waveforms show membrane potentials at key sites when (left column) an action potential propagates in one input cell (at site 1), (center column) an action potential propagates in another input cell (at site 2), and (right column) action potentials propagate in both input cells. Potentials in the target cell are shown in the dendrite (at site 3) and at the trigger zone (at site 4).

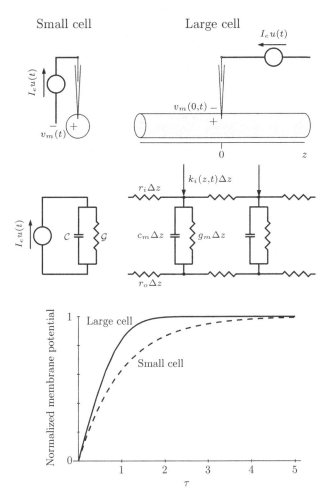

Figure 3.37 A comparison between recording arrangements (upper panel), electrical network models (middle panel), and step responses (lower panel) for small and large cells. The small spherical cell is impaled with a micropipette, and the incremental membrane potential $v_m(t)$ is recorded. The large cylindrical cell is impaled at $z = 0$, and the incremental membrane potential $v_m(0, t)$ is recorded at that location. The electrical network model for the small cell is a lumped-parameter circuit with total conductance $\mathcal{G}$ and total capacitance $\mathcal{C}$. The electrical network model for the large cell is a distributed network model, or cable, with resistances per unit length of r_i and r_o and membrane conductance and capacitance per unit length of g_m and c_m, respectively. The step responses of both cells are shown in normalized coordinates, $\hat{v}_m(\tau)/\hat{v}_m(\infty)$ for the small cell and $\hat{v}_m(0, \tau)/\hat{v}_m(0, \infty)$ for the large cell.

3.5 Summary: A Comparison of Small and Large Cells

The relation between the properties of electrically small and large cells is summarized in Figure 3.37. For small cells, the potential across the membrane is independent of position and depends on only one independent variable—time. For a cell whose membrane is electrically uniform and can be represented by a simple conductance and capacitance in parallel, the electrical network model of the cell is a lumped-parameter network consisting of a conductance and a capacitance in parallel. The relation between an external current source and the membrane potential is that of a first-order linear ordinary differential equation with constant coefficients (Equation 3.4). The response to a step of current across the membrane is a simple exponential increase in membrane potential whose time dependence is governed by one parameter—the time constant.

For a large cell, the potential across the membrane depends upon position along the cell as well as on time. For a cell whose membrane is electrically uniform and can be represented by a simple conductance and capacitance in parallel, the electrical network model of the cell is a distributed-parameter model, a cable, consisting of distributed conductances and capacitances. The relation between the external current, which can itself be distributed in space, and membrane potential is that of a second-order linear partial differential equation with constant coefficients—a modified diffusion equation (Equation 3.15). The response of such a network to a step of current depends upon both time and position and is given by the error functions shown in Figure 3.29. The spatial and temporal dependence are governed by two parameters—the time constant and the space constant. The time constant is a property of the membrane and is independent of cell dimensions; the space constant depends upon the cell dimensions.

A comparison of responses at the stimulating electrodes of the small and those of the large cell reveals differences in response kinetics. In normalized coordinates, it is readily seen that the response for the small cell is

$$\hat{v}_m(\tau) = \mathcal{R}I_e \left(1 - e^{-\tau}\right) u(\tau), \tag{3.54}$$

where $\mathcal{R} = 1/\mathcal{G}$ is the total resistance of the membrane of the cell, $\tau_M = \mathcal{R}C$, and $\tau = t/\tau_M$. The response for the large cell can be obtained from an equation identical to Equation 3.51 except for a substitution of r_i for r_o, because in Figure 3.37 the external current is applied to the internal rather than the external conductor (see Problem 2.6). With that substitution, we can evaluate Equation 3.51 at $\lambda = z/\lambda_C = 0$ to obtain

$$\hat{v}_m(0, \tau) = \frac{\lambda_C r_i}{2} I_e \text{erf}\left(\sqrt{\tau}\right) u(\tau). \tag{3.55}$$

As shown in Figure 3.37, if the small and large cells have the same membrane time constant τ_M, then the kinetics for charging the membrane potential of the large cell are appreciably faster than for charging the membrane of the small cell.

A given cell can be considered an electrically small cell if its dimensions are appreciably smaller than its space constant. Under these conditions, a lumped-parameter network model can be used to represent the relation between incremental voltage and current. The simplest such model that accounts for electrophysiological properties of cells is a parallel combination of a conductance and a capacitance. When the cell dimensions are not appreciably smaller than the cell space constant, then the cell is an electrically large cell and the relation between current and voltage is that given by a distributed-parameter network. The simplest such model is the conductance-capacitance cable model, which accounts quite well for the spread of potential along cylindrical cells such as invertebrate axons and muscle cells.

Exercises

3.1 Define both the space constant and the time constant.

3.2 Does the time constant of a cylindrical cell depend on its dimensions? Does the space constant of a cylindrical cell depend on its dimensions?

3.3 The space constant decreases as the specific membrane conductance is increased. Give a physical explanation of this result.

3.4 Give a quantitative explanation of what is meant by an *infinitesimal electrode*.

3.5 For each of the following statements, assume that the electrical properties of a patch of the membrane of the cell can be represented as a parallel resistance and capacitance. Assume that the cell has a cylindrical shape, with a radius that is small compared to the length of the cell. Determine if each assertion is true or false, and give a reason for your choice.

　a. For an electrically small cell, the membrane potential in response to a step of current through the membrane is an exponential function of time.

　b. For an electrically small cell, the steady-state value of the membrane potential in response to a step of current applied through the membrane at one position along the cell is a constant that is independent of position along the cell.

　c. For an electrically large cell, the membrane potential in response to a step of current applied through the membrane at one position along the cell is an exponential function of time.

　d. For an electrically large cell, the steady-state value of the membrane potential in response to a step of current applied through the membrane at one position along the cell is an exponential function of longitudinal position along the cell.

e. For an electrically large cell, the steady-state value of the membrane potential in response to a step of current applied through the membrane at one position along the cell is a Gaussian function of position along the cell.

3.6 What are the distinctions among the conductance variables G_m, $\mathcal{G}_m$, and g_m?

3.7 Physically, what does the characteristic conductance of an infinite cable represent?

3.8 Explain why a cell is an electrically small cell if its dimensions are small compared to the space constant.

3.9 Explain why the intersection of the vertical dashed line with the abscissa in Figure 3.20 gives the space constant of the muscle cell.

3.10 Figure 3.13 shows all the voltage and current variables as a function of position along an infinite cylindrical cell in response to a constant current applied extracellularly to the cell.

a. Explain why the external longitudinal current, $\hat{\imath}_o(\lambda)$, is discontinuous at $\lambda = 0$, while the internal longitudinal current, $\hat{\imath}_i(\lambda)$, is continuous at $\lambda = 0$.

b. Explain why the internal potential, $\hat{v}_i(\lambda)$, approaches zero as r_i/r_o goes to zero.

c. Both the internal potential, $\hat{v}_i(\lambda)$, and the external potential, $\hat{v}_o(\lambda)$, approach the same linear dependence on λ for large values of λ. Explain why the magnitude of the slope of this line increases as r_i/r_o increases.

3.11 Figure 3.31 shows that the time required for the step response of a linear cable to reach $1/2$ its final value is a linear function of position along the cable. Does this result imply that the step response propagates at constant velocity along the cable? Explain.

3.12 Given two cells, one an electrically small cell and the other an electrically large cell with the same membrane time constant, is the step response of the electrically small cell slower, faster, or the same as the step response of the electrically large cell?

$I_m(t)$

$V_m(t)$ ◯ Source

Figure 3.38 Space-clamped axon (Problem 3.2).

3.13 Equations 3.5 and 3.52 both express the membrane potential response of a cell to a step of current. They are mathematically quite different. Using words—no mathematics—explain the physical bases of the differences in these results.

Problems

3.1 A squid axon is placed in a large volume of seawater so that you may assume that $r_o \ll r_i$. The following data are given: resistivity (specific resistance) of squid axoplasm, $\rho_i = 30\,\Omega \cdot$ cm; diameter of axon, 500 μm; space constant, $\lambda_C = 6$ mm; thickness of membrane, $d = 50$ Å; capacitance per unit area of membrane, $C_m = 1\mu F/cm^2$.

a. Find the conductance of the axon per unit length, g_m.

b. Find the conductance of the axon per unit area, G_m.

c. Find λ_C for unmyelinated axons whose membranes have specific properties (i.e., G_m and ρ_i) that are identical to those of the squid axon, but whose diameters are 1 mm, 0.1 mm, 0.01 mm, and 0.001 mm.

d. Find τ_M for the squid axon and for the same axons considered in part c.

3.2 A giant nerve fiber is arranged with external and internal electrodes that effectively short circuit the resistance of the outer and inner conductors (Figure 3.38). These electrodes eliminate spatial variations of the voltages and currents (a *space clamp*). Assume that $V_m(t) = V_m^o$ for $t < 0$ and that the changes in $V_m(t)$ from the resting value, V_m^o, are small enough that the nerve membrane can be represented as a circuit with constant capacitance C_m, constant resistance $\mathcal{R}_m$, and resting potential V_m^o.

a. Assume that for $t < 0$, $V_m = V_m^o$ and that the electrodes are connected to a current source such that $I_m(t) = Iu(t)$. Determine $V_m(t)$ for $t \geq 0$.

b. If the source is an ideal voltage source such that $V_m(t) = V_m^o + Vu(t)$, determine the membrane current $I_m(t)$.

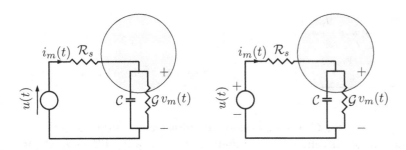

Figure 3.39 Effects of series resistance (Problem 3.3). The unit step function is $u(t)$.

c. If the source is represented as an ideal voltage source identical to that in part b in series with a resistance R_s, determine and sketch $V_m(t)$ and $I_m(t)$ for $R_s = 10^{-2}R_m$, $R_s = R_m$, and $R_s = 10^2 R_m$.

3.3 In the model of a small cell shown in Figure 3.4, the resistances of the electrode, the external medium, and the cytoplasm of the cell were ignored. This problem is concerned with the effects of such a series resistance. As shown in Figure 3.39, a simple representation of this resistance is as a resistance, R_s, in series with the membrane.

a. For an external current that is a unit step (left panel of Figure 3.39), find the change in potential across the membrane, $v_m(t)$. What is the effect of the series resistance on the potential across the membrane?

b. For an external voltage that is a unit step (right panel of Figure 3.39), find the potential across the membrane. What is the effect of the series resistance on the potential across the membrane?

3.4 Constant currents I_1 and I_2 are applied to the exteriors of axons 1 and 2 (Figure 3.40), respectively, and the resulting time-independent changes in membrane potential are $v_{m1}(z)$ and $v_{m2}(z)$, respectively. I_1 and I_2 are adjusted so that $v_{m1}(0) = v_{m2}(0) = 10$mV. This change in potential is sufficiently small that the membrane voltage-current characteristic may be assumed to be linear. You may also assume that $r_o \ll r_i$ for both axons and that r_o is the same for both axons. Data on axons 1 and 2 are given in Table 3.3, where a is the axon radius, ρ_i is the cytoplasmic resistivity, and G_m is the specific membrane conductance.

a. Let $v_{m1}(-0.1)$ and $v_{m2}(-0.1)$ be the membrane potential changes at $z = -0.1$cm for the two axons. Determine the value of the ratio $A = v_{m1}(-0.1)/v_{m2}(-0.1)$.

b. Determine the value of the ratio $B = I_1/I_2$.

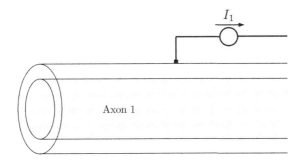

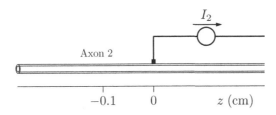

$-0.1 \quad 0 \qquad z \text{ (cm)}$

Figure 3.40 External stimulation of two axons (Problem 3.4).

Table 3.3 Data on two axons (Problem 3.4).

Axon no.	a (μm)	ρ_i ($\Omega \cdot$cm)	G_m (S/cm^2)
1	100	100	5×10^{-3}
2	10	100	$(1/8) \times 10^{-3}$

3.5 An axon is stimulated by an externally applied constant current source, I_e, and the potential V_e across the current source is measured as shown in Figure 3.41. Assume that the external current is applied to the axon through cylindrically symmetric infinitesimal electrodes separated by the distance l. You may assume that the cable model applies.

a. Find and sketch $v_m(z)$.

b. Find and sketch $i_o(z)$, the external longitudinal current.

c. Find the resistance between the electrodes, $\mathcal{R}_e = V_e/I_e$.

3.6 The first accurate estimates of the electrical resistance of a biological membrane were obtained by the experimental arrangement shown in Figure 3.42. A squid axon was sealed at both ends and suspended between

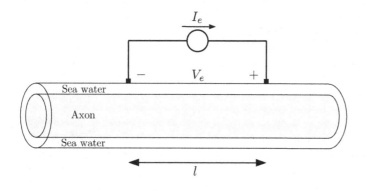

Figure 3.41 Arrangement of electrodes (Problem 3.5).

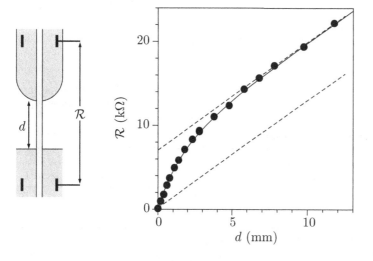

Figure 3.42 Method of measuring membrane resistance (left) for an axon, along with results (right) (adapted from Cole and Hodgkin, 1939, Figure 3) (Problem 3.6).

two seawater baths separated by oil. The oil gap between the baths, d, could be varied. The resistance $\mathcal{R}$ between baths was measured as a function of d, with the results shown in Figure 3.42.

a. Derive a relation between $\mathcal{R}$ and d assuming the axon can be represented by the cable model.

b. Using the results from part a, estimate the values of r_i, r_o, and g_m from the measurements shown in Figure 3.42.

3.7   This problem is concerned with the time-independent (i.e., $\partial v_m/\partial t = 0$) spread of potential in dendrites. You may assume that a dendrite can be represented as a linear cable with the following cable parameters:

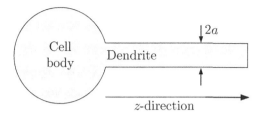

Figure 3.43 Schematic diagram of a cell body and a dendrite (Problem 3.7).

c_m is the membrane capacitance per unit length,

g_m is the membrane conductance per unit length,

r_o is the resistance per unit length of the outer conductor,

r_i is the resistance per unit length of the inner conductor.

a. Consider a dendrite that is oriented in the z-direction as shown in Figure 3.43. Let $z = 0$ be at the junction of the dendrite with the cell body. Assume that the electrical properties of the dendrite are uniform along its length and that the dendrite is infinitely long. The cell body is uniformly depolarized, so that $V_m^{cb} = V_1 + V_m^o$. Determine and sketch the incremental membrane potential in the dendrite, $v_m(z) = V_m(z) - V_m^o$.

b. Determine the incremental longitudinal current inside the dendrite, $i_i(z)$.

c. Determine the ratio $v_m(z)/i_i(z)$ at $z = 0$.

d. By definition, the characteristic conductance of the dendrite is $G_c = \sqrt{g_m/(r_o + r_i)}$. Assume that $r_o \ll r_i$, and determine an expression for G_c that involves the dimensions and the specific properties of the cytoplasm and membrane only. That is, determine G_c in terms of a, G_m, and ρ_i, where

   a is the radius of the dendrite,

   G_m is the conductance per unit area of membrane,

   ρ_i is the resistivity of the inner conductor (cytoplasm).

e. Now consider a dendrite that is bifurcated into two (infinitely long) branches, each of radius b, as shown in Figure 3.44. At the bifurcation, the inside longitudinal current divides between the two branches; the membrane potential is the same in all three portions of the dendrite. Assume that $r_o \ll r_i$ and that G_m and ρ_i are the same in the cell body and in all the dendrite branches. Determine the relation between the radii a and b such that the relation of voltage to current in the cell

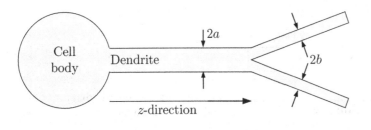

Figure 3.44 Schematic diagram of a cell body and a branched dendrite (Problem 3.7).

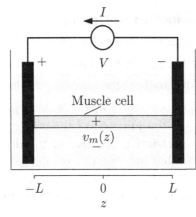

Figure 3.45 Arrangement of electrodes for measuring the resistance of a muscle (Problem 3.8).

body is the same for an infinite dendrite with radius a (as shown in Figure 3.43) or a branched dendrite (as shown in Figure 3.44).

3.8   A skeletal muscle cell is placed in a bath between two electrodes (Figure 3.45) that deliver a constant current, I; the potential measured across these electrodes is V. This current causes an incremental membrane potential, $v_m(z)$, that is sufficiently small that the cable model applies. The external resistance per unit length is r_o, and the internal resistance per unit length is r_i; the space constant is λ_C. Because the muscle membrane has a high resistance, you may assume that at the electrodes the incremental internal longitudinal current is zero, i.e., $i_i(-L) = i_i(+L) = 0$.

a. The dependence of membrane potential on distance has the form

$$v_m(z) = A \sinh(z/\lambda_C) \qquad \text{or} \qquad v_m(z) = A \cosh(z/\lambda_C).$$

Determine which form is correct. Explain your reasons carefully.

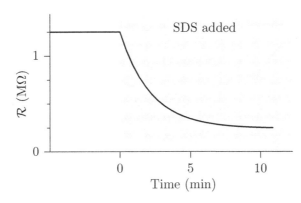

Figure 3.46 Effect of the membrane detergent SDS on resistance of the muscle (Problem 3.8).

b. Find A in terms of the I, r_o, r_i, and λ_C.

c. Determine the external longitudinal current $i_o(z)$.

d. Find an expression for the resistance of the muscle cell as viewed from the electrodes, i.e., find $\mathcal{R} = V/I$. Determine $\mathcal{R}$ for a small cell for which $L \ll \lambda_C$ and for a large cell for which $L \gg \lambda_C$. Explain these results.

e. The resistance $\mathcal{R}$ is measured continuously as SDS, a membrane detergent, is added to the bath to disrupt the membrane of the muscle cell, with the result shown in Figure 3.46. You may assume that SDS makes the membrane resistance zero. On the basis of the results shown in Figure 3.46, what can you conclude about the ratio of the muscle length to the muscle space constant $2L/\lambda_C$? Explain.

3.9 Assume that the incremental membrane voltage, $v_m(z, t)$, of an unmyelinated axon is described by the homogeneous cable equation,

$$\lambda_C^2 \frac{\partial^2 v_m(z, t)}{\partial z^2} = v_m(z, t) + \tau_M \frac{\partial v_m(z, t)}{\partial t},$$

and that at $t = 0$ the voltage is nonzero in a region near $z = 0$ such that

$$\int_{-a}^{a} v_m(z, 0)\, dz = \frac{Q_o}{c_m},$$

as shown in Figure 3.47, where c_m is the membrane capacitance per unit length. Determine the total charge $Q(t)$, where

$$Q(t) = \int_{-\infty}^{\infty} c_m v_m(z, t)\, dz \quad \text{for } t > 0.$$

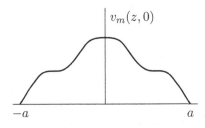

Figure 3.47 Initial voltage along a cable (Problem 3.9).

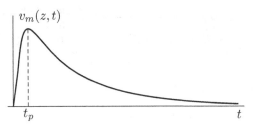

Figure 3.48 Muscle potential response (Problem 3.10).

3.10 This problem involves a comparison of a prediction based on the cable model with measurements on a muscle fiber.

a. Assume that the muscle fiber can be represented as a linear cable (action potentials are prevented from occurring by pharmacological means) and that an impulse of current is delivered to the muscle fiber at $z = 0$ with an arrangement of stimulating electrodes as shown in in Figure 3.22. At each point along the muscle fiber (except $z = 0$), the membrane potential deviation, $v_m(z, t)$, has the form shown in Figure 3.48, where t_p is the time of occurrence of the maximum value of $v_m(z, t)$. Show that for any point z,

$$\left(\frac{z}{\lambda_C}\right)^2 = 4\left(\frac{t_p}{\tau_M}\right)^2 + 2\left(\frac{t_p}{\tau_M}\right).$$

b. Figure 3.49 shows measurements of the end-plate potential of a muscle fiber of the quantity

$$4\left(\frac{t_p}{\tau_M}\right)^2 + 2\left(\frac{t_p}{\tau_M}\right)$$

plotted versus z^2. The end-plate potential is generated by stimulating the motonerve fiber that innervates the muscle.

i. Compare predictions of the cable model with these measurements.

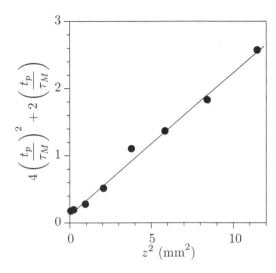

Figure 3.49 Measurements of potential response properties from a muscle fiber (adapted from Fatt and Katz, 1951, Figure 11) (Problem 3.10).

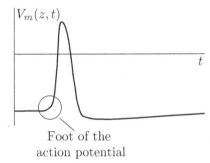

Foot of the action potential

Figure 3.50 Foot of the action potential (Problem 3.11).

ii. Estimate the space constant of the muscle fiber.

iii. How long does it take the peak of the potential to travel one space constant? You may express your answer in terms of τ_M.

3.11 During the initial portion or *foot* of the action potential (Figure 3.50), the deviation of the membrane potential from its resting value is sufficiently small that the electrical characteristics of the membrane can be represented by a linear electrical network consisting of a resistance and a capacitance in parallel. Assume that an action potential is propagated a constant velocity v in the positive z-direction down an unmyelinated axon whose space constant is λ_C and whose membrane time constant is τ_M.

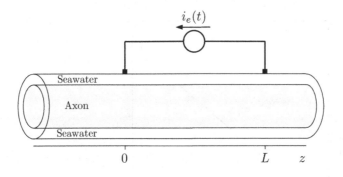

Figure 3.51 Arrangement for stimulating an axon (Problem 3.12).

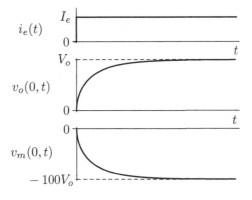

Figure 3.52 Change in potentials caused by a current step (Problem 3.12).

a. Show that during the foot of the action potential, the membrane potential can be expressed as

$$V_m(z, t) = V_m^o + Ae^{\alpha(t - z/v)},$$

where A is a constant.

b. Determine the value of α in terms of λ_C, τ_M, and v.

3.12 An axon of radius 200 μm is placed in oil and stimulated with an external current source applied between locations $z = 0$ and $z = L$ as shown in Figure 3.51. The distance L is large compared with the space constant of the cell. The current is a step of amplitude I_e that produces a change in membrane potential, $v_m(z, t)$, and in external potential, $v_o(z, t)$. The potentials at $z = 0$ are shown in Figure 3.52. The steady-state value of the external potential $v_o(0, \infty) = V_o$, and the steady-state value of the membrane potential $v_m(0, \infty) = -100V_o$. Figure 3.53 shows the dependence of V_o on I_e. The resistivity of the cytoplasm of this cell is $\rho_i = 25\Omega \cdot$ cm.

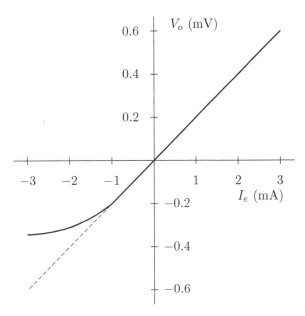

Figure 3.53 Steady-state voltage-current characteristic (Problem 3.12).

a. For $I_e = 1$ mA, find the value of the space constant, λ_C, and sketch $v_o(z, \infty)$ versus z in response to the current step. (Omit the term that occurs in the interpolar region of the form $r_i r_o / (r_i + r_o) I_e z$.) Indicate the values of all key amplitudes and spatial variables on your sketch.

b. For $I_e = 2$ mA, $v_o(z, \infty)$ has one of the forms shown in Figure 3.54. (The term that occurs in the interpolar region of the form $r_i r_o / (r_i + r_o) I_e z$ has been omitted.) Select the correct form of $v_o(z, \infty)$, and briefly explain the reasons for your choice.

3.13 A cylindrical unmyelinated axon of radius a is placed in a large volume of isotonic seawater, and the conduction velocity of the action potential, v_1, is measured. The osmotic pressure of the seawater is then doubled by increasing the salt concentration, and the conduction velocity, v_2, is measured. You may assume that at the time of the second measurement of the conduction velocity (1) osmotic equilibrium has been reached, (2) no net transport of any solute has occurred, (3) the specific electrical properties of the membrane (the capacitances and conductances per unit area) are unchanged, and (4) the length of the axon remains unchanged.

a. Explain which cellular variables will change as a result of the change in solution osmotic pressure and how these variables would affect the conduction velocity.

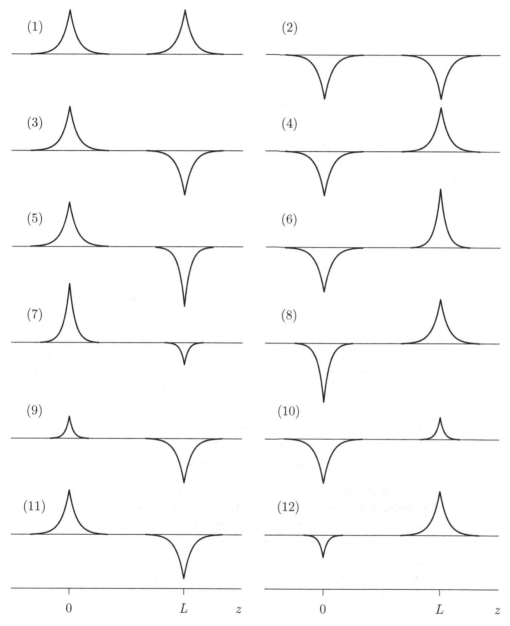

Figure 3.54 Spatial distributions of extracellular potential (Problem 3.12).

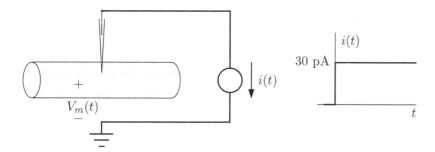

Figure 3.55 Stimulation of and recording from a cell (Problem 3.14).

b. List the equations you require to determine the value of v_2. For each of these equations, give a brief statement indicating the assumptions on which the equation is based.

c. Find v_2 in terms of v_1.

3.14 A cylindrical cell with a uniform membrane is stimulated with a current source, $i(t)$, as shown in Figure 3.55. The amplitude of the current stimulus, 30 pA, is sufficiently low that you may assume that the cable model applies. The cable parameters of this cell are $\tau_M = 2$ ms and $\lambda_C = 5$ mm. The specific membrane capacitance is 1 μF/cm^2. The cell's largest dimension is 10 μm, its surface area is 6×10^{-6}cm^2, and its resting potential is -50 mV.

a. Determine the simplest appropriate equivalent network model of this cell that relates $i(t)$ to $V_m(t)$. Give *all* the parameter values of your network. State any approximations that you make.

b. Find and sketch $V_m(t)$.

3.15 A cylindrical cell whose diameter is 500 μm and whose length is $L = 4$ cm has the following cable parameters: $g_m = 100$ μS/cm, $c_m = 150$ nF/cm, $r_i = 10$ kΩ/cm, and $r_i \gg r_o$.

a. Determine the cell space constant and the membrane time constant.

b. The cell is impaled with a micropipette in the center of its length, and the membrane potential is measured as shown in the upper panel of Figure 3.56. Determine the potential across the membrane at $z = 0$, i.e., determine $v_m(0, t)$.

c. Axial electrodes are used to record the potential across the membrane as shown in the lower panel of Figure 3.56. The axial electrodes have a resistance per unit length of 5 Ω/cm. Determine the potential across the membrane at $z = 0$, i.e., determine $v_m(0, t)$.

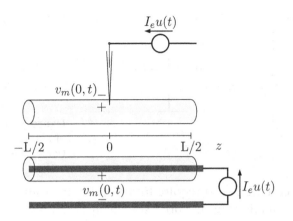

Figure 3.56 Comparison of potential recorded with a micropipette and an axial electrode from a cell (Problem 3.15).

3.16 A large invertebrate axon is immersed in oil, and five different arrangements of electrodes for delivering current stimuli and for measuring potential responses are attached to the axon as shown in Figure 3.57. The stimulus current, a brief positive pulse at $t = 0$, is the same for each arrangement of electrodes. The pulse has a duration that is much shorter than the membrane time constant of the cell and a strength that is low enough that the cell's voltage response remains in its linear range of operation. For each of the different arrangements, parts a–e of Figure 3.57, determine the waveform of $v(t)$ from among those shown in Figure 3.58. If none apply, answer "None." Explain the basis of your choice in each case. In each case, $v(t)$ is the deviation of the potential from its resting value.

3.17 An axon of radius a is placed in a bath of seawater. The conduction velocity of the action potential in seawater is v. The axon space constant is λ_C, and the membrane time constant is τ_M. The seawater surrounding the axon is replaced by a solution that consists of 78% seawater and 22% distilled water by volume. It is assumed that (1) the axon is cylindrical in both solutions, (2) the change in length of the axon is negligible, (3) the volume of the axon is equal to that of the intracellular water, (4) the quantity of intracellular ions remains constant, (5) the replacement does not affect the electrical characteristics of a unit area of membrane, and (6) the extracellular resistance per unit length is negligible (i.e., $r_o \ll r_i$). Determine each of the following, and explain your reasoning:

a. The conduction velocity of the action potential in the diluted sea water in terms of v.

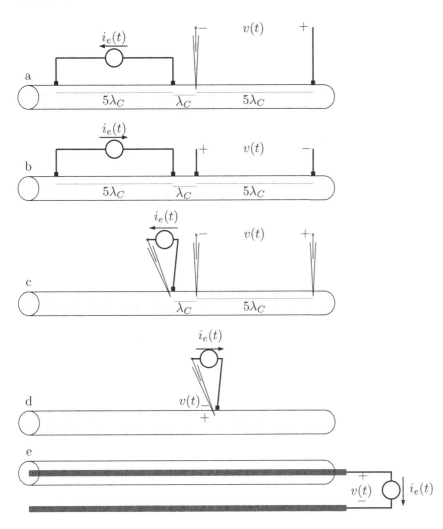

Figure 3.57 Electrode arrangements for measuring the potential produced by an unmyelinated fiber in response to a pulse of current (Problem 3.16). The space constant of the axon is λ_C. In arrangement d, the potential is recorded at the same longitudinal position as that at which current is delivered. In part e, the electrodes are much longer than λ_C.

    b. The axon space constant in the diluted seawater in terms of λ_C.

    c. The membrane time constant in the diluted seawater in terms of τ_M.

3.18 An electrically large cylindrical cell, that can be represented by a linear cable model, is stimulated electrically by an extracellular electrode that imposes an incremental potential $v_o(z, t)$ on the outside of the cell.

    a. Show that for this case the cable equation has the form

$$\lambda_C^2 \frac{\partial^2 v_m(z, t)}{\partial z^2} = v_m(z, t) + \tau_M \frac{\partial v_m(z, t)}{\partial t} - \lambda_C^2 \frac{\partial^2 v_o(z, t)}{\partial z^2}.$$

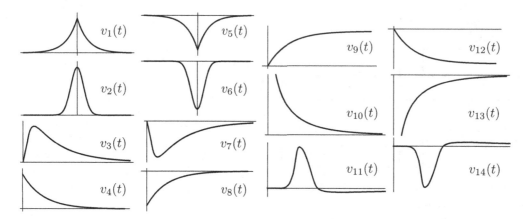

Figure 3.58 Waveforms of voltage versus time (Problem 3.16). The horizontal axis corresponds to $v(t) = 0$, and the vertical axis to $t = 0$.

b. A time independent potential $v_o(z) = (1/2)z^2 u(z)$ is imposed on the cell, where $u(z)$ is the unit step function. Determine and sketch $V_m(z)$.

3.19 A cylindrical cell is stimulated with a constant current as shown in Figure 3.9. Show that the parallel combination of the extracellular and intracellular resistances per unit length can be estimated from a measurement of the spatial gradient in the extracellular potential in the interpolar region, i.e., show that

$$\frac{r_i r_o}{r_i + r_o} = -\frac{1}{I_e} \lim_{z \to \infty} \frac{dv_o(z)}{dz}.$$

3.20 This problem concerns the estimation of cable parameters from measurements of the extracellular potential of frog muscle fibers in response to rectangular pulses of external current using a system similar to that shown schematically in Figure 3.1. Two characteristics of the response to the onset of the rectangular wave were measured. Let the response to the onset (step response) be $v_o(z, t)$, where z is position along the fiber measured from the location of the current stimulus and in the extrapolar region and t is time after the onset of the current step. Figure 3.59 shows measurements of the steady-state value of the response $v_o(z, \infty)$ versus z and of the time $(t_{1/2})$ it takes the response to reach half its steady-state amplitude versus z. The time at which the amplitude reaches half

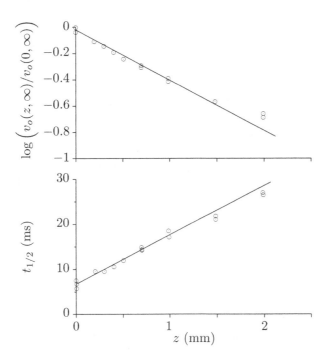

Figure 3.59 Estimation of cable parameters from measurements on muscle fibers (Problem 3.20). Measurements of the steady-state value of the membrane potential (upper panel) and of the time it takes for the step response to reach half its steady-state value (lower panel) are both plotted versus position along the muscle fiber (adapted from Katz, 1948, Figure 9). The muscle fibers were from M. extensor digitorum IV of *Rana temporaria*.

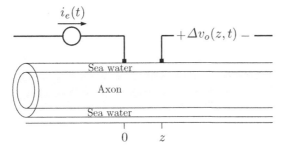

Figure 3.60 Schematic diagram of the measurement of extracellular potentials (Problem 3.21). $\Delta v_o(z, t)$ is the difference between the extracellular potential measured at z and that measured at a distant electrode (not shown).

its steady-state value occurs when $v_o(z, t_{1/2}) = v_o(z, \infty)/2$. Estimate the cell space constant λ_C and the membrane time constant τ_M from these measurements.

3.21 An unmyelinated fiber of 50 μm diameter is placed in oil. A thin layer of seawater clings to the fiber, and the extracellular potential response to an external current $i_e(t)$ is measured as indicated schematically in Figure 3.60. The extracellular potential $\Delta v_o(z, t)$ is a function of time

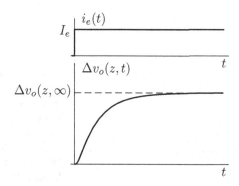

Figure 3.61 Response of a cell (lower panel) to a step of current (upper panel) as a function of time at a single location (Problem 3.21). The steady-state value of the step response is $\Delta v_o(z, \infty)$ and is shown with a dashed line.

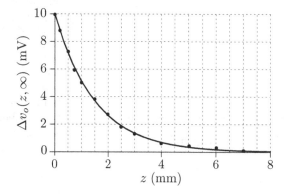

Figure 3.62 Measurement of steady-state value of the extracellular potential in response to a step of current (Problem 3.21).

and position along the fiber. Two different types of measurements are obtained for this fiber.

Measurement #1 In response to a brief pulse of current, the monophasic extracellular action potential is measured and found to have a peak amplitude of 40 mV. The transmembrane action potential has a peak amplitude of 120 mV relative to the resting potential.

Measurement #2 The stimulus current is a step $i_e(t) = I_e u(t)$, where $u(t)$ is the unit step function. The amplitude of the current I_e is reduced until the relation between the current amplitude and the amplitude of the voltage response is linear. It is found that for a current amplitude of $I_e = 0.5\ \mu A$, the extracellular voltage response is related linearly to the current, and all further measurements are obtained at this current level. The time dependence of the voltage response $\Delta v_o(z, t)$ at a location z is shown in Figure 3.61. The steady-state value of the extracellular potential is shown as a function of position in Figure 3.62.

a. Determine the value of the cell space constant, λ_C.

b. Determine the value of the extracellular longitudinal resistance per unit length, r_o.

c. Determine the value of the intracellular longitudinal resistance per unit length, r_i.

d. Determine the value of the membrane conductance per unit length, g_m.

e. Determine the value of the specific membrane conductance (the conductance per unit area of membrane), G_m.

f. Determine the resistivity of the cytoplasm, ρ_i.

References

Books and Reviews

Aidley, D. J. (1989). *The Physiology of Excitable Cells.* Cambridge University Press, Cambridge, England.

Cole, K. S. (1968). *Membranes, Ions, and Impulses.* University of California Press, Berkeley, CA.

Florey, E. (1966). *An Introduction to General and Comparative Animal Physiology.* W. B. Saunders, Philadelphia.

Jack, J. J. B., Noble, D., and Tsien, R. W. (1975). *Electric Current Flow in Excitable Cells.* Clarendon, Oxford, England.

Johnston, D. and Wu, S. M. S. (1995). *Foundations of Cellular Neurophysiology.* MIT Press, Cambridge, MA.

Katz, B. (1966). *Nerve, Muscle, and Synapse.* McGraw-Hill, New York.

Lighthill, M. J. (1960). *Introduction to Fourier Analysis and Generalised Functions.* Cambridge University Press, Cambridge, England.

Rall, W. (1977). Core conductor theory and cable properties of neurons. In Brookhart, J. M. and Mountcastle, V. B., eds., *Handbook of Physiology,* sec. 1; *The Nervous System,* vol. 1, *Cellular Biology of Neurons,* pt. 1, 39-97. American Physiological Society, Bethesda, MD.

Segev, I., Rinzel, J., and Shepherd, G. M., eds. (1995). *The Theoretical Foundation of Dendritic Function.* MIT Press, Cambridge, MA.

Weiss, T. F. (1996). *Cellular Biophysics,* vol. 1, *Transport.* MIT Press, Cambrdige, MA.

Original Articles

Asami, K., Hanai, T., and Koizumi, N. (1976). Dielectric properties of yeast cells. *J. Membr. Biol.,* 28:169-180.

Cao, B. J. and Abbott, L. F. (1993). A new computational method for cable theory problems. *Biophys. J.,* 64:303-313.

Cole, K. S. (1941). Rectification and inductance in the squid giant axon. *J. Gen. Physiol.*, 25:29–51.

Cole, K. S. and Baker, R. F. (1941a). Longitudinal impedance of the squid giant axon. *J. Gen. Physiol.*, 24:771–788.

Cole, K. S. and Baker, R. F. (1941b). Transverse impedance of the squid giant axon during current flow. *J. Gen. Physiol.*, 24:535–549.

Cole, K. S. and Curtis, H. J. (1936). Electric impedance of nerve and muscle. In *Cold Spring Harbor Symposia on Quantitative Biology*, vol. 4, 73–89. Long Island Biological Society, Cold Spring Harbor, NY.

Cole, K. S. and Curtis, H. J. (1941). Membrane potential of the squid giant axon during current flow. *J. Gen. Physiol.*, 24:551–563.

Cole, K. S. and Hodgkin, A. L. (1939). Membrane and protoplasm resistance in the squid giant axon. *J. Gen. Physiol.*, 22:671–687.

Curtis, H. J. and Cole, K. S. (1938). Transverse electric impedance of the squid giant axon. *J. Gen. Physiol.*, 21:757–765.

Falk, G. and Fatt, P. (1964). Linear electrical properties of striated muscle fibres observed with intracellular electrodes. *Proc. R. Soc. London, Ser. B*, 160:69–123.

Fatt, P. (1964). An analysis of the transverse electrical impedance of striated muscle. *Proc. R. Soc. London, Ser. B*, 159:606–651.

Fatt, P. and Katz, B. (1951). An analysis of the end-plate potential recorded with an intracellular electrode. *J. Physiol.*, 115:320–370.

Fricke, H. (1925). The electric capacity of suspensions with special reference to blood. *J. Gen. Physiol.*, 9:137–152.

Fricke, H. and Morse, S. (1925). The electric resistance and capacity of blood for frequencies betwen 800 and 4 1/2 million cycles. *J. Gen. Physiol.*, 9:153–167.

Gage, P. W. and McBurney, R. N. (1973). An analysis of the relationship between the current and potential generated by a quantum of acetylcholine in muscle fibers without transverse tubules. *J. Membr. Biol.*, 12:247–272.

Grosse, C. and Schwan, H. P. (1992). Cellular membrane potentials induced by alternating fields. *Biophys. J.*, 63:1632–1642.

Hanai, T., Asami, K., and Koizumi, N. (1979). Dielectric theory of concentrated suspensions of shell-spheres in particular reference to the analysis of biological cell suspensions. *Bull. Inst. Chem. Res.*, 57:297–305.

Hanai, T., Haydon, D. A., and Taylor, J. (1964). An investigation by electrical methods of lecithin-in-hydrocarbon films in aqueous solution. *Proc. R. Soc. London, Ser. A*, 281:377–391.

Hodgkin, A. L. (1947). The membrane resistance of a non-medullated nerve fibre. *J. Physiol.*, 106:305–318.

Hodgkin, A. L. and Rushton, A. H. (1946). The electrical constants of a crustacean nerve fibre. *Proc. R. Soc. London, Ser. B*, 133:444–479.

Katz, B. (1948). The electrical properties of the muscle fibre membrane. *Proc. R. Soc. London, Ser. B*, 135:506–534.

Major, G. (1993). Solutions for transients in arbitrarily branching cables: III. Voltage clamp problems. *Biophys. J.*, 65:469–491.

Major, G. and Evans, J. D. (1994). Solutions for transients in arbitrarily branching cables: IV. Nonuniform electrical parameters. *Biophys. J.*, 66:615–634.

Major, G., Evans, J. D., and Jack, J. J. B. (1993a). Solutions for transients in arbitrarily branching cables: I. Voltage recording with a somatic shunt. *Biophys. J.*, 65:423-449.

Major, G., Evans, J. D., and Jack, J. J. B. (1993b). Solutions for transients in arbitrarily branching cables: II. Voltage clamp theory. *Biophys. J.*, 65:450-468.

Pauly, H. and Schwan, H. P. (1966). Dielectric properties and ion mobility in erythrocytes. *Biophys. J.*, 6:621-639.

Rall, W. (1959). Branching dendritic trees and motoneuron membrane resistivity. *Exp. Neurol.*, 1:491-527.

Rall, W. (1960). Membrane potential transients and membrane time constant of motoneurons. *Exp. Neurol.*, 2:503-532.

Rall, W. (1962a). Electrophysiology of a dendritic neuron model. *Biophys. J.*, 2:145-167.

Rall, W. (1962b). Theory of physiological properties of dendrites. *Ann. N.Y. Acad. Sci.*, 96:1071-1092.

Rall, W. (1964). Theoretical significance of dendritic trees for neuronal input-output relations. In Reiss, R. F., ed., *Neural Theory and Modeling*, ch. 4, 73-97. Stanford University Press, Stanford, CA.

Rall, W. (1969a). Distributions of potential in cylindrical coordinates and time constants for a membrane cylinder. *Biophys. J.*, 9:1509-1541.

Rall, W. (1969b). Time constants and electrotonic length of membrane cylinders and neurons. *Biophys. J.*, 9:1483-1508.

Rall, W. (1989). Cable theory for dendritic neurons. In Koch, C. and Segev, I., eds., *Methods in Neuronal Modeling*, 9-62. MIT Press, Cambridge, MA.

Redwood, W. R., Takashima, S., Schwan, H. P., and Thompson, T. E. (1972). Dielectric studies on homogeneous phosphatidylcholine vesicles. *Biochim. Biophys. Acta*, 255:557-566.

Schwan, H. P. (1957). Electrical properties of tissue and cell suspensions. *Adv. Biol. Med. Phys.*, 5:147-209.

Schwan, H. P. (1963). Determination of biological impedances. In Nastuk, W. L., ed., *Physical Techniques in Biological Research*, vol. 6, 323-407. Academic Press, New York.

Schwan, H. P. (1965). Biological impedance determinations. *J. Cell. Comp. Physiol.*, 66, Part II:5-11.

Schwan, H. P., Takashima, S., Miyamoto, V. K., and Stoeckenius, W. (1970). Electrical properties of phospholipid vesicles. *Biophys. J.*, 10:1102-1119.

Spach, M. S. and Heidlage, J. F. (1992). A multidimensional model of cellular effects on the spread of electrotonic currents and on propagating action potentials. *CRC Crit. Rev. Biomed. Eng.*, 20:141-169.

Takashima, S. (1978). Frequency domain analysis of asymmetry current in squid axon membrane. *Biophys. J.*, 22:115-119.

Takashima, S. (1979). Admittance change of squid axon during action potentials. *Biophys. J.*, 26:133-142.

Takashima, S., Asami, K., and Takahashi, Y. (1988). Frequency domain studies of impedance characteristics of biological cells using micropipet technique. *Biophys. J.*, 54:995-1000.

Takashima, S. and Schwan, H. P. (1974). Passive electrical properties of squid axon membrane. *J. Membr. Biol.*, 17:51-68.

Takashima, S. and Yantorno, R. (1977). Investigation of voltage-dependent membrane capacity of squid giant axons. *Ann. N.Y. Acad. Sci.*, 303:306–321.

Taylor, R. E. (1965). Impedance of the squid axon membrane. *J. Cell. Comp. Physiol.*, 66, Part II: 21–25.

Thomson, W. (1855). On the theory of the electric telegraph. *Proc. R. Soc. London, Ser. A*, 7:382–399.

Van Pelt, J. (1992). A simple vector implementation of the Laplace-transformed cable equations in passive dendritic trees. *Biol. Cybern.*, 68:15–21.

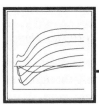

The Hodgkin-Huxley Model

Finally there was the difficulty of computing the action potentials from the equations which we had developed. We had settled all the equations and constants by March 1951 and hoped to get these solved on the Cambridge University computer. However, before anything could be done we learnt that the computer would be off the air for 6 months or so while it underwent a major modification. Andrew Huxley got us out of that difficulty by solving the differential equations numerically using a hand-operated Brunsviga. The propagated action potential took about three weeks to complete and must have been an enormous labour for Andrew. But it was exciting to see it come out with the right shape and velocity and we began to feel that we had not wasted the many months that we had spent in analysing records.
—Hodgkin, 1977

The computations . . . were done by hand. This was a laborious business: a membrane action potential took a matter of days to compute, and a propagated action potential took a matter of weeks. But it was often quite exciting. For example, when calculating the effect of a stimulus close to the threshold value, one would see the forces of accommodation—inactivation of the sodium channel, and the delayed rise of potassium permeability—creeping up and reducing the excitatory effect of the rapid rise of sodium permeability. Would the membrane potential get away into a spike, or die in a subthreshold oscillation? Very often my expectations turned out to be wrong, and an important lesson I learned from these manual computations was the complete inadequacy of one's intuition in trying to deal with a system of this degree of complexity.
—Huxley, 1964

4.1 Introduction

4.1.1 Historical Perspective

Prior to the first intracellular measurements of action potentials in animal cells (Hodgkin and Huxley, 1939; Curtis and Cole, 1940), most of the key properties of electrically excitable cells were already well known. Although the results were based on a variety of relatively indirect methods of measurement, as described in Section 1.2 (see in particular Figure 1.11), such properties as the all-or-none character of the action potential, its sharp threshold, refractoriness, accommodation, and so on were all familiar to physiologists.

With the growth of knowledge of the properties of electrically excitable cells, phenomenological theories to account for these properties were formulated. These so-called *two-factor theories* attempted to describe the relation between current stimulation and the occurrence of action potentials in terms of hypothetical threshold and excitation variables. However, these theories had only limited success in explaining the electrically excitable properties, and they yielded little insight into membrane mechanisms. With the advent of intracellular recording, the electrical properties of membranes could be investigated directly.

The culmination of this effort was the systematic measurements and the theoretical formulations of Hodgkin and Huxley (Hodgkin et al., 1952; Hodgkin and Huxley, 1952a, 1952b, 1952c, 1952e), for which they were awarded the Nobel Prize in 1963. Their theory is one of the most successful mathematical theories in biology and greatly accelerated research in neurobiology. The theory explained the properties of the electrically excitable squid giant axon in terms of the measured relations of the membrane potential and the membrane current. The primitive entities of this theory were a set of hypothetical transmembrane ionic channels.

Hence, this research focused the attention of neurobiologists on the identification and elucidation of the properties of these ionic channels. Since the 1970s, electrophysiological techniques have been developed to record the ionic current through such isolated single channels, and molecular biological techniques have been developed to isolate the channel macromolecules. These topics are discussed in Chapter 6. In this chapter we shall focus on the Hodgkin-Huxley model of the giant axon of the squid. In the succeeding chapters we shall indicate the extent to which these concepts apply to other electrically excitable cells.

4.1.2 Key Notions Leading to the Hodgkin-Huxley Model

We shall now briefly review the concepts that led to the experiments that gave rise to the Hodgkin-Huxley model.

4.1.2.1 Ionic Basis of Resting Membrane Potential

By the late 1940s, the notion that membranes were selectively permeable to different ions was well established. It was appreciated that the generation of the resting membrane potential could be understood by an equivalent network in which each species of ion flows through an independent channel (Weiss, 1996, Chapter 7, Figure 7.26). The driving force for the nth ion is the electrochemical potential difference, $V_m - V_n$. The ease with which ions pass through the membrane is represented by the conductance G_n. At rest, the potassium conductance, G_K, is much larger than the sodium conductance, G_{Na}. Hence, the resting potential is near V_K. In general, the resting potential is determined by a weighted sum of the Nernst potentials of the permeant ions plus a contribution from electrogenic ion pumps. The weighting factor for the Nernst potential of each ion equals the fraction of the total membrane conductance of that ion.

4.1.2.2 Membrane Conductance Changes during an Action Potential

The measurements of membrane impedance and cable properties showed that in the linear range of operation of a membrane, its electrical characteristics could be represented approximately by a network consisting of parallel combination of a conductance and a capacitance (see Figure 3.3). The early measurements were made by placing a cell, such as the giant axon of the squid, between two electrodes that were attached to the two arms of a Wheatstone bridge. This arrangement allowed the electrical impedance to be measured as discussed in Chapter 3. Typically, the impedance between the electrodes was measured over a range of frequencies, and the membrane impedance was inferred from these measurements. A key experiment was to record the impedance of the membrane during the occurrence of an action potential. It was found (Cole and Curtis, 1939) that during the occurrence of an action potential, the membrane capacitance was essentially constant,[1] but that the membrane conductance showed a large transient increase. The waveshapes of the membrane potential and the conductance are shown in Figure 4.1. The conductance increase lags behind the increase in membrane potential and decays somewhat more slowly than does the membrane potential.

---

1. More recent measurements (Takashima, 1979) have revealed that the capacitance does in fact change a little during an action potential. The significance of these findings will be described in Chapter 6.

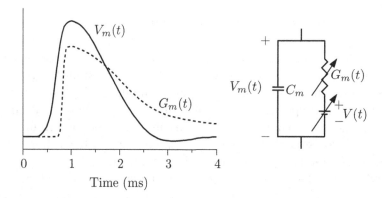

Figure 4.1 Waveforms of action potential and membrane conductance recorded simultaneously (adapted from Cole, 1968, Figure 2:18). The measurements shown are based on the original measurements (Cole and Curtis, 1939). During the action potential, there is a change in the membrane conductance (indicated by the arrow through the conductance in the equivalent network), as well as in the source of membrane potential, but no appreciable change in the membrane capacitance.

4.1.2.3   *Dependence of the Peak of the Action Potential on Sodium Concentration*

From the 1900s to the late 1940s, the dominant theory for resting and action potentials was the Bernstein theory. According to the Bernstein theory, which was based on extracellular recordings from cell populations, at rest the membrane was permeable to potassium ions only. Hence, the resting potential of a cell equaled the potassium equilibrium potential (Weiss, 1996, Chapter 7, Section 7.5). During an action potential, the membrane transiently lost its selective permeability for potassium; hence, the membrane potential approached zero. The first intracellular measurements of action potentials (Hodgkin and Huxley, 1939; Curtis and Cole, 1940) showed that the peak of the action potential of the squid axon exceeds ("overshoots") a potential of zero, thus refuting the Bernstein theory.[2]

---

2. It is curious that the notion of an overshoot of the action potential was much discussed among neuroscientists, including Bernstein, prior to the formulation of the Bernstein hypothesis. However, all the earlier discussions were based on indirect measurements (of the compound action potential of whole nerves and muscles). Bernstein's theoretical notions had no room for an overshoot, and the overshoot was indeed unexpected when the first intracellular action potentials were recorded in the 1940s (Grundfest, 1965).

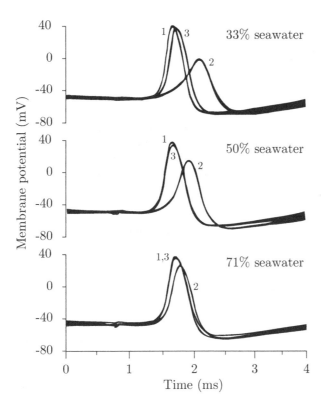

Figure 4.2 Effect of extracellular sodium concentration on the membrane potential of a giant axon of a squid (adapted from Hodgkin and Katz, 1949a, Figure 4). In each panel, traces 1 and 3 are the action potentials obtained in normal seawater before and after trace 2, which was obtained in seawater with reduced sodium content. Isotonic solutions were obtained by using mixtures of seawater and a glucose solution. In the three panels, the solution contained 33, 50, and 71% seawater.

By 1949, Hodgkin and Katz had clearly shown that the action potential was sensitive to the concentration of sodium ions. Figure 4.2 shows that when a squid axon is placed in a seawater solution with reduced sodium concentration, the action potential waveform is changed dramatically and reversibly. In particular, a decrease in the extracellular concentration of sodium reduces the peak value of the action potential. The value of the resting potential is not changed appreciably. As shown in Figure 4.3, at a high external sodium concentration the peak of the action potential approaches the Nernst equilibrium potential for sodium. This result strongly suggested that during an action potential there is a transient increase in the permeability of the membrane to sodium so that the sodium conductance greatly exceeds the potassium conductance, thus reversing the conditions that obtain in a resting membrane. For this reason, the peak of the action potential approaches the sodium equilibrium potential. Later this notion that the rising phase of the action potential is caused by a transient increase of sodium conductance was further supported by experiments in which the intracellular concentration was raised

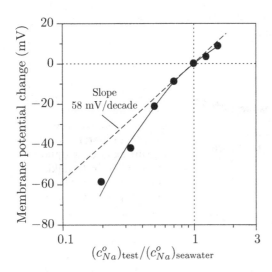

Figure 4.3 Dependence of the peak of the action potential on extracellular sodium concentration (adapted from Hodgkin and Katz, 1949a, Figure 7). The change in peak value of the action potential is plotted versus the logarithm of the extracellular concentration of sodium in the test solution normalized to the sodium concentration in normal seawater. If the peak of the action potential equaled the sodium equilibrium potential, the measured points would lie on the dashed line. The points are averages obtained in one to six experiments.

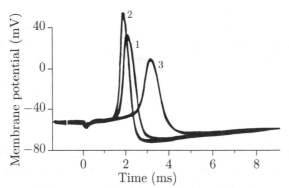

Figure 4.4 Effect of intracellular sodium concentration on the membrane potential of a giant axon of a squid (adapted from Baker et al., 1961, Figure 5). Curve 2 was obtained with an intracellular solution that contained isotonic potassium sulfate. In curves 1 and 3, one-quarter and one-half of the potassium ions, respectively, were replaced by sodium.

(Figure 4.4). An increase in intracellular sodium concentration decreases the sodium equilibrium potential and decreases the peak of the action potential.

In similar experiments, it has been shown that a modest change in extracellular potassium concentration results in a relatively small change in the resting potential and in the peak of the action potential, but in a large change in the undershoot of the action potential (Figure 4.5). A decrease in extracellular potassium concentration makes the undershoot more negative. For the range of potassium concentration shown in the figure, the changes in potential for a population of four axons was about 7 mV for the resting potential, 6 mV for the peak of the action potential, and 16 mV for the undershoot of the action potential (Hodgkin and Katz, 1949a). The effect of a large increase

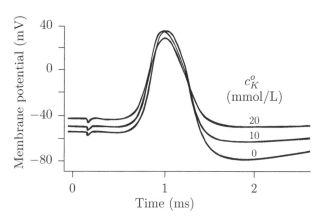

Figure 4.5 Effect of extracellular potassium concentration on the membrane potential of a giant axon of a squid (adapted from Hodgkin and Katz, 1949a, Figure 12). Three solutions were used in which the potassium concentration of seawater was 0, 10, and 20 mmol/L.

in the potassium concentration on the action potential could not be studied, because no action potential can be elicited in a solution with a high external potassium concentration. These results show that potassium also affects the action potential, but in a manner that is characteristically different from the effect of sodium.

These results show that the Bernstein hypothesis is fundamentally wrong. Instead of losing its selective permeability during an action potential, the membrane transiently reverses its selective permeability, as we shall see. The membrane changes from being more permeable to potassium than to sodium at rest to being transiently more permeable to sodium than to potassium.

4.1.3 Mathematical Description of the Hodgkin-Huxley Model

The Hodgkin-Huxley model evolved from the three important observations available in the late 1940s. First was the notion that the resting potential could be explained largely on the basis of a semipermeable membrane whose permeability to potassium exceeded that to other ions. This property established the notion that the membrane potential was a consequence of both the membrane permeabilities and the Nernst equilibrium potentials of the permeant ions. Second, during an action potential the conductance of the membrane changed transiently. Finally, the effect of changes in sodium and potassium concentration on the action potential indicated that the permeability of the membrane to sodium and potassium increased markedly. Before we describe the experiments that led to a description of the mechanism underlying these permeability changes, we shall briefly describe the overall structure of the Hodgkin-Huxley model of a patch of membrane and of an axon.

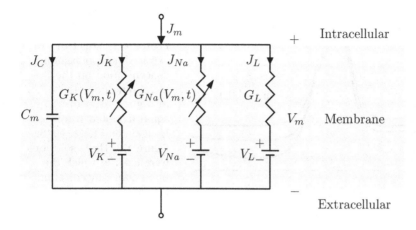

Figure 4.6 The Hodgkin-Huxley model of a patch of membrane relates the membrane current density to the membrane potential. The Js are current densities (A/cm^2), the Vs are potentials (V), C_m is the specific capacitance of the membrane (F/cm^2), and the Gs are specific ionic conductances (S/cm^2).

4.1.3.1   The Equations

The Hodgkin-Huxley model of a patch of membrane is shown in Figure 4.6. The network relates the membrane current density, J_m, to the membrane potential, V_m. There are four branches of current density. One branch is the capacitance current, which represents the displacement current through the membrane. The other three branches represent ionic currents due to sodium, potassium, and leakage. The leakage current represents the sum of membrane currents other than those flowing through the sodium and potassium branches and includes currents through the cut end of a dissected axon; hence, the name *leakage.* The sodium and potassium branches are each represented by a conductance in series with the Nernst equilibrium potential for that ion. The important new notion is that the sodium and potassium conductances are explicitly dependent on membrane potential and time, which is signified by an arrow through the conductance. The leakage branch is represented by a Thévenin equivalent network with a constant conductance and an equivalent battery.

From Figure 4.6, and using Kirchhoff's current law, the membrane current density, J_m, can be expressed as

$$J_m = J_C + J_{ion},$$

where the ionic current density is

$$J_{ion} = J_K + J_{Na} + J_L, \tag{4.1}$$

which can be written as

$$J_m = J_C + J_K + J_{Na} + J_L. \tag{4.2}$$

Substitution of Ohm's law for each of the conductances and the constitutive law for the membrane capacitance into Equation 4.2 yields

$$J_m = C_m \frac{\partial V_m}{\partial t} + G_K(V_m, t) \, (V_m - V_K)$$
$$+ G_{Na}(V_m, t) \, (V_m - V_{Na}) + G_L(V_m - V_L). \tag{4.3}$$

V_{Na} and V_K are the Nernst equilibrium potentials for sodium and potassium, respectively, defined as (Weiss, 1996, Chapter 7)

$$V_{Na} = \frac{RT}{F} \ln \frac{c_{Na}^o}{c_{Na}^i} \text{ and } V_K = \frac{RT}{F} \ln \frac{c_K^o}{c_K^i},$$

where R is the molar gas constant, T, is absolute temperature, F, is Faraday's constant, and c_{Na}^i and c_{Na}^o are the molar concentrations of sodium intracellularly and extracellularly, respectively.

The core conductor equation (Equation 2.25) relates the coupling of current and voltage along a cylindrical cell as follows:

$$\frac{\partial^2 V_m}{\partial z^2} = 2\pi a (r_o + r_i) J_m. \tag{4.4}$$

If we substitute Equation 4.3 into 4.4, we can eliminate the membrane current density to obtain

$$\frac{1}{2\pi a (r_o + r_i)} \frac{\partial^2 V_m}{\partial z^2} = C_m \frac{\partial V_m}{\partial t} + G_K(V_m, t) \, (V_m - V_K)$$
$$+ G_{Na}(V_m, t) \, (V_m - V_{Na}) + G_L(V_m - V_L). \tag{4.5}$$

Equation 4.5 describes the potential along an axon whose membrane electrical characteristics are described by the Hodgkin-Huxley model and combines the core conductor model and the Hodgkin-Huxley model as depicted in Figure 4.7.

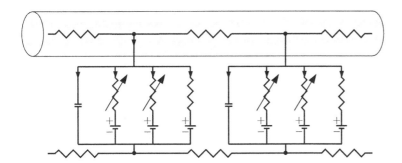

Figure 4.7 Model of an axon according to the Hodgkin-Huxley model of a patch of membrane.

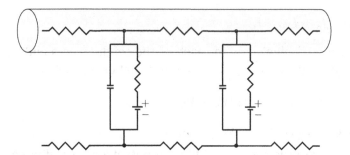

Figure 4.8 Cable model of an axon.

4.1.3.2 The Hodgkin-Huxley Model as a Model of Resting Potential and Cable Properties

The Hodgkin-Huxley model of an axon, defined by Figure 4.7 and Equation 4.5, is a generalization of previous models we have investigated. Note that at rest, $J_m = 0$ and $V_m = V_m^o$ by definition. Since the membrane potential is constant, the current through the membrane capacitance is zero. Hence, both $J_m = 0$ and $\partial^2 V_m / \partial z^2 = 0$. Therefore, Equation 4.3 reduces to the an equation for the resting potential as a sum of weighted Nernst potentials for the permeant ions (Weiss, 1996, Equation 7.76).[3] Thus, for the axon at rest, the Hodgkin-Huxley model reduces to a model of resting potential (Weiss, 1996, Chapter 7). The Hodgkin-Huxley model also reduces to the cable model described in Chapter 3, provided that the membrane conductances are constant. If the conductances in Equation 4.5 are assumed to be constant,[4] the three ionic channels can be combined into one equivalent network as shown in Figure 4.8, which is the equivalent network of the cable model described in Chapter 3. Thus, the Hodgkin-Huxley model is a generalization of the models we have discussed previously. This model incorporates both the resting and the cable properties of the axon.

---

3. Mechanisms for the active transport of cations are not included; these could easily be added, but have little effect on the excitation and conduction of action potentials on the brief time scale for which the Hodgkin-Huxley model is valid. Furthermore, the direct effect of active transport on the resting potential is small for the giant axon of the squid.

4. As we shall see, the equivalent network of the membrane for small perturbations in membrane potential implied by the Hodgkin-Huxley model is more complex than the simple parallel combination of a conductance and a capacitance assumed in the cable model. Furthermore, the network model implied by the Hodgkin-Huxley model more accurately reflects the measured membrane impedance than does a parallel conductance and capacitance network model.

4.1.3.3 The Importance of Ionic Conductances

The membrane potential predicted by the Hodgkin-Huxley model can be computed from Equation 4.5 (if suitable boundary conditions are specified) if the functions $G_K(V_m, t)$ and $G_{Na}(V_m, t)$ are known. Thus, the essence of the Hodgkin-Huxley model involves the specification of the ionic conductances, G_K and G_{Na}. These were obtained in a series of experiments in which the conductances were measured as a function of V_m and t.

4.2 Revelation of Ionic Mechanisms by the Voltage-Clamp Technique

4.2.1 Experimental Techniques

The essence of the Hodgkin-Huxley model is contained in the ionic conductances, G_K and G_{Na}. It is therefore important to understand the principles of the experimental techniques used to measure these conductances (Cole, 1949; Marmont, 1949; Hodgkin et al., 1949; Hodgkin et al., 1952; Hodgkin and Huxley, 1952a, 1952b, 1952c, 1952e). The technical problem is to measure the Gs that are embedded in a distributed nonlinear cable. As illustrated schematically in Figure 4.9, the principle of the technique involves first preventing the normal propagation of an action potential. This objective can be achieved experimentally by inserting a wire (with a low resistance per unit length) inside the axon and providing a good conductor, such as seawater, at the exterior of the axon. This technique is called the *space-clamp technique.* In an ideally space-clamped axon, all membrane variables become independent of z. In particular, the potential across the membrane is independent of z, and, if driven by an external current, the axon can generate an action potential—called a *membrane action potential* to distinguish it from the *propagated action potential*—simultaneously along the entire length of axon contained within the space-clamp electrodes. A space-clamped axon driven by a current is called *current clamped.* The space clamp provides an important conceptual simplification, since the description of the electrical properties of the axon changes from that of a distributed-parameter network to that of a lumped-parameter network. The differential equations describing the axon change from partial differential equations to ordinary differential equations. That is, the space clamp technique effectively changes the cell from an electrically large cell to an electrically small cell. The efficacy of the space clamp has already been discussed in Chapter 2 (Figure 2.15).

The space-clamp technique eliminates the dependence of membrane variables on the spatial dimension, z. Thus, as seen in Figure 4.9, this technique enables a direct investigation of the relation between V_m and J_m. Since, as we shall see, the membrane is voltage controlled rather than current controlled, it is far simpler to study the V_m-J_m relation by constraining the membrane potential with a voltage source and recording the membrane current than to do the reverse. This technique is called the *voltage-clamp technique.* In this technique, a voltage source is placed across the membrane as shown schematically in Figure 4.10, and the membrane current is measured. Since the membrane resistance of the axon varies over a wide range and can become quite low, a good voltage source (i.e., one with a low source impedance) is required. This condition can be achieved most easily by means of a feedback amplifier as shown in Figure 4.11. The membrane potential, $V_m(t)$, is subtracted from an external voltage source, or *command voltage,* and the difference, or *error signal,* is amplified (through an amplifier with a high input impedance and a gain of K) and fed back to the input as shown. Note that $V_o(t) = V_m(t) = K(V_c(t) - V_m(t))$. Therefore, $V_m(t) = (K/(K + 1))V_c(t)$. Thus, for large values of K, the membrane potential, $V_m(t)$, can be made to follow the command voltage, $V_c(t)$. The membrane current, $J_m(t)$, that results from maintaining the membrane potential equal to $V_c(t)$ is measured. This feedback-amplifier technique is a method of achieving a low-impedance voltage source to control the membrane potential of the axon. Two electrodes are inserted into the axon to reduce the effect on the measurement of the membrane potential of junction potentials due to current through the electrode. One electrode is used to measure the potential, the other to carry the membrane current. To minimize the effect of junction potentials, the reference electrodes for current and potential are also separate (not shown in Figure 4.11).

If the membrane potential of a space-clamped axon is constrained to be a step of voltage or a *step voltage clamp*, then (except at the discontinuity in $V_m(t)$) $\partial V_m/\partial t = 0$; therefore, the capacity current is zero. Thus, for such an axon $\partial V_m/\partial z = \partial V_m/\partial t = 0$, and the relation between $J_{ion}(t)$ and $V_m(t)$ can be measured as indicated in Figure 4.9. With proper techniques, the transient

Figure 4.9 Illustration of techniques for recording ionic currents from an axon. The top panel shows the model of an axon. The object is to measure properties of the ionic currents. The space clamp is achieved by shorting out the inside and outside conductors. The inside conductor is shorted out by placing an axial wire in the axon's cytoplasm. The capacitance current is eliminated by using the voltage-clamp method, in which the membrane potential is maintained constant by external circuitry. This arrangement allows measurement of the ionic current. The final step is to separate the ionic current into its constituent ionic current components.

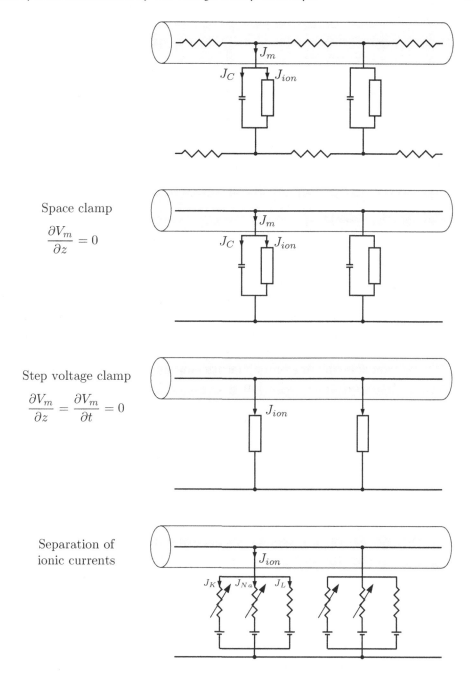

Space clamp

$$\frac{\partial V_m}{\partial z} = 0$$

Step voltage clamp

$$\frac{\partial V_m}{\partial z} = \frac{\partial V_m}{\partial t} = 0$$

Separation of
ionic currents

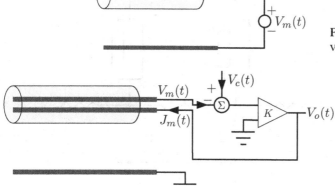

Figure 4.10 Schematic diagram of a voltage-clamp measurement.

Figure 4.11 Schematic diagram of voltage-clamped axon using a feedback amplifier.

pulse of capacity current that occurs at the change in the membrane potential is complete by the time the ionic current begins to change. Thus, it is feasible to separate in time the membrane ionic current components from the capacity current. The ionic current can be separated into sodium and potassium components by changes in the composition of the solutions on the two sides of the axonal membrane or by pharmacological agents that specifically block passive transport of a specific ion (see Chapter 6).

4.2.2 Membrane Current Components

In response to depolarizing and hyperpolarizing steps of membrane potential, the membrane current density in a voltage-clamped membrane patch has the form shown in Figure 4.12. In response to a hyperpolarizing step, there is a large, brief (ca. 50 μs) surge of inward current plus a persistent current with a small amplitude that is just barely discernible at the scale shown in Figure 4.12. In response to a depolarizing step, the initial large, brief surge is outward. In addition, there is a transient early inward current plus a persistent late outward current. These features of the $J_m(t)$ waveform can be attributed to distinct current components.

4.2.2.1 Capacitance Current

The large, brief surge of current is a capacity current resulting from a step of voltage applied across the membrane capacitance. The time course of this

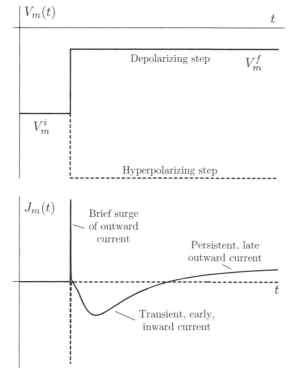

Figure 4.12 Schematic diagram showing the components of the membrane current in response to both depolarizing (solid line) and hyperpolarizing (dashed line) steps of membrane potential.

current is predictable from the value of the capacitance plus the measured resistance in series with the membrane (as explored in Problem 4.4), as well as from the bandwidth of the electronic stimulation and measurement devices (Figure 4.13). In a step voltage-clamp experiment, the capacitance current charges the membrane capacitance to the final voltage, V_m^f.

4.2.2.2 Ionic Currents

Identification of Ionic Current Components

The small, persistent current (not discernible in Figure 4.12) that is seen most easily as a small inward current in response to a hyperpolarizing step, but is also present as a small outward current for a depolarization, results from components of the membrane conductance that do not change with time, which include the leakage current. As we shall see, the transient early current in Figure 4.12, which is inward provided that $V_m^f < V_{Na}$, is normally carried predominantly by sodium ions, and the persistent late current, which is outward provided that $V_m^f > V_K$, is normally carried predominantly by potassium

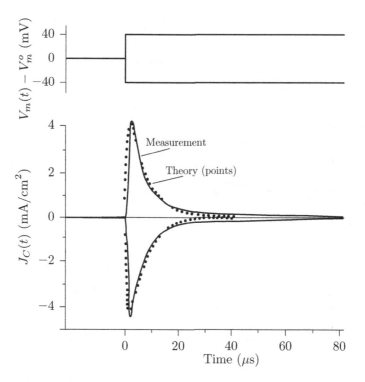

Figure 4.13 Capacitance current density recorded in a squid giant axon in response to depolarization and hyperpolarization voltage steps of ±40 mV (adapted from Hodgkin et al., 1952, Figure 16). The measurements (solid lines) are compared with calculation (points) based on a simple model of the membrane and of the stimulating and recording arrangement.

ions. Since the capacity current and the leakage current are, to a first-order approximation, related linearly to the potential change, these two components can be subtracted from the membrane current to more clearly reveal the other two ionic currents.[5]

A series of voltage-clamp measurements for different values of V_m^f (Figure 4.14) reveals that the ionic current depends in a complex manner on the membrane potential and on time. The transient early component of current is inward for small voltage steps and outward for large voltage steps. This component reverses polarity from inward to outward as V_m^f increases. The potential V_m^f at which the early current component reverses is called the *reversal potential*. The reversal of the early component is seen more clearly in Figure 4.15. This behavior of the transient early current is consistent with its

---

5. In fact, a careful examination of these types of measurements reveals that there are systematic differences in the waveshapes of the capacitance currents for potential steps of equal magnitude in the depolarization and hyperpolarization directions. The significance of these differences will be taken up in Chapter 6.

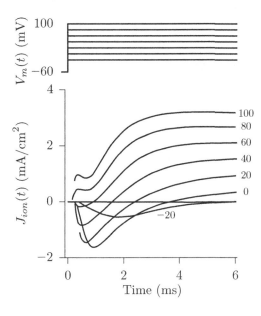

Figure 4.14 Ionic currents under voltage clamp recorded from a Chilean squid, *Dosidicus gigas* (adapted from Armstrong, 1969, Figure 2b). Each current trace corresponds to one voltage trace. The voltage traces each start at a holding potential of −60 mV and step to a final potential that ranges from −20 to 100 mV; the final voltage corresponding to each current trace is indicated near each current trace. The temperature was 6.6°C.

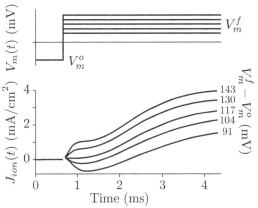

Figure 4.15 Ionic currents under voltage clamp recorded from a giant axon of a squid, *Loligo forbesi* (adapted from Hodgkin et al., 1952, Figure 14). The temperature was 3.5°C. The transient early current is clearly inward for $V_m^f - V_m^o = 104$ mV and outward for 130 mV. For a change in potential of 117 mV, the early current is not discernible.

being carried by sodium ions. Note that a current carried by sodium would obey the following:

$$J_{Na} = G_{Na} \left(V_m^f - V_{Na} \right). \tag{4.6}$$

For $V_m^f < V_{Na}$, $J_{Na} < 0$, i.e., J_{Na} is inward. For $V_m^f = V_{Na}$, $J_{Na} = 0$. For $V_m^f > V_{Na}$, $J_{Na} > 0$, i.e., J_{Na} is outward. Therefore, a current component carried by sodium ions should reverse at the sodium equilibrium potential. By changing

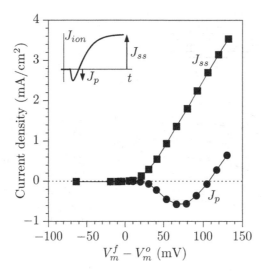

Figure 4.16 Peak and steady-state currents measured from a giant axon of a squid under voltage clamp plotted versus membrane potential (adapted from Hodgkin et al., 1952, Figure 13). The inset shows the measurement method. The ionic current density, J_{ion}, is measured in response to a voltage step to potential V_m^f. J_p is the peak of the transient early current density component, and J_{ss} is the steady-state value of the persistent late current density component. J_p and J_{ss} are plotted versus $V_m^f - V_m^o$. The temperature was 3.8°C.

the sodium concentration either inside or outside the cell, V_{Na} can be changed and the experiment can be repeated to find the reversal potential for the early current component. It is found that the reversal potential for the early component of current equals V_{Na}. This reversal potential is strong evidence that the early component of current is carried by sodium ions.

The strong voltage dependence of the ionic current components is illustrated in a current-voltage characteristic (Figure 4.16). In this figure, the peak values of the transient early component and of the persistent late component of the membrane ionic current density are plotted versus the membrane potential. Over the potential range shown, the persistent late current is a nonlinear function that increases monotonically as the membrane potential is increased. The transient current is inward for membrane potentials below the sodium equilibrium potential (about 110 mV above the resting potential), zero at about the sodium equilibrium potential, and outward for larger potentials.

Separation of Ionic Current Components by Ion Substitution

The observation that the putative sodium current is zero for membrane potentials at the sodium equilibrium potential can be used to separate the ionic currents into sodium and potassium components. So, for example, the ionic currents can be measured for voltage steps with the same final value, but different external sodium concentrations. As shown in Figure 4.17, changing the external sodium concentration changes the Nernst equilibrium potential for sodium. Three cases are shown: one with $V_{Na} > V_m^f$, one with $V_{Na} < V_m^f$, and a third with $V_{Na} = V_m^f$. For the latter case, there is no sodium current; hence,

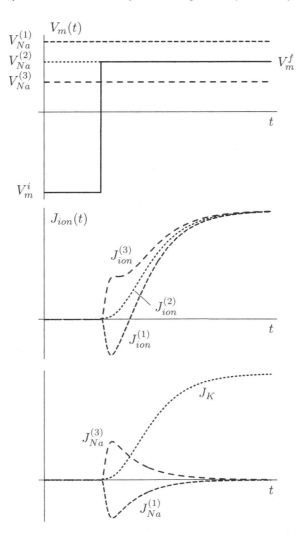

Figure 4.17 Illustration of a method for determining the sodium and potassium currents from the ionic current. It is assumed that the leakage current has been removed from the records. The ionic current is measured for the same voltage step in three solutions that have different external sodium concentrations. Solutions with different sodium concentrations lead to different Nernst equilibrium potentials for sodium, as shown. For $V_m^f = V_{Na}$, $J_{Na} = 0$, so that the ionic current equals the potassium current. Therefore, $J_K = J_{ion}^{(2)}$, which is the same for all three solutions. Thus, $J_{Na}^{(3)} = J_{ion}^{(3)} - J_{ion}^{(2)}$, and $J_{Na}^{(1)} = J_{ion}^{(1)} - J_{ion}^{(2)}$.

the ionic current is due entirely to potassium, and $J_{ion}^{(2)} = J_K$. If the potassium current is independent of the sodium concentration, subtraction of the potassium current obtained at $V_{Na} = V_m^f$ from the currents obtained for other values of V_{Na} gives the sodium currents for the corresponding sodium concentrations. This method was used by Hodgkin and Huxley to separate ionic currents into sodium and potassium components. As illustrated schematically in Figure 4.17, it was found that the sodium current is transient, whereas the potassium current is persistent.

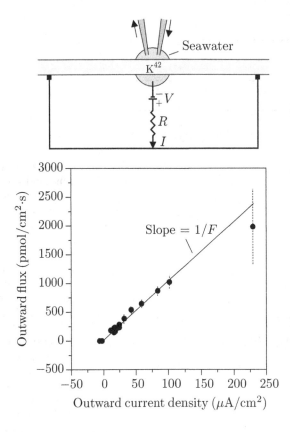

Figure 4.18 Comparison of persistent outward current with potassium efflux in cuttlefish (*Sepia officinalis*) axons (adapted from Hodgkin and Huxley, 1953, Figure 4). The upper panel shows a schematic diagram of the stimulating and recording arrangement, and the lower panel shows the results. The points are the mean flux determined from the count rate and the vertical line segments are the standard error of the mean of the count rate and give an indication of the expected variation of the mean flux.

Direct Determination of Ionic Species Responsible for Ionic Current Components

The method of determining the sodium and potassium currents by varying the ion concentrations indicates which of these ions is likely to contribute to a given feature of the total ionic current, but it does not determine the sodium and potassium currents directly. Direct evidence that the persistent current is carried by potassium ions was obtained by comparing the flux of labeled potassium with the persistent current component (Hodgkin and Huxley, 1953). In experiments on cuttlefish, an axon was soaked in seawater containing radioactive potassium ions ($^{42}K^+$). Thus, a portion of the internal potassium was labeled. The axon was placed in oil, and a bubble of seawater was placed along a portion of the axon as shown schematically in Figure 4.18. A current was passed through the membrane via the bubble for a period of time, and the seawater from the bubble was sampled by means of a pair of pipettes—one to remove the seawater and the other to replace it with fresh seawater—and counted to determine the efflux of radioactive potassium.

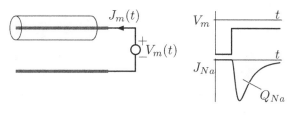

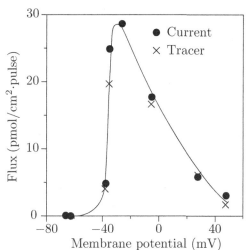

Figure 4.19 Measurement of sodium ions transferred per pulse of membrane potential as a function of the membrane potential from squid (*Dosidicus gigas*) axons (adapted from Atwater et al., 1969, Figure 1B). The upper panel shows a schematic diagram of the electrical recording method. The total charge density carried by sodium is Q_{Na}. As indicated, it is the area of $J_{Na}(t)$. The filled circles were obtained by integrating the sodium current, and the $\times$s were obtained from counting the radioactive tracers in the perfusion solution.

Thus, both the current and the flux of potassium through the membrane were determined through the known surface area of membrane in the bubble. The measurement was repeated for different values of current. Figure 4.18 shows the relation between the efflux of potassium (above the resting value) and the outward current density. The slope of the line shown in the figure is the reciprocal of Faraday's constant (F). This result demonstrates that "the steady outward current associated with depolarization is mainly carried by potassium ions" (Hodgkin and Huxley, 1953).

An experiment on the relation between the putative sodium transferred by the current measured in a voltage-clamp experiment and the efflux of radioactively labeled sodium has given a similar result (Figure 4.19). In this experiment, the axon was bathed in radioactive seawater containing $^{22}Na^+$, and the internal perfusion fluid was collected and sampled. Then a periodic sequence of pulses of membrane potential was applied under voltage-clamp conditions. The increment in internal sodium concentration per pulse was determined from the radioactive tracer. In addition, the total charge density resulting from the transient early current was measured by separating the sodium and potassium currents with pharmacological blockers and integrating the transient early current, as shown schematically in Figure 4.19, to yield the total charge density Q_{Na} due to the putative sodium current. As shown in Figure 4.19, the increment in sodium concentration per unit area of membrane and per pulse of voltage computed from the current records fits reasonably well with that obtained directly from the radioactive tracer. The ratio of flux per pulse determined electrically to that determined by the tracer averaged over all experiments and membrane potentials was 0.92 ± 0.15 (average $\pm$ one standard deviation). Therefore, the transient early component of the ionic current is carried mainly by sodium ions.

Other Methods of Separating Ionic Currents

The results shown in Figures 4.18 and 4.19 indicate that the transient early current is indeed carried by sodium and that the persistent late current is carried by potassium. As we shall see in more detail in Chapter 6, methods of separating ionic current components are now available that are more convenient than the ion substitution experiments initially used by Hodgkin and Huxley. Since 1952, it has been found that there are ionic species to which the membrane is relatively impermeable and there exist pharmacological agents that specifically block either the sodium current or the potassium current. Tetrodotoxin (TTX), a potent neurotoxin isolated from puffer fish, selectively blocks the early current component normally carried by sodium and does not affect the later component normally carried by potassium. On the other hand, tetraethylammonium chloride (TEA) selectively blocks the late current normally carried by potassium ions without affecting the early sodium current. Figure 4.20 indicates the results that can be achieved using experimental protocols in which both impermeant ions and pharmacological blockage are used to isolate the ionic current components. In the upper panel, TTX was used to block sodium currents, so the ionic current is due almost entirely to potassium. In the lower panel, cesium was substituted for the intracellular potassium, and the seawater contained no potassium ions. Since, as we shall see, the membrane is relatively impermeant to cesium, there is no outward current carried by cesium, so the potassium current, which is normally outward, is blocked. Therefore, the early current carried by sodium ions is shown in isolation.

There are now several techniques for separating ionic currents into components carried by different species of ions, and the different separation techniques give consistent results. Furthermore, the success of the pharmacological methods shows that the sodium and potassium currents are

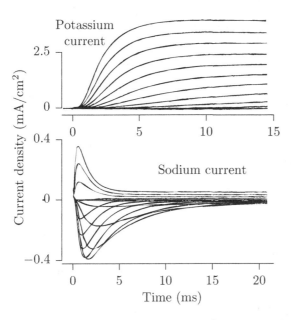

Figure 4.20 Superimposed traces of potassium (upper panel) and sodium (lower panel) currents measured under voltage clamp from giant axons of squids (adapted from Keynes and Aidley, 1991, Figures 4.12 and 4.13). Potassium currents were obtained from an axon bathed in artificial seawater that contained 1 μmol/L TTX and which was dialyzed intracellularly with 350 mmol/L of KF for voltage steps from a holding potential of -70 mV to potentials that varied from -60 to $+40$ mV in 10 mV steps. Sodium currents were obtained from an axon bathed in K-free artificial seawater that contained 103 mmol/L NaCl and 421 mmol/L Tris buffer and dialyzed intracellularly with 330 mmol/L CsF plus 20 mmol/L NaF for voltage steps from a holding potential of -70 mV to potentials that varied from -40 to $+80$ mV in 10 mV steps. The temperature was 4–5°C.

due to mechanisms that are not only kinetically, but also pharmacologically, distinct.

Inactivation of the Sodium Current

The sodium current in response to a step of depolarization is transient. Its magnitude increases rapidly to a maximum and then decreases much more slowly. This behavior results from two mechanisms, one of which increases the sodium current when the membrane is depolarized (called *activation*), and the other of which decreases the current magnitude when the membrane is depolarized (called *inactivation*). Hodgkin and Huxley investigated the properties of inactivation of the sodium current with a two-step paradigm in voltage-clamp experiments. As shown in Figure 4.21, the membrane potential in such

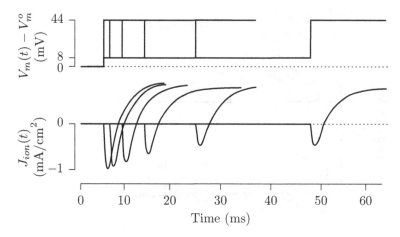

Figure 4.21 Superimposed traces of ionic currents measured under voltage clamp from a giant axon of a squid illustrating the time course of inactivation (adapted from Hodgkin and Huxley, 1952c, Figure 1). The membrane potential was depolarized 8 mV and held at this level for different time intervals, then stepped to a total depolarization of 44 mV as shown in the upper panel. The resulting ionic currents are shown in the lower panel. The temperature was 5°C.

experiments consists of a step to one level followed by a step to another level. The response to the second step is a measure of the inactivation caused by the first step. As indicated in Figure 4.21, the magnitude of the peak of the early response to the second depolarization decreases as the duration of the initial depolarization of 8 mV is increased. The inactivation is nearly complete in 25 ms for the conditions shown. Note that the steady-state value of the outward current is relatively independent of the duration of the initial depolarization. These results indicate that the sodium current inactivates, but the potassium current does not.

Figure 4.22 shows the results of the experiment shown in Figure 4.21, along with results obtained for different values of the initial membrane potential. When the membrane is initially depolarized, the response to the second step decreases exponentially with time to a steady-state value, i.e., the response is inactivated. When the membrane is initially hyperpolarized, this response increases exponentially with time to a larger value, i.e., the inactivation of the response is removed by the initial hyperpolarization. In general, an initial depolarization of the membrane decreases this response, and an initial hyperpolarization increases this response. The time constant of inactivation is a nonmonotonic function of the membrane potential; it is maximal near a depolarization of zero and is small both for large depolarizations and for large hyperpolarizations.

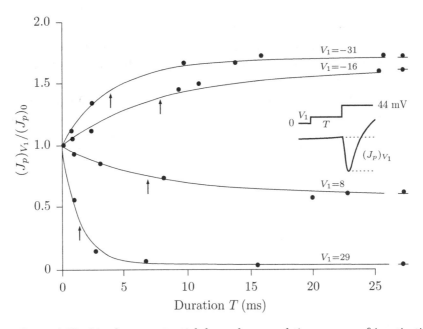

Figure 4.22 Membrane potential dependence and time course of inactivation of the sodium current of a giant axon of a squid (adapted from Hodgkin and Huxley, 1952c, Figure 3). The membrane potential was depolarized to the potential V_1 for different time intervals and then stepped to a total depolarization of 44 mV as shown in the insets. The peak value of the early current in response to the second depolarization, $(J_p)_{V_1}$, normalized to its value when $V_1 = 0$ is plotted versus the duration of the first depolarization for different values of the membrane potential used in the first depolarization. The points are the measurements, and the lines are exponential functions of time. Each arrow indicates the value of the time constant for the indicated curve. The points on the far right were obtained with long-duration pulses to allow the inactivation to approach its final value.

4.2.3   Ionic Conductances

4.2.3.1   *Fitting the Measured Data*

The ionic conductances can be obtained directly from the ionic currents measured under voltage clamp,

$$G_K(V_m, t) = \frac{J_K(V_m, t)}{V_m - V_K} \tag{4.7}$$

and

$$G_{Na}(V_m, t) = \frac{J_{Na}(V_m, t)}{V_m - V_{Na}}. \tag{4.8}$$

The ionic conductances can be calculated as a function of time for several different values of V_m (Figure 4.23) from measurements of the sodium and potassium currents as a function of time for several values of V_m. Both the sodium conductance and the potassium conductance are positive quantities. The sodium conductance in response to a step of depolarization rises rapidly along an S-shaped curve to a maximum and then decays exponentially to zero with a slower time course. The potassium conductance rises along an S-shaped curve to a persistent final value. The rate of rise of the potassium conductance is roughly similar to the rate of decay of the sodium conductance. Both the time course and the amplitude of these conductance changes depend upon membrane potential.

Hodgkin and Huxley fit the ionic conductances with powers of variables satisfying first-order kinetic equations as follows:

$$G_K(V_m, t) = \overline{G}_K n^4(V_m, t),\tag{4.9}$$

and

$$G_{Na}(V_m, t) = \overline{G}_{Na} m^3(V_m, t) h(V_m, t),\tag{4.10}$$

where $\overline{G}_K$ and $\overline{G}_{Na}$ are constants and the variables n, m, and h are each solutions of first-order ordinary differential equations whose coefficients are instantaneous functions of membrane potential, V_m, only. Let x be a generic (in)activation factor that represents any one of the variables n, m, or h and satisfies a first-order kinetic equation that can be written in two convenient forms. One form, which is expressed in terms of rate constants, is

$$\frac{dx}{dt} = \alpha_x(1 - x) - \beta_x x.\tag{4.11}$$

The other form, which is in terms of a time constant and a final value, is

$$\tau_x \frac{dx}{dt} + x = x_\infty,\tag{4.12}$$

where α_x, β_x, τ_x, and x_∞ are functions of membrane potential only, i.e., $\alpha_x(V_m)$, $\beta_x(V_m)$, $\tau_x(V_m)$, and $x_\infty(V_m)$. The relations between the final value of x, x_∞, and the time constant τ_x in Equation 4.12 and the rate constants in Equation 4.11 are $x_\infty = \alpha_x/(\alpha_x + \beta_x)$ and $\tau_x = 1/(\alpha_x + \beta_x)$.

Thus, the time dependence of x results from the first-order kinetic equation and the voltage dependence results from the parameters x_∞ and τ_x. Under step voltage-clamp conditions, the solution has the form

$$x(t) = x_\infty - (x_\infty - x_0)e^{-t/\tau_x}, \qquad t \ge 0,\tag{4.13}$$

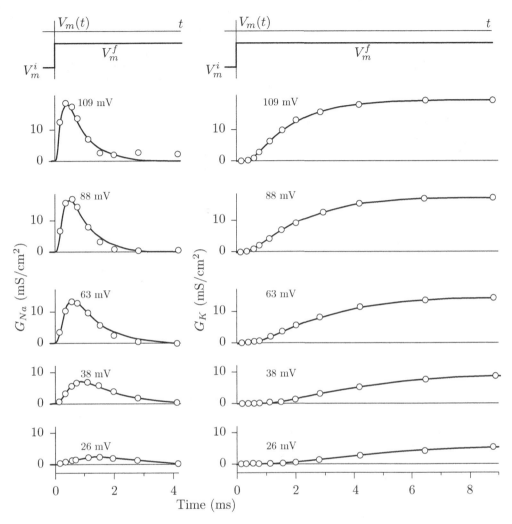

Figure 4.23 Comparison of measured (points) and calculated (lines) sodium and potassium conductances obtained from voltage-clamp experiments on the giant axon of the squid (adapted from Hodgkin, 1964, Figure 29). The upper panels show the membrane potential as a function of time, and the lower panels show the conductances. The calculations are based on the Hodgkin-Huxley model of the conductances. The parameter next to each curve is the displacement of the potential from its initial value $(V_m^f - V_m^i)$. The temperature was 6-7°C.

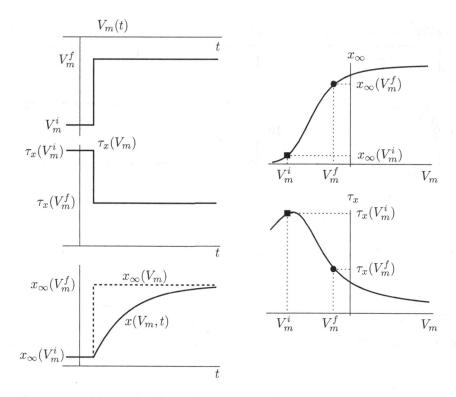

Figure 4.24 Time and voltage dependence of the generic activation factor x. The membrane potential is a step from V_m^i to V_m^f. The values of x_∞ and τ_x at V_m^i (filled square) and V_m^f (filled circle) are shown in the right panels. Both x_∞ and τ_x are step functions of time as shown in the left panels. These parameters determine the exponential time course of x, which starts at $x_\infty(V_m^i)$ and approaches $x_\infty(V_m^f)$ exponentially with time constant $\tau_x(V_m^f)$.

where x_∞ is the final value of $x(t)$ and depends upon the final value of the voltage, V_m^f; x_o is the initial value of $x(t)$ and depends upon the initial value of the voltage, V_m^i; and τ_x is the time constant and depends upon V_m^f. Figure 4.24 indicates how x depends upon both time and membrane potential. The potential is assumed to have been at its initial value, V_m^i, for a sufficient time so that x has reached its steady-state value. The potential is then stepped to its final value, V_m^f. Both x_∞ and τ_x step instantaneously from their initial values to their final values. However, x follows first-order kinetics, so it starts at $x_\infty(V_m^i)$ and approaches $x_\infty(V_m^f)$ exponentially with the time constant $\tau_x(V_m^f)$.

To completely characterize G_K and G_{Na} under voltage-clamp conditions, it is necessary to determine the dependence on V_m of n_∞, m_∞, h_∞, τ_n, τ_m, and τ_h or all the rate constants—α_m, β_m, α_h, β_h, α_n, and β_n. Hodgkin and Huxley measured the voltage dependence of all the rate constants and fit the rate constants with analytic expressions as follows:

$$\alpha_m = \frac{-0.1(V_m + 35)}{e^{-0.1(V_m+35)} - 1}, \tag{4.14}$$

$$\beta_m = 4e^{-(V_m+60)/18}, \tag{4.15}$$

$$\alpha_h = 0.07e^{-0.05(V_m+60)}, \tag{4.16}$$

$$\beta_h = \frac{1}{1 + e^{-0.1(V_m+30)}}, \tag{4.17}$$

$$\alpha_n = \frac{-0.01(V_m + 50)}{e^{-0.1(V_m+50)} - 1}, \tag{4.18}$$

$$\beta_n = 0.125e^{-0.0125(V_m+60)}, \tag{4.19}$$

where V_m is expressed in mV and all the αs and βs are expressed in 1/ms. The time constants and equilibrium values can be defined in terms of the rate constants by comparing Equation 4.11 to Equation 4.12, as follows:

$$
\begin{aligned}
\tau_m &= \frac{1}{\alpha_m + \beta_m}, \quad \text{and} \quad m_\infty = \frac{\alpha_m}{\alpha_m + \beta_m}, \\
\tau_h &= \frac{1}{\alpha_h + \beta_h}, \quad \text{and} \quad h_\infty = \frac{\alpha_h}{\alpha_h + \beta_h}, \\
\tau_n &= \frac{1}{\alpha_n + \beta_n}, \quad \text{and} \quad n_\infty = \frac{\alpha_n}{\alpha_n + \beta_n}.
\end{aligned}
\tag{4.20}
$$

The Hodgkin-Huxley model comprises Equations 4.5, 4.9, 4.10, 4.11, and 4.14–4.19 plus the following numerical parameters: $\overline{G}_{Na} = 120$, $\overline{G}_K = 36$, and $G_L = 0.3$ mS/cm^2; $C_m = 1$ μF/cm^2; $c_{Na}^o = 491$, $c_{Na}^i = 50$, $c_K^o = 20.11$, and $c_K^i = 400$ mmol/L; $V_L = -49$ mV; and temperature is 6.3°C.

The dependences on membrane potential of the time constants and equilibrium values are shown in Figure 4.25. Both n_∞ and m_∞ increase monotonically as V_m increases. Therefore, these factors make the conductances increase as the membrane potential increases. Hence, they are called *activation factors*. In contrast, h_∞ decreases monotonically as V_m increases. Therefore, it makes the sodium conductance decrease as the membrane potential increases. Hence, h is called an *inactivation factor*. Note that τ_m is about a factor of 10

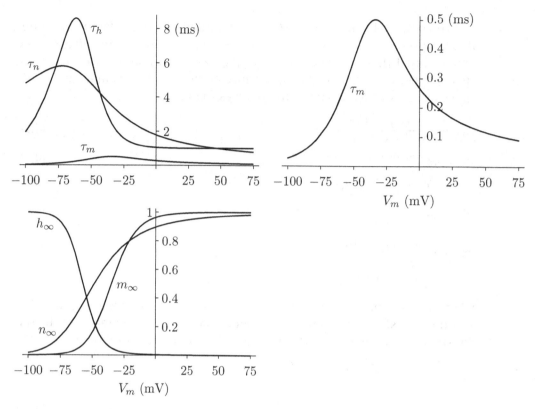

Figure 4.25 Dependence on membrane potential of parameters of activation and inactivation factors. Because it is much smaller than either τ_n or τ_h, τ_m is also shown on a separate scale.

smaller than either τ_n or τ_h. As we shall see, this difference in time constants is crucial in the excitation of an action potential.

Figure 4.26 shows how the model fits the conductance data. Consider the potassium conductance first. A rectangular pulse of depolarization results in a rectangular pulse change in both n_∞ and τ_n, and n follows n_∞ with first-order kinetics with a time constant appropriate to the voltage. This voltage dependence of τ_n can be seen in the n waveform, where the rise of n has about half the time constant of the decay. G_K is proportional to n^4, which accounts for the S-shaped onset of the potassium conductance. This onset can be seen by examining G_K, which is

$$G_K(V_m, t) \approx \overline{G}_K \left(n_\infty - (n_\infty - n_o)e^{-t/\tau_n} \right)^4.$$

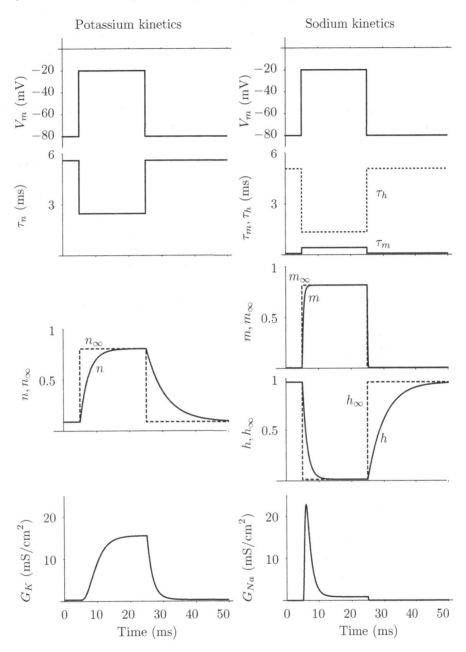

Figure 4.26 Hodgkin-Huxley model of the sodium (right panel) and potassium (left panel) conductances in response to a 20 ms pulse of membrane potential from −80 to −20 and back to −80 mV. The activation and inactivation factors as well as their time constants and equilibrium values are shown in separate panels.

For $t/\tau_n \ll 1$, the exponential can be approximated by the first two terms of a Taylor's series expansion about $t = 0$ to give

$$G_K(V_m, t) \approx \overline{G}_K \left(n_\infty - (n_\infty - n_o)(1 - t/\tau_n)\right)^4 = \overline{G}_K \left(n_o + (n_\infty - n_o)t/\tau_n\right)^4 ;$$

this shows that the conductance increases initially as a fourth-order parabola.

The sodium conductance is a bit more complicated, since it depends upon two factors. A rectangular pulse of membrane potential results in rectangular pulses of the parameters m_∞, τ_m, h_∞, and τ_h. Also, m follows m_∞, and h follows h_∞. The values of the time constants in each segment of constant potential depend upon the value of the potential. The depolarization makes m increase rapidly and h decrease slowly. Hence, G_{Na} shows a rapid onset and a slower decline. Because $G_{Na} \propto m^3 h$, the sodium conductance has an S-shaped onset. An argument similar to the one given for the potassium conductance shows that the sodium conductance increases initially as a third-order parabola. As can be seen in Figure 4.23, this model of the conductances fits the measurements rather well. Note also that G_{Na} is small at rest and after a large depolarization has been on for some time. However, G_{Na} is small at rest because m is small, whereas G_{Na} is small during a large depolarization because h is small. The implications of this distinction are critical for understanding the excitation of action potentials. When the sodium conductance is small because m is small and h is large, as is the case at rest, the sodium conductance is capable of a rapid increase in response to a depolarization. When the sodium conductance is small because m is large and h is small, as is the case during a large depolarization, the sodium conductance is not capable of a rapid increase in response to a depolarization; the sodium conductance is inactivated.

4.2.3.2 *Properties of the Conductances*

Passivity

Because the Gs are nonnegative quantities, i.e., $G_n \geq 0$, they are passive chord conductances. This property can be seen directly by examining the power flow into this element (see also Appendix 4.1). The power density flow, $\mathcal{P}_n$, into the conductance G_n is the product of the current density through the conductance and the potential across the conductance, which is

$$\mathcal{P}_n(t) = J_n(t)\,(V_m(t) - V_n) = G_n(t)\,(V_m(t) - V_n)^2 . \tag{4.21}$$

Therefore, since $G_n \geq 0$, Equation 4.21 implies that $\mathcal{P}_n \geq 0$, i.e., the power flow is always into the conductance and energy is always dissipated by the conductance. Hence, these Gs are purely dissipative elements, which is why they are called *passive chord conductance.*

Voltage Control

The Gs depend explicitly on the membrane potential, V_m, and not, for example, on J_m. Thus, the membrane potential has two effects on the current carried by an ion. For example, for potassium

$$J_K = \underbrace{G_K(V_m, t)}_{\text{gating}} \underbrace{(V_m - V_K)}_{\text{driving force}}. \tag{4.22}$$

The first effect results because the membrane potential acts as a driving force on potassium ions. That is, a change in potential causes a change in potassium current directly by changing the electrochemical potential difference $(V_m - V_K)$, which drives the potassium ions across the membrane. The second effect results from the effect of the potential on the gating of potassium. That is, a change in potential also changes the potassium conductance of the membrane, $G_K(V_m, t)$, which changes the potassium current. The same argument applies to sodium. It might be useful to invoke a hydraulic analogy. Imagine a water faucet whose valve position depends upon the hydraulic pressure in the line feeding the faucet. Thus, an increase in hydraulic pressure will directly drive more water through the valve, but if the increase in hydraulic pressure also opens the valve, there will be a further increase in water flow. Alternatively, we could imagine an inactivating valve in which the increased hydraulic pressure closed the valve.

Voltage Dependence of Maximum Conductance

Both the magnitude of the conductance and its time course depend upon the membrane potential. A clue to the molecular mechanisms underlying the conductances, which will be explored further in Chapter 6, is contained in the measurements of the voltage dependence of the maximum conductance, as shown in Figure 4.27. The results show that the maximum sodium and potassium conductances grow monotonically with membrane potential. For small values of membrane potential, the conductances grow exponentially with membrane potential and correspond to an e-fold change in conductance in 4–5 mV. For large depolarizations, the conductances saturate.

Memory

Both G_K and G_{Na} have memory in the sense that the present values of G_K and G_{Na} depend upon previous values. This property can be seen from the definitions of the variables n, m, and h. The present value of each of these variables depends upon one initial condition. Hence, since G_K depends upon n

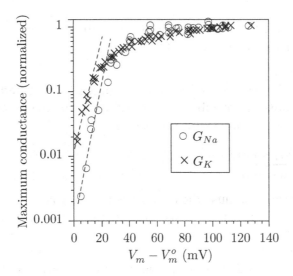

Figure 4.27 Measurements of the voltage dependence of the maximum sodium and potassium conductances during a voltage clamp for the giant axon of the squid (adapted from Hodgkin and Huxley, 1952b, Figures 9 and 10). Results are pooled from those obtained from five axons. The maximum sodium and potassium conductance was measured for each value of membrane potential during a voltage clamp. These maximum conductances were normalized to the maximum conductance obtained at $V_m - V_m^o = 100$ mV. The ordinate scale is logarithmic. The dashed lines correspond to exponential dependence of conductance on membrane potential, with an e-fold change in conductance for a 4 mV (for the sodium conductance) and a 5 mV (for the potassium conductance) change in potential.

only, G_K depends upon only one initial condition. Similarly, G_{Na} depends upon two "state" variables, m and h, and is determined by two initial conditions.

Continuity

In response to a discontinuous change in membrane potential—such as occurs in a voltage-clamp experiment—the factors n, m, and h change continuously in time (Figures 4.24 and 4.26). Therefore, since the conductances are products of these factors, the conductances also change continuously in time. However, the ionic currents are products of the conductances and the deviation of the membrane potential from the equilibrium potential. Hence, if the membrane potential is discontinuous, the ionic currents will be discontinuous functions of time even though the conductances are continuous functions of time.

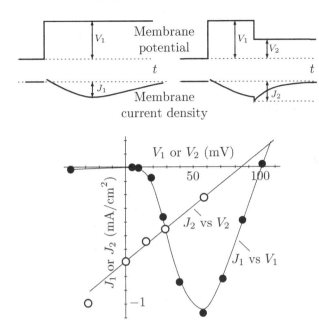

Figure 4.28 Measurements of the instantaneous current-voltage relation for the giant axon of the squid (adapted from Hodgkin and Huxley, 1952a, Figure 6). The upper panels show the membrane potential and the resulting ionic currents as a function of time. The upper left panel shows a single-step voltage-clamp profile; the upper right panel shows a two-step voltage clamp. The lower panel shows the current-voltage characteristics defined in the upper panels. The temperature is $4°$C. All the voltages in this figure are deviations of the membrane potential from its resting value.

Ohmic Properties

In response to a step of membrane potential, the ionic current shows a time-varying waveform. Why is it reasonable to represent this variation by a time-varying conductance rather than some other equivalent electrical network, e.g., one consisting of capacitances and inductances? One way to address this question is to measure the current-voltage characteristic instantaneously. The instantaneous current-voltage relation for a conductance is quite different from that for a capacitance or an inductance. The results of such an experiment are shown in Figure 4.28. Two current-voltage relations are shown: one is based on the peak early current just as in Figure 4.16, and the other is based on the instantaneous relation between the current and the voltage in response to the second step. While the former is grossly nonlinear, the latter is linear. That is, the instantaneous change in current is proportional to the change in potential.[6] Thus, the element obeys Ohm's law; therefore, the ele-

---

6. In normal seawater, the instantaneous relation between current and voltage is linear for the giant axon of the squid, but it is nonlinear in seawater whose sodium content has been reduced. Nevertheless, the current-voltage relation is still that of a conductance—a nonlinear conductance.

ment can be represented as a conductance and not as either a capacitance or an inductance.

4.3 Synthesis of the Hodgkin-Huxley Model

In the previous sections, the techniques for measuring the electrical properties of the squid giant axon under voltage-clamp conditions were explained. The model devised by Hodgkin and Huxley was based on such measurements and resulted in the model given in Equations 4.5–4.19. Under voltage-clamp conditions, no action potential occurs. In this section, we shall show that when the voltage-clamp conditions are removed, the same equations produce action potentials. In the following section, we shall examine key properties of electrically excitable cells and compare these properties with the predictions of the Hodgkin-Huxley model.

The Hodgkin-Huxley model is analytically quite complex when the membrane potential is not constant. Hence, we shall examine properties of the equations using two methods. Some properties are discussed by examining numerical solutions of the Hodgkin-Huxley equations. This procedure can give information on the accuracy with which the equations account for membrane phenomena for complex situations. Access to a simulation of these equations allows the performance of simulation experiments, which can greatly enhance intuition about these equations. We shall also use a variety of approximate analytical methods to obtain mathematical and physical insight into the model. These approximate results can, of course, be checked by more accurate numerical calculations.

4.3.1 Propagated Action Potential

4.3.1.1 Waveform and Conduction Velocity

The Hodgkin-Huxley equations for the squid giant axon are a set of four coupled nonlinear partial differential equations plus a set of algebraic equations (Equations 4.5, 4.9–4.11, and 4.14–4.19). Solutions of these equations obtained with a digital computer (Cooley and Dodge, 1966) are shown in Figure 4.29. Solutions for $V = V_m - V_m^o$ are shown as a function of t in response to a simulated suprathreshold current applied intracellularly to the axon at $z = 0$ starting at $t = 0$ with a duration of 0.2 ms. The solution is symmetrical about $z = 0$, but only the portion for $z > 0$ is shown in the figure. The initial conditions are

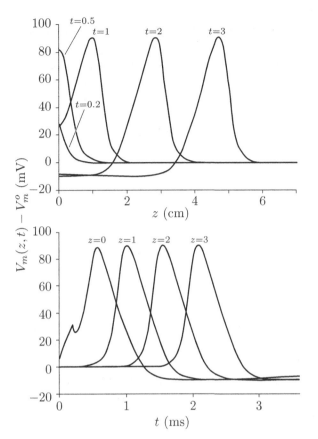

Figure 4.29 Solution of the Hodgkin-Huxley equations for a simulated suprathreshold pulse of current (adapted from Cooley and Dodge, 1966, Figure 2). Solutions are shown as a function of z (upper panel) at five instants of time (in ms) and as a function of t (lower panel) at four locations (in cm).

that $V = 0$ for $t < 0$. The solution corresponds to a pulse of potential that travels away from $z = 0$ with a constant velocity shortly after the initial starting transient is complete. The conduction velocity estimated from these calculations is about 18.7 m/s. This velocity is in the range of the normal propagation velocity for a squid axon. Thus, the equations predict that a propagated action potential will occur, as well as the value of the conduction velocity. Note also that there is a local response at the stimulation site ($z = 0$ in the lower part of Figure 4.29) that does not appear at the more distant locations. This local response is the membrane potential response to the current pulse, but it propagates by the cable equation and is attenuated so that it is not discernible at the more distant locations.

The solution of these equations is time consuming (see the quotes at the beginning of this chapter). The solutions shown in Figure 4.29 were performed on an IBM 7094 before 1966 and took many minutes to perform. If it is assumed that Equation 4.5 can be satisfied by a traveling wave with constant propagation velocity, v, a considerable savings in computation time can be obtained, because the partial differential equation for V_m can be reduced to an ordinary differential equation, which can be solved more rapidly. As shown in Chapter 2, if V_m represents a wave traveling at constant velocity in the positive z-direction, $V_m(z, t) = f(t - z/v)$, and V_m satisfies the wave equation,

$$\frac{\partial^2 V_m(z, t)}{\partial z^2} = \frac{1}{v^2} \frac{\partial^2 V_m(z, t)}{\partial t^2}. \tag{4.23}$$

By substituting the wave equation into Equation 4.5, we obtain the ordinary differential equation

$$\frac{1}{2\pi a(r_o + r_i)v^2} \frac{d^2 V_m}{dt^2} = C_m \frac{dV_m}{dt} + \overline{G}_K n^4(V_m, t) (V_m - V_K)$$
$$+ \overline{G}_{Na} m^3(V_m, t) h(V_m, t) (V_m - V_{Na})$$
$$+ G_L (V_m - V_L). \tag{4.24}$$

Equation 4.24 was first solved with a hand-operated calculator (Hodgkin and Huxley, 1952e). A solution of Equation 4.24 can be obtained by guessing a value of v and seeking a stable solution (one for which V_m is bounded) by integrating the differential equation numerically starting from $t = 0$. At some point in time, the trial solution for V_m may begin to diverge (as shown by the dashed lines in Figure 4.30). A new value of v is chosen, and the calculation is repeated. By this trial-and-error method, a stable solution (solid line in Figure 4.30) can be found and the value of v that satisfies the equation can also be determined ($v \approx 18.74$ m/s). A stable solution of Equation 4.24 that corresponds to a propagated action potential occurs for only one value of v.

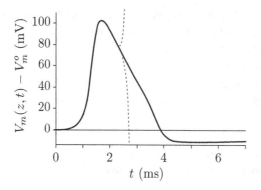

Figure 4.30 A propagated action potential computed according to Equation 4.24. The dashed lines show divergent solutions. The solid line shows a stable solution that corresponds to $v = 18.74$ m/s (adapted from Fitzhugh and Antosiewicz, 1959, Figure 5).

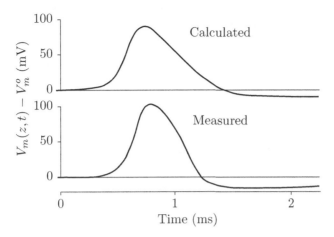

Figure 4.31 Comparison of waveforms of a computed propagated action potential (upper trace) and a measured (lower trace) action potential from a squid axon (adapted from Hodgkin and Huxley, 1952e, Figure 15).

This value of v agrees with the computation of Cooley and Dodge and falls in the range of values obtained from measurements on squid axons.

The waveshape of the computed propagated action potential is shown in Figure 4.31 (upper trace) and may be compared with data from a squid axon (lower trace). The waveshapes, including the time courses and amplitudes, agree reasonably well, although the falling phase of the measured action potential is more rapid than the falling phase of the calculated action potential.

4.3.1.2 Underlying Variables

We can take a deeper look at the events that underlie the propagated action potential by examining all the Hodgkin-Huxley variables (Figure 4.32). Four time intervals can be identified. During interval I, a propagating action potential that invades the membrane from a neighboring patch of membrane produces an outward membrane current that is carried predominantly by the capacitance current. The outward current charges the membrane capacitance and increases the membrane potential. However, the change in potential is sufficiently small that the variables m, n, and h and the ionic conductances are not changed significantly from their resting values. During this time interval, the change in membrane potential can be accounted for by the cable model (Chapter 3). During interval II, the membrane potential increases rapidly, which increases m rapidly, which increases G_{Na} and drives V_m toward V_{Na}.

It is this regenerative action, described in further detail in Figure 4.33, that accounts for the onset of the action potential; an increase in membrane

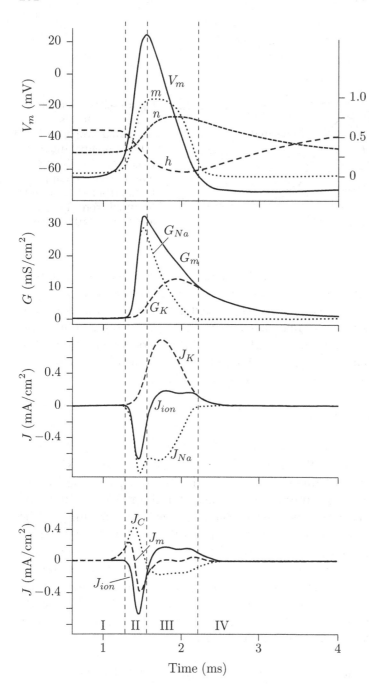

Figure 4.32 Relations among membrane potential, current components, ionic conductances, and activation factors during a propagated action potential (adapted from Cooley and Dodge, 1965, Figure 2.4). The dashed vertical lines separate four time intervals, which are numbered just above the time axis.

Depolarization mechanism – positive feedback

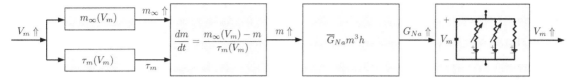

Repolarization mechanisms – negative feedback

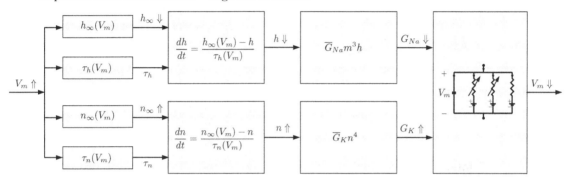

Figure 4.33   Illustration of factors responsible for regenerative depolarization at the onset of the action potential (upper panel) and at the onset of repolarization (lower panel). During the onset of depolarization (upper panel), an increase in V_m causes an increase in m_∞, which results in an increase in m. Since the kinetics of m are rapid, the change in m is much faster than the change in n and in h. The increase in m results in an increase in G_{Na}. The change in G_{Na} in the network model increases V_m further and drives V_m toward V_{Na}. Thus, there is a positive feedback effect or a regenerative action; an increase in V_m leads to an increase in V_m. As this process proceeds, the factors that repolarize the membrane (lower panel) also occur, but more slowly. During the onset of repolarization, the increase in V_m causes h_∞ to decrease and n_∞ to increase. Therefore, h decreases to decrease G_{Na}, and n increases to increase G_K. These changes in the network model result in a decrease in V_m and drive V_m toward V_K. Thus, there is a negative feedback effect; an increase in V_m causes the repolarization mechanisms to reduce V_m.

potential increases the sodium conductance, which increases the membrane potential. During this interval, the ionic current is inward and is carried predominantly by sodium ions. Toward the end of interval II, n has increased appreciably and h has decreased appreciably. That these variables change more slowly than m is a critical factor in the generation of the action potential.

At the beginning of interval III, the decrease in h results in a decrease in G_{Na} and the increase in n results in an increase in G_K, both of which result in a decrease in V_m toward V_K, as is shown in detail in Figure 4.33. During

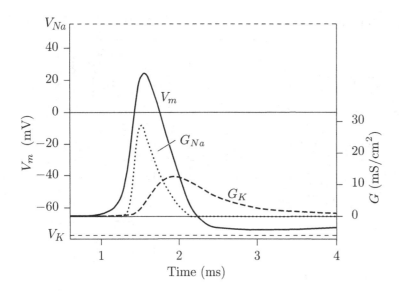

Figure 4.34 Comparison of membrane potential and ionic conductances during a propagated action potential computed from the Hodgkin-Huxley model (adapted from Cooley and Dodge, 1966, Figure 2). $V_K = -72$ mV, and $V_{Na} = +55$ mV.

interval III, h is small and n is large. Both these factors tend to prevent a rapid depolarization of the membrane. During this interval, no action potential can be generated, and the membrane is refractory. During interval IV, G_{Na} has decreased to near its resting level, but G_K has a magnitude that exceeds its resting value. Hence, the membrane potential approaches V_K more closely than it does at rest. This result accounts for the undershoot of the action potential.

The relation between the membrane potential and the ionic conductances is shown in more detail in Figure 4.34. At rest, $G_K \gg G_{Na}$ and V_m is near V_K (see Table 4.1). If the membrane is depolarized, G_{Na} increases rapidly. This increase in G_{Na} drives the membrane potential toward V_{Na}. At the peak of the action potential, $G_{Na} > G_K$; hence, the peak value V_m^p is nearer V_{Na} than V_K. Then G_{Na} decreases and G_K increases. Both factors tend to drive V_m toward V_K. During the falling phase of the action potential, G_K is larger than its resting value; therefore, V_m approaches V_K and leads to the undershoot of the action potential. Thus, the shape of the action potential can be seen to result from changes in ionic conductances.

4.3.1.3 Membrane Conductance

Figure 4.35 shows measurements of the membrane potential and the membrane conductance compared with calculations of the same two quantities. For

Table 4.1   Values of the sodium and potassium conductances at the resting potential (V_m^o) and at the peak of the action potential (V_m^p) for squid axon (Hodgkin and Huxley, 1952e).

Ion	$G_i(V_m^o)$ (mS/cm^2)	$G_i(V_m^p)$ (mS/cm^2)	$G_i(V_m^p)/G_i(V_m^o)$
K$^+$	0.25	5	20
Na$^+$	3×10^{-3}	30	10^4

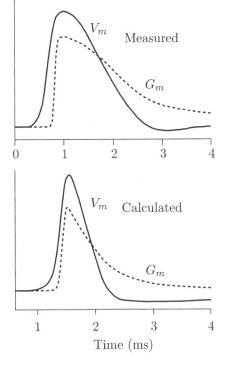

Figure 4.35 Comparison between measurements and calculations of the waveforms of the action potential and the membrane conductance. The measurements are reproduced from Figure 4.1 and the calculations from Figure 4.32. The measurements were made in the temperature range 2–4°C, and the calculations correspond to a temperature of 6.3°C. This difference accounts, in part, for the difference in time course between the measurements and the calculations.

both measured and calculated waveforms, the conductance increase lags behind the potential, rises more rapidly than the potential, peaks at about the same time as the potential, and outlasts the potential change. Thus, these major features of the relation between the measured membrane conductance and the membrane potential are reproduced by the Hodgkin-Huxley model. The peak conductance change predicted by the model for the calculations shown was 32 mS/cm^2, which is in the range of the measured change in conductance

(19–68 mS/cm^2) and close to the average value of 36 mS/cm^2 (Cole and Curtis, 1939).

4.3.1.4 *Net Transfer of Ions during an Action Potential*

The results in Figure 4.34 can also be used to compute the net transfer of Na$^+$ and K$^+$ during an action potential, and these computations fit with measurements obtained using radioactive tracers (see Problem 4.6). These calculations match the measurements, which shows that a few pmol/cm^2 are transferred across the membrane per action potential (Weiss, 1996, Table 7.4).

4.3.2 Membrane Action Potential

4.3.2.1 *Waveform*

Many of the properties of the action potential of a squid axon can be observed in a space-clamped squid axon that gives rise to a membrane action potential. Computation of the membrane action potential can be done much more rapidly than the computation of a propagated action potential, since one need only solve a nonlinear ordinary differential equation rather than a nonlinear partial differential equation.

In a space-clamped axon, a brief shock of membrane current delivered to the axon can be used to vary the initial membrane potential to any value and to examine the resulting membrane potential change following the shock (see Figure 4.36). The initial membrane potential change is $\Delta V_m = Q_e/C_m$,

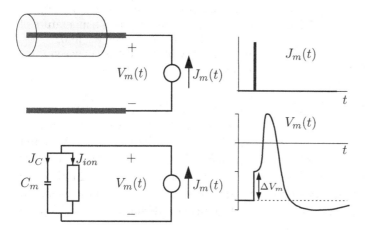

Figure 4.36 Schematic diagram indicating the method of recording a membrane action potential from a space-clamped axon. The left panel shows the recording arrangement (upper) and an equivalent network (lower) in which the ionic current is isolated from the capacitance current. The right panel shows a schematic diagram of the response (lower) to a brief pulse of membrane current (upper).

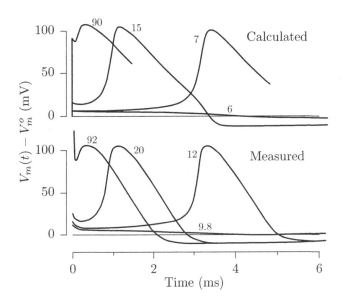

Figure 4.37 Comparison of measured and computed membrane action potentials (adapted from Hodgkin and Huxley, 1952e, Figure 12). The numbers attached to the curves are the charge densities (in nC/cm^2) delivered by a brief pulse of current. For a membrane capacitance of 1 μF/cm^2, these numbers also equal the initial potential change in mV. Both the measurements and the computations are for a temperature of 6°C.

where Q_e is the charge per cm^2 of membrane delivered by the shock and C_m is the capacitance per cm^2 of membrane. The membrane potential change corresponding to this space-clamped axon can also be computed directly from the Hodgkin-Huxley equation by setting $\partial^2 V_m / \partial z^2 = 0$ to obtain

$$0 = C_m \frac{dV_m}{dt} + \overline{G}_K n^4 (V_m, t)\,(V_m - V_K)$$

$$+ \overline{G}_{Na} m^3 (V_m, t) h(V_m, t)\,(V_m - V_{Na}) + G_L\,(V_m - V_L)\,, \tag{4.25}$$

which can be solved for the initial condition $V_m(0) = V_m^o + \Delta V_m$. Results of both the experiments and calculations are compared in Figure 4.37. Both results show that for large-amplitude shocks membrane action potentials are generated and for small-amplitude shocks no action potential occurs. These observations and calculations embody the all-or-none principle. Thus, the Hodgkin-Huxley model accounts for the all-or-none property of the axon.

A direct comparison of the waveform of the membrane action potential measured in a squid giant axon and that computed from the Hodgkin-Huxley model is given in Figure 4.38. The main difference in shape occurs in the falling phase of the membrane action potential, as was the case for the propagated action potential.

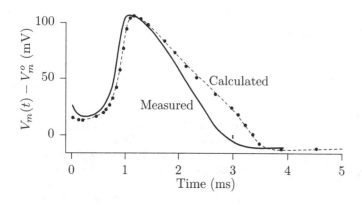

Figure 4.38 Comparison of waveforms of measured (solid line) and calculated (dashed line and points) membrane action potentials (adapted from Fitzhugh and Antosiewicz, 1959, Figure 1). The dashed-line curve was obtained by solving the equations with a digital computer; the dots are the original computations of Hodgkin and Huxley (Hodgkin and Huxley, 1952e).

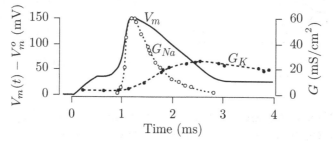

Figure 4.39 Measurements of the action potential (solid line) and the sodium (dotted line) and potassium (dashed line) conductances. Measurements were obtained from a space-clamped squid giant axon at a temperature of 9.5°C (adapted from Bezanilla et al., 1970, Figure 10). The resting potential of the axon was −78 mV.

4.3.2.2 Ionic Conductances

The sodium and potassium conductances during a membrane action potential have been estimated by eliciting a membrane action potential and then rapidly switching on a voltage-clamp system during the action potential. The ionic current was measured in response to a voltage step. By clamping to the sodium or to the potassium equilibrium potential, it was possible to estimate the potassium and sodium currents and conductances, respectively. The results shown in Figure 4.39 indicate that the major features of the sodium and potassium conductances predicted by the Hodgkin-Huxley model are consistent with the measurements. The sodium conductance lags behind and rises more rapidly than the potential, peaks near the peak of the potential, and declines more rapidly than the potential. The potassium conductance lags behind the sodium conductance and rises more slowly than the potential. The potassium conductance increase outlasts the peak of the membrane potential.

4.4 Explanation of the Electrical Excitability of the Giant Axon of the Squid

In this section we shall examine general properties of electrically excitable cells as revealed by the Hodgkin-Huxley model of the giant axon of the squid. Our purpose is to explore the implications of the model of such attributes of electrically excitable cells as the all-or-none property with its inherent threshold property, as well as refractoriness, the strength-duration relation, accommodation, anode-break excitation, repetitive activity, subthreshold oscillations, and the effect of temperature.

4.4.1 Threshold

An essential feature of the all-or-none property of the action potential is the notion of a threshold for eliciting an action potential. In this section, we shall first show that the Hodgkin-Huxley equations have a thresholdlike behavior in response to electrical stimulation, as does an axon. Then we shall look more closely at those features of the electrical characteristics of excitable membranes that give rise to sharp threshold characteristics.

The range of current levels from subthreshold to suprathreshold is exceedingly narrow. The upper six traces in Figure 4.40 were all measured with current stimuli that were identical within three significant figures. Action potentials occurred in response to three of these current pulses, and local responses occurred in response to the other three current pulses. These measurements show that a change in current strength of less than 1 part in 100 spans the entire threshold region. Within this threshold region, the responses of an axon are probabilistic. Since the Hodgkin-Huxley model is a deterministic model, the stochastic behavior of electrically excitable cells near threshold cannot be explained with this model.

Calculations using the Hodgkin-Huxley model reveal that the equations exhibit an even narrower threshold region, as indicated in Figure 4.41. The figure shows pairs of calculations, one suprathreshold and one subthreshold, in a decreasingly narrow range of current. For example, calculations A and A′ correspond to current strengths that differ by 1 part in 10^8, yet trace A yields an action potential and A′ does not. Traces B and B′ differ by 1 part in 10^{14}. Therefore, the Hodgkin-Huxley model does not really exhibit a dis-

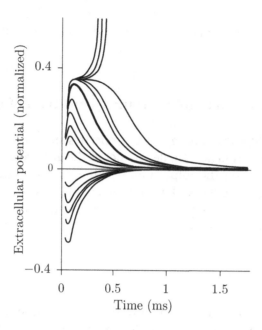

Figure 4.40 Membrane potential in response to near-threshold current pulses recorded from single axons of the crab *Carcinus maenas* (adapted from Hodgkin, 1938, Figure 8). The axon was placed in oil, and extracellular stimulating and recording electrodes were applied to its surface. The ordinate scale is the extracellular potential divided by the amplitude of the propagated action potential, which was 40 mV. The shock strengths, normalized to the value that just gave an action potential, are 1.00 (top six traces), 0.96, 0.85, 0.71, 0.57, 0.43, 0.21, −0.21, −0.43, −0.57, −0.71, and −1.00.

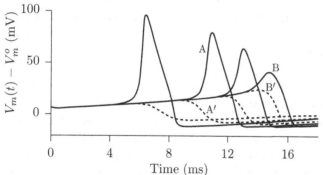

Figure 4.41 Computations of membrane potential for near-threshold current pulses (adapted from Fitzhugh and Antosiewicz, 1959, Figure 2). Pairs of computations, one suprathreshold (solid line) and one subthreshold (dashed line), are shown. The difference between the shock levels of the two pairs decreases from left to right.

crete threshold, but rather exhibits an exceedingly narrow threshold region. As the range of current narrows to near threshold, the occurrence of the peak of the action potential is further delayed in time. The calculations do not really obey the all-or-none principle, since intermediate-size action potentials can be generated by the equations. In theory, some of these intermediate-size action potentials can propagate (Huxley, 1959). These phenomena exhibited by the equations occur for very precisely specified current stimuli. The inherent noise in the membranes prevents these phenomena (intermediate-

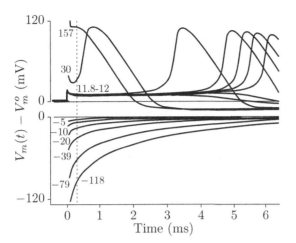

Figure 4.42 Membrane potential response to current pulses in a space-clamped squid axon (adapted from Hodgkin et al., 1952, Figure 9). The numbers indicate the shock strength, which is the charge density delivered by the brief current stimulus in nC/cm^2. The temperature is 6°C. The dashed vertical line marks a time 0.29 ms after the shock.

size action potentials) from being observed in real cells. The presence of membrane noise throws the membrane potential out of this precisely determined threshold region into either a full-blown action potential or into a local response.

Which property of the membrane gives rise to a threshold? In order to answer this question, we shall examine in more detail the membrane potential change following a brief shock of membrane current as shown in Figure 4.42. Note that there are seven curves for shock strengths in the range 11.8–12 nC/cm^2; five are suprathreshold, and two are subthreshold. For stimuli in this narrow threshold range, the response shows a long plateau at the beginning before either initiation of an action potential or completion of a local response. What is happening during this interval of time when the membrane is "deciding" whether to produce an action potential? We shall examine the electrical properties of the membrane during this critical time period.

Consider an instant of time after the shock is completed, but before the peak of the action potential (indicated by the dashed vertical line in Figure 4.42). After completion of the shock, the membrane current is zero, i.e., $J_m = 0$. Therefore, $J_C + J_{ion} = 0$; hence,

$$J_{ion} = -J_C = -C_m \frac{dV_m}{dt}. \tag{4.26}$$

At any instant of time, t_o, it is possible to measure $V_m(t_o)$ and $(dV_m/dt)_{t_o}$. With Equation 4.26, $J_{ion}(t_o)$ can be calculated. Hence, it is possible to plot $J_{ion}(t_o)$ versus $V_m(t_o)$. Different values of $V_m(t_o)$ were obtained by delivering

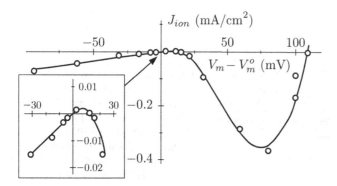

Figure 4.43 Relation between ionic current density and membrane potential for a space-clamped squid axon (the same axon that gave the measurements shown in Figure 4.42) at time 0.29 ms after the application of the shock (adapted from Hodgkin et al., 1952, Figure 10).

shocks of different strengths, which delivered different amounts of charge. The curve relating $J_{ion}(t_o)$ to $V_m(t_o)$ depends on the value of t_o chosen. We shall examine this characteristic at a time, t_o, that occurs just after the shock is completed and prior to the time of occurrence of the peak of the action potential (Figure 4.43). During this interval, the dynamics of the model result in either an action potential or a local response. The J_{ion}-V_m characteristic is sigmoidal and crosses the V_m-axis at three values of the membrane potential— one at the resting value of the membrane potential ($V_m - V_m^o = 0$), a second near the Nernst equilibrium potential for sodium ($V_m - V_m^o = 110$ mV), and a third at a potential about 15 mV above the resting membrane potential. In order to explore the significance of these points, we shall use a schematic diagram of the J_{ion}-V_m characteristic (Figure 4.44) in which the scales have been distorted to simplify the exposition.

Certain points on the curve in Figure 4.44 have the property that if the membrane potential is displaced to one of those points, the membrane potential will stay at this value. In particular,

if $J_{ion}(t_o) = 0$, then $\left(\dfrac{dV_m}{dt} \right)_{t_o} = 0,$ (4.27)

and the membrane potential $V_m(t_o)$ will remain at such an equilibrium value. These equilibrium points have been circled in Figure 4.44. Two types of equilibrium points are indicated in the figure: stable equilibrium points and unstable equilibrium points. A stable equilibrium point is one to which the potential will return if it is displaced from that point by an incremental change in membrane potential. To decide which of the three equilibrium points are stable, we note that Equation 4.26 shows that for $J_{ion} > 0$, $dV_m/dt < 0$, and for $J_{ion} < 0$, $dV_m/dt > 0$. An arrow has been placed on each part of the curve to indicate the direction of change of the membrane variables with increas-

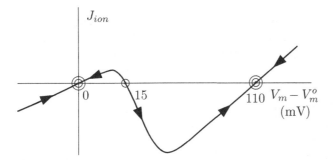

Figure 4.44   Schematic diagram of the relation between ionic current density and membrane potential for a space-clamped squid axon. The diagram is a schematic version of Figure 4.43 to allow for identification of equilibrium points at which the J_{ion}-V_m contour crosses the axis for which $J_{ion} = 0$. Stable equilibrium points are indicated with two concentric circles, the unstable equilibrium point with a single circle.

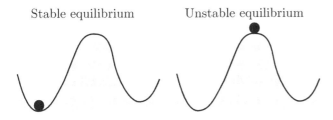

Figure 4.45   Schematic diagram showing the mechanical stability of a sphere subject to gravity at both a stable and an unstable equilibrium point.

ing time. For instance, for $V_m > 110$ mV, $J_{ion}(t_o) > 0$, which implies that $dV_m/dt < 0$; therefore, the membrane potential decreases in this part of the curve. It can be seen that the equilibrium points with the double circles are stable, since a small deviation in membrane potential to either side of the equilibrium point will result in a return of the membrane potential to its equilibrium value. From Figure 4.43, it is clear that two stable equilibrium points occur, one at the resting potential ($V_m(t_o) - V_m^o = 0$) and another at $V_m(t_o) - V_m^o = 110$ mV, which is near the sodium equilibrium potential. On the other hand, the equilibrium at $V_m(t_o) - V_m^o = 15$ mV is unstable, since a slight decrease in the membrane potential will cause it to return to the resting value, and a slight increase in the membrane potential will cause it to increase toward the sodium equilibrium potential. Thus, $V_m(t_o) - V_m^o = 15$ mV is a threshold potential.

The difference between stable and unstable equilibrium points is made quite clear with the mechanical analog shown in Figure 4.45. Consider a sphere subject to gravity and moving on a surface that contains local maxima and minima. Assume that there is friction between the sphere and the surface. Both the minima and the maxima are equilibrium points in that if the sphere is carefully placed on such a point, it will remain there forever. However, the minima are stable equilibrium points, because a small displacement of the sphere from a minimum results in a return of the sphere to the equilibrium point. At the maximum, an infinitesimal displacement of the sphere causes it to leave this unstable equilibrium point.

We return to the analysis of stable and unstable equilibria in Figure 4.44. Which property makes two of these stable equilibrium points and the other an unstable equilibrium point? We note that the unstable equilibria occur at points for which the slope of the J_{ion}-V_m characteristic is negative; stable equilibria occur when this slope is positive. The slope of the characteristic has the dimensions of a conductance, and we call it a *slope conductance* to distinguish it from the *chord conductance* on which we have focused until now. The definitions of the slope conductance, g, and the chord conductance, G, are

$$G \equiv \frac{J_{ion}(V_m)}{V_m} \quad \text{and} \quad g \equiv \frac{dJ_{ion}(V_m)}{dV_m}. \tag{4.28}$$

Both of these definitions are illustrated in Figure 4.46. Thus, unstable equilibria occur at points for which

$$g = \frac{dJ_{ion}}{dV_m} < 0 \quad \text{and} \quad J_{ion} = 0, \tag{4.29}$$

and stable equilibria occur at points for which

$$g = \frac{dJ_{ion}}{dV_m} > 0 \quad \text{and} \quad J_{ion} = 0. \tag{4.30}$$

The stability criterion can be seen directly by expanding the function $J_{ion}(V_m)$ about an equilibrium point (at which the potential is V_m^e) for small perturbations in V_m, which we designate as v_m. The expansion has the form

$$J_{ion} = J_{ion}^e + \left[\frac{dJ_{ion}}{dV_m} \right]_{V_m^e} v_m + \cdots.$$

If we keep only the first two terms in this expansion and equate incremental quantities, we obtain

$$j_{ion} = \left[\frac{dJ_{ion}}{dV_m} \right]_{V_m^e} v_m = g v_m. \tag{4.31}$$

Using Equation 4.26 for small perturbations of V_m, we obtain

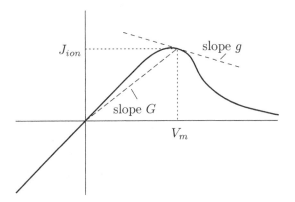

Figure 4.46 Relation between the chord conductance (G) and the slope conductance (g).

$$gv_m = -C_m \frac{dv_m}{dt}, \tag{4.32}$$

whose solution is

$$v_m = v_m(t_o)e^{-(t-t_o)/\tau} \quad \text{for } t \geq t_o, \tag{4.33}$$

where $\tau = g/C_m$. Provided that $\tau > 0$ (i.e., $g > 0$), a perturbation in V_m will result in a convergence of V_m to its equilibrium value. If $\tau < 0$ (i.e., $g < 0$), a perturbation in V_m will result in a divergence of V_m from its equilibrium value. Therefore, the essential character of a thresholdlike property in the relation of J_{ion} to V_m is that the J_{ion}-V_m characteristic crosses the $J_{ion} = 0$ axis with a negative slope.

Since the sodium conductance rises rapidly during the rising phase of the action potential, it is reasonable to inquire whether the sodium conductance alone can account for a threshold point in the J_{ion}-V_m characteristic. Note that if we assume that $J_{ion} = J_{Na}$, then $J_{Na} = G_{Na}(V_m - V_{Na})$. Since the chord conductance $G_{Na} \geq 0$, there is only one equilibrium point, and at that point $V_m = V_{Na}$. To determine if this point is stable, we need to examine the slope conductance at this point. We can write the following:

$$g = \left(\frac{dJ_{Na}}{dV_m}\right)_{V_m=V_{Na}} = \left[\left(\frac{dG_{Na}}{dV_m}\right)(V_m - V_{Na}) + G_{Na}\right]_{V_m=V_{Na}} \tag{4.34}$$

$$= [G_{Na}]_{V_m=V_{Na}}.$$

Hence, for $V_m = V_{Na}$, $g = G_{Na} > 0$, and the equilibrium point is necessarily stable. We conclude that a model in which a single ion is transported passively ($G_{Na} > 0$) according to an equation of the form

$$J_{Na} = G_{Na}(V_m - V_{Na})$$

cannot yield a thresholdlike behavior. We shall see that the requirement is that two ions be transported passively and that one of these ions have a sufficiently nonlinear dependence of chord conductance on membrane potential.

To further examine the conductance of the membrane, we assume that during the rising phase of the action potential, the time constant of $m(t)$ is short compared to the time course of the action potential and that the time constants of $h(t)$ and $n(t)$ are long. Specifically, we assume that m is at its final value at each potential, so $m(V_m, t) = m_\infty(V_m(t))$, and that both n and h remain at their resting values, so $n(V_m, t) = n_\infty(V_m^o)$ and $h(V_m, t) = h_\infty(V_m^o)$. These assumptions are qualitatively consistent with the Hodgkin-Huxley model as illustrated in Figures 4.25 and 4.32, and they allow us to see essential phenomena without the necessity of laborious calculations. Using these assumptions, the approximate form of the dependence of ionic current on membrane potential at a point t_o after the initial depolarization can be plotted as in Figure 4.47. We note that $G_{Na}(V_m)$ is a nonlinear function of V_m and that $J_{Na} = G_{Na}(V_m - V_{Na})$ has a negative slope conductance region. The condition for a negative slope conductance can be derived by differentiating J_{Na} with respect to V_m as follows:

$$g_{Na} = \frac{dJ_{Na}}{dV_m} = \frac{dG_{Na}}{dV_m}(V_m - V_{Na}) + G_{Na}. \tag{4.35}$$

Hence, if $g_{Na} < 0$ and $V_m < V_{Na}$, then

$$\frac{dG_{Na}}{dV_m} > \frac{G_{Na}}{V_{Na} - V_m}. \tag{4.36}$$

Thus, for $V_m < V_{Na}$, if the rate of increase of G_{Na} with V_m is sufficiently great, $g_{Na} < 0$. Note also that the occurrence of a negative slope conductance region depends upon the value of V_{Na}.

Although the J_{Na}-V_m characteristic has a negative slope conductance region, the characteristic does not cross the $J_{Na} = 0$ line with a negative slope conductance. However, the J_{ion}-V_m characteristic—where the ionic current is the sum of sodium, potassium, and leakage current (Equation 4.1)—crosses the $J_{ion} = 0$ line with a negative slope conductance at one point, the "threshold."

In summary, a threshold can be achieved with a model of the membrane that contains three parallel branches: a membrane capacitance branch and two

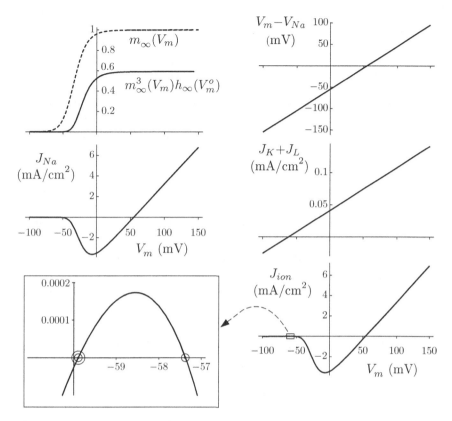

Figure 4.47 Components of ionic current as a function of membrane potential under the assumption that m changes rapidly and n and h do not change from their resting values. The upper left panel shows the variables that make up the sodium conductance: $G_{Na}(V_m, t) \propto m_\infty^3(V_m(t))h_\infty(V_m^o)$. The upper right panel shows $V_m - V_{Na}$ plotted versus V_m. The product of G_{Na} and $V_m - V_{Na}$ is the sodium current, J_{Na}, shown in left center panel. $J_K + J_L$ is shown in the center right panel. The total ionic current is $J_{ion} = J_{Na} + J_K + J_L$ and is plotted versus V_m in the lower right panel. The small boxed region in this panel is shown on an expanded scale in the lower left panel.

ionic branches. One ionic branch need have only a constant chord conductance, whereas the other ionic branch requires a sufficiently nonlinear chord conductance (according to the criterion in Equation 4.36).

The previous analysis gives some insight into how threshold points can occur in networks with nonlinear conductors (also see Appendix 4.1). However, the analysis cannot precisely account for the threshold inherent in the Hodgkin-Huxley model, because the analysis was based entirely on the J_{ion}-V_m characteristic at one point in time, t_o, and this characteristic changes

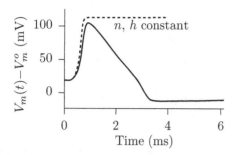

Figure 4.48 Calculations of the membrane action potential for an impulse of membrane current that leads to an initial depolarization of 20 mV (Fitzhugh, 1969). The solid line shows the solution of the Hodgkin-Huxley equations. The dashed line shows the solution of the same equations, except that h and n are constant. The rising phase of the potential is similar for both calculations.

with time during an action potential. So, for example, the difference between the threshold potential and the resting potential predicted by this analysis is less than 3 mV (Figure 4.47), whereas the measured value is about 15 mV (Figure 4.43). We shall now consider a more precise, albeit still approximate, analysis that takes into account the time variation of the membrane voltage-current characteristics in an approximate manner.

We shall assume once again that the sodium activation factor $m(t)$ has a faster time course than either the sodium inactivation factor $h(t)$ or the potassium activation factor $n(t)$. But this time we shall include the time course of the first-order kinetics of m. We shall examine the behavior of the Hodgkin-Huxley equations of a space-clamped axon under the assumption that $h(t)$ and $n(t)$ do not change at all, but remain at their initial values, i.e., $h(t) = h_\infty(V_m^o)$ and $n(t) = n_\infty(V_m^o)$. Therefore, Equation 4.25 becomes

$$0 = C_m \frac{dV_m}{dt} + \overline{G}_K n_\infty^4(V_m^o)\,(V_m - V_K)$$

$$+ \overline{G}_{Na} m^3(V_m, t) h_\infty(V_m^o)\,(V_m - V_{Na}) + G_L\,(V_m - V_L), \tag{4.37}$$

where

$$\tau_m(V_m) \frac{dm(t)}{dt} + m(t) = m_\infty(V_m). \tag{4.38}$$

Equations 4.37 and 4.38 can be solved for different initial conditions $V_m(0+)$, each of which results from a different current strength. We can compare these solutions to those obtained from the normal Hodgkin-Huxley equations as shown in Figure 4.48 for a single suprathreshold value of $V_m(0+)$. These results demonstrate that the rising phase of the action potential is predominantly due to the rapid rise in the sodium activation factor $m(t)$.

Insight can be gained by examining these solutions in the phase plane shown in Figure 4.49. In such a phase-plane plot, the state of Equations 4.37 and 4.38 is characterized by a state point in the m-V plane whose coordi-

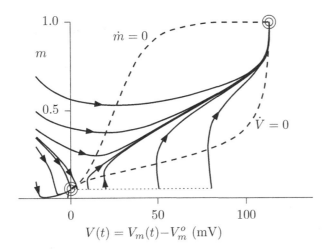

$$V(t) = V_m(t) - V_m^o \ (\text{mV})$$

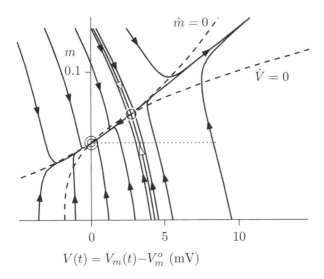

$$V(t) = V_m(t) - V_m^o \ (\text{mV})$$

Figure 4.49   Phase-plane trajectories of the Hodgkin-Huxley equations under the assumption that h and n are constant (adapted from Fitzhugh, 1969, Figure 4-4). Here $m(t)$ is plotted versus $V = V_m(t) - V_m^o$. The lower part of the figure shows details of the trajectory near $V = 0$ on expanded scales. The isoclines are indicated with dashed lines. The separatrix is indicated with open arrows. The phase-plane trajectories are indicated with filled arrows. Stable equilibria are indicated with two concentric circles, unstable equilibria with single circles.

nates are $m(t)$ and $V(t) = V_m(t) - V_m^o$. As t changes, the state point moves in this plane and traces a phase trajectory. The direction of motion of the state point along the trajectory is indicated by an arrow. Each phase trajectory corresponds to a different initial state (initial condition). The dotted horizontal line corresponds to the locus of initial states for which $m(0) = m_\infty(V_m^o)$ and $V(0)$ is varied, i.e., the locus of initial states that result from an impulse of current passed through a membrane that is initially at rest. The dashed curves, called *isoclines*, correspond to trajectories for which either $\dot{m} = dm/dt = 0$ or for which $\dot{V} = dV/dt = 0$. The $\dot{m} = 0$ isocline can be gotten directly from Equation 4.38 and is

$$m(t) = m_\infty\,(V_m(t)) \tag{4.39}$$

The $\dot{V} = 0$ isocline is obtained directly from Equation 4.37 and can be written as

$$m(t) = \left(\frac{-\overline{G}_K n_\infty^4(V_m^o)\,(V_m - V_K) - G_L\,(V_m - V_L)}{\overline{G}_{Na} h_\infty(V_m^o)\,(V_m - V_{Na})} \right)^{1/3} \tag{4.40}$$

There are three points at which the isoclines intersect. At these points, both $\dot{m} = 0$ and $\dot{V} = 0$. Hence, these three points are equilibrium points, and the system can remain at these points if it is not disturbed. One point corresponds to the resting state at which $V = 0$; another point corresponds to a potential near the Nernst equilibrium potential for sodium. The arrows on the phase trajectories point toward these points, so these points are stable equilibria. The arrows on the phase trajectories do not approach the third point of intersection (seen in the lower panel in Figure 4.49). This point is a threshold point and lies at the intersection of three curves: the two isoclines plus a trajectory indicated by open arrows in Figure 4.49 and called a *separatrix*. The separatrix divides the phase trajectories into two groups; trajectories with initial values starting on one side of the separatrix all eventually converge to one of the stable equilibrium points, whereas those that start on the other side of the separatrix all converge to the other stable equilibrium point. Initial states that lie on the separatrix approach the unstable equilibrium point, but with a phase velocity, $(\dot{m}^2 + \dot{V}^2)^{0.5}$, that converges to zero. Such a singular point in phase space is called a *saddle point*.

This analysis gives a more precise description of threshold for the Hodgkin-Huxley model than our previous analysis. Furthermore, this more precise analysis indicates that for pulses of current that are of brief duration compared to all the time constants of the Hodgkin-Huxley model (τ_m, τ_h, τ_n, and $\tau_M = R_m C_m$), the membrane threshold *does* occur at a fixed value of the membrane potential. That point in phase space is at the intersection of the separatrix and the locus of initial values of V_m. If the duration of the pulse is not brief compared to all these time constants, the threshold value does *not* in general correspond to a fixed membrane potential. However, even this more precise analysis is only approximate. If h and n are allowed to vary, the threshold behavior cannot be explained exactly by examining the state of the membrane in the m-V phase plane; the equations must be described in the four-dimensional (m-h-n-V) phase plane.

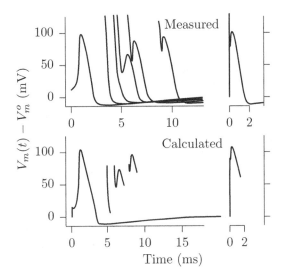

Figure 4.50 Comparison of measured and calculated responses to two pulses of membrane current illustrating refractory properties (adapted from Hodgkin and Huxley, 1952e, Figure 20). One pulse, with a shock strength of 15 nC/cm^2, occurs at $t = 0$, and the second pulse, with a shock strength of 90 nC/cm^2, occurs at various times after the first pulse. The response to 90 nC/cm^2 alone is shown on the right. The responses to the second pulses presented at various times after the first are shown superimposed. The measurements were obtained at 9°C, and the calculations are for 6°C. The difference in time scales is appropriate to the the difference in temperatures.

4.4.2 Refractoriness

For an interval of time, called the *absolute refractory period,* immediately after an action potential has been elicited by a stimulus, no second action potential can be elicited, no matter how intense the stimulus. Following the absolute refractory period is a *relative refractory period* during which a second action potential can be elicited but the threshold is elevated so that a higher strength of current is required. Therefore, as an action potential travels along an axon, it leaves a refractory wake behind. The threshold for eliciting an action potential in this wake is raised. Thus, refractoriness limits the rate at which action potentials can be propagated along an axon. In this section, we shall discuss the mechanisms responsible for the refractory properties of axons.

The membrane potential of a space-clamped squid axon and of the space-clamped Hodgkin-Huxley model are shown in Figure 4.50 in response to two pulses of membrane current. The first pulse occurs at $t = 0$ and is sufficient to elicit an action potential. The second pulse, which delivers six times the charge of the first pulse to the membrane, elicits no action potential if it occurs less than 5 ms after the first pulse. For interpulse intervals of 5 to 10 ms, the second pulse elicits an action potential of reduced amplitude. For interpulse intervals exceeding 10 ms, a full action potential occurs in response to the second pulse. The model exhibits behavior that is similar to that of the axon.

Figure 4.51 shows calculations based on the Hodgkin-Huxley model of a space-clamped squid giant axon in response to two current pulses. The first

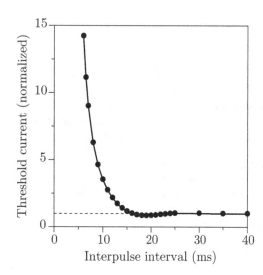

Figure 4.51 Calculations of the threshold of a second current pulse as a function of time after a first suprathreshold current pulse based on the Hodgkin-Huxley model of a space-clamped squid giant axon. Both pulses had a duration of 0.5 ms. The threshold of a single pulse was $I_{1th} = 13.377$ $\mu A/cm^2$. The first pulse was set at twice the threshold value. For each interpulse interval, the amplitude of the second pulse I_{2th} was determined so as to just elicit an action potential. The threshold of the second pulse is normalized to the threshold of a single pulse I_{1th}, i.e., the normalized threshold is I_{2th}/I_{1th}. The standard parameters of the Hodgkin-Huxley model were used at a temperature of 6.3°C.

current pulse is suprathreshold, and the threshold current amplitude of the second pulse is shown as a function of the time interval between the two pulses. For short interpulse intervals, no second action potential occurs; this is the absolute refractory period. For interpulse intervals of 6–16 ms, a second action potential is elicited, but it has a high threshold. The threshold decreases rapidly as the interpulse interval increases in the range of 6–16 ms. In the range of 16–24 ms, the threshold is lower than that for a single current pulse by as much as 17%. For interpulse intervals exceeding 24 ms, the threshold for the second pulse is equal to that for a single pulse.

The Hodgkin-Huxley model can reveal the basis of refractoriness. Figure 4.52 shows the time variation of the two factors responsible for refractoriness. During the peak and repolarization phase of the action potential, h is reduced and G_K is increased above the resting value. The decreased value of h prevents a large, rapid increase in G_{Na}, which is the basis of the action potential. The increased value of G_K tends to hold the membrane potential near the Nernst equilibrium potential for potassium.

The effect of the inactivation factor on threshold can be examined approximately in the m-V phase plane (Figure 4.53). Here we assume once again that the pulse durations are shorter than any of the time constants in the model and that h and n are constant. Hence, with these assumptions, the behavior of the Hodgkin-Huxley model is characterized in the m-V phase plane. Shown in Figure 4.53 are the $\dot{m} = 0$ isocline and several isoclines for $\dot{V} = 0$, each corresponding to a different value of $h = h_\infty$. As h_∞ decreases, the threshold, which is the intersection of the $\dot{m}$ and $\dot{V}$ isoclines, occurs at higher values

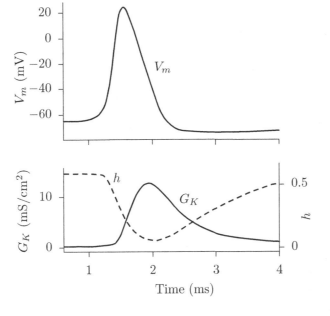

Figure 4.52 The time course of V_m, h, and G_K during a propagated action potential (adapted from Cooley and Dodge, 1965, Figure 2.4).

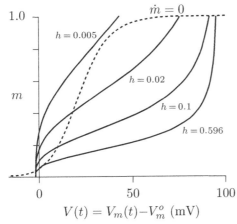

Figure 4.53 Phase-plane plot of m versus V showing effect of $h = h_\infty$ on location of threshold (adapted from Fitzhugh, 1960, Figure 4). The solid lines are the V isoclines for different values of h_∞. The dashed line is the m isocline.

of V, i.e., the threshold potential increases. Therefore, if the membrane were depolarized for some time, thus decreasing the value of h_∞, this inactivated membrane would have a higher threshold. This condition corresponds to the relative refractory period. If h_∞ is reduced sufficiently (to 0.005), the isoclines intersect at one point only, which corresponds to the resting potential. There is no intersection corresponding to threshold, and all phase trajectories terminate at the stable equilibrium point at rest. Thus, no action potential occurs for a membrane that is sufficiently inactivated. This condition corresponds to the absolute refractory period.

4.4.3 The Strength-Duration Relation

Suppose a rectangular pulse of current density of amplitude J and duration T is applied to an axon and that for each value of T, J is adjusted to the threshold excitation value J_{th}, i.e., the minimum value of J that elicits an action potential. Measurements on electrically excitable cells (e.g., Figure 1.13) have shown that as T increases, J_{th} decreases. This relation between current strength and duration to just elicit an action potential is called the *strength-duration relation*. The strength-duration relation was studied extensively at the end of the nineteenth century and the beginning of the twentieth century using measurements of the compound action potential of a population of fibers in a nerve (Katz, 1939). Current pulses of various waveforms were found to obey a similar relation, although we shall discuss only rectangular current pulses. It was found that the strength-duration relation has the following asymptotic behavior:

$$\lim_{T \to 0} J_{th}(T) = \frac{Q_{th}}{T} \quad \text{and} \quad \lim_{T \to \infty} J_{th}(T) = J_R, \tag{4.41}$$

where Q_{th} and J_R are constants that have units of charge density and current density, respectively.

Equation 4.41 shows that for very brief pulses of current, an action potential is elicited when the charge delivered by the current exceeds a threshold value of Q_{th}. This result is consistent with measurements and theoretical analyses that we discussed earlier in this section. We showed that the charge per unit area delivered by an impulse of membrane current density to a space-clamped axon, Q, results in an initial membrane potential $V_m = V_m^o + Q/C_m$, where C_m is the capacitance of the membrane per unit area. We have also shown that for brief current pulses the membrane behaves as if the threshold for eliciting an action potential corresponds to a fixed membrane potential, which we call V_{th}. Combining these two results shows that for brief current pulses—brief compared to all the time constants of the Hodgkin-Huxley model so that the current pulses can be regarded as impulses—the charge delivered at threshold is constant and has the value $Q_{th} = C_m V_{th}$. Equation 4.41 also shows that for long-duration current pulses, the current amplitude required to elicit an action potential approaches a constant value, J_R, called the *rheobase*, which is independent of duration T.

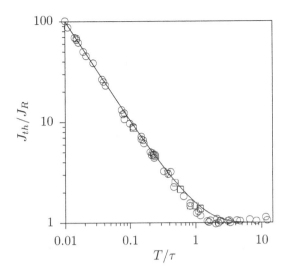

Figure 4.54 The strength-duration relation for space-clamped squid axons plotted in normalized logarithmic coordinates. The open circles are the measurements on six squid axons (adapted from Cole, 1955, Figure 10); the line is the function $J_m/J_R(T/\tau) = 1/(1 - e^{-T/\tau})$, which is based on the simple theory described in the text; and the open squares and triangles (of which there are ten) are based on calculations using the Hodgkin-Huxley equations (adapted from Cooley et al., 1965, Figure 5). The open squares are based on calculations of the equations for a propagating action potential; the triangles are for a membrane action potential.

It is convenient to express the strength-duration relation in normalized variables. Therefore, we define $\tau = Q_{th}/J_R$ and express Equation 4.41 in normalized variables $\hat{J} = J_{th}/J_R$ and $\hat{T} = T/\tau$. Although the strength-duration relation depends strongly on temperature, the strength-duration relation expressed in these normalized variables is relatively insensitive to temperature. The asymptotic behavior of the strength-duration relation can then be expressed in these normalized variables as follows:

$$\lim_{\hat{T} \to 0} \hat{J}(\hat{T}) = \frac{1}{\hat{T}} \quad \text{and} \quad \lim_{\hat{T} \to \infty} \hat{J}(\hat{T}) = 1. \tag{4.42}$$

Measurements of the relation of $\hat{J}$ to $\hat{T}$ are shown in Figure 4.54. These results demonstrate that the axons exhibit the asymptotic strength-duration relations shown in Equation 4.42. The parameters τ and J_R vary from axon to axon and depend upon temperature, but at 20°C the values for the squid giant axon are $\tau \approx 1$ ms and $J_R \approx 10\,\mu A/cm^2$ (Guttman and Barnhill, 1966). Thus, the value of τ is close to the value of the membrane time constant, τ_M. These values show that for $C_m \approx 1\,\mu F/cm^2$, $Q_{th} \approx 10\,nC/cm^2$ and $V_{th} \approx 10\,mV$. Also shown in Figure 4.54 is the theoretical relation

$$\hat{J} = \frac{1}{1 - e^{-\hat{T}}}, \tag{4.43}$$

which is based on a very simple model of threshold.

Before we examine the predictions of the Hodgkin-Huxley model, we shall explore this simpler model, which gives the correct asymptotic behavior of

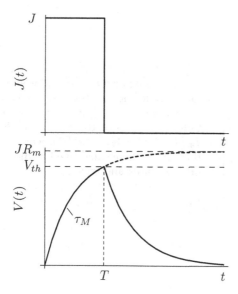

Figure 4.55 The relation of $V(t)$ and $J(t)$ for a simple model of subthreshold response.

the strength-duration relation and approximates the behavior of the relation over the whole range of durations. Suppose the subthreshold behavior of the membrane is approximated by an RC network so that the relation of membrane current density, J, to the change in membrane potential, V, is given by

$$J(t) = C_m \frac{dV}{dt} + \frac{V}{R_m}. \tag{4.44}$$

If $J(t)$ is a rectangular pulse of current as shown in Figure 4.55, the change in membrane potential is

$$V(t) = J_{th}R_m(1 - e^{-t/\tau_M}) \quad \text{for } 0 \leq t \leq T, \tag{4.45}$$

where $\tau_M = R_m C_m$. Now suppose that when the membrane potential reaches a threshold value, V_{th}, an action potential occurs. The amplitude of the current pulse for each duration is adjusted so that the membrane potential just reaches threshold. Therefore,

$$V_{th} = J_{th}(T)R_m(1 - e^{-T/\tau_M}). \tag{4.46}$$

For $T \to \infty$, Equation 4.45 gives the asymptotic value $V_{th} = J_{th}(\infty)R_m$. But by definition of the rheobase, $J_{th}(\infty) = J_R$, and

$$J_{th}(T) = \frac{J_R}{1 - e^{-T/\tau_M}}, \tag{4.47}$$

which in normalized coordinates equals Equation 4.43. Thus, a simple model in which the membrane potential charges up to a fixed potential threshold to elicit an action potential adequately accounts for the asymptotic behavior of the strength-duration relation. However, this model deviates somewhat from the measurements for current pulse durations that are close to the membrane time constant. That is, for T/τ near one the measurements of J_{th}/J_R (Figure 4.54) are systematically smaller than predicted by Equation 4.47.

Calculations based on the Hodgkin-Huxley model for a space-clamped axon as well as the model for an unclamped axon are also shown in Figure 4.54. The calculations based on the Hodgkin-Huxley model fit the measurements somewhat better than does the empirical relation based on a fixed threshold. This result is particularly apparent near $T/\tau \approx 1$. Hence, the Hodgkin-Huxley model adequately accounts for the strength-duration relation.

4.4.4 Accommodation

If a sufficiently slowly increasing current is applied to an electrically excitable cell, no action potential is elicited even if the current amplitude greatly exceeds that required to elicit an action potential with a current step (Figure 1.15). This phenomenon, called *accommodation*, has been studied since the late nineteenth century (Katz, 1939). More recently the phenomenon has been studied in space-clamped squid axons (Hagiwara and Oomura, 1958; Guttman and Barnhill, 1968) by stimulating these axons with ramp membrane currents, i.e., $J_m(t) = atu(t)$, where a is the slope of the ramp. Measurements show that if the slope is below some threshold value a_{th}, no action potential occurs even though the current greatly exceeds that required to elicit an action potential with a current step.

Does the Hodgkin-Huxley model exhibit the phenomenon of accommodation? The answer to this question is a bit subtle. The Hodgkin-Huxley model with the standard numerical parameters does not exhibit accommodation. Although the peak current required to elicit an action potential increases as the slope of a current ramp is decreased, the model inevitably gives an action potential if the computation is carried out for a sufficient time interval. Analysis of the model in the phase plane is consistent with this observation (Fitzhugh and Antosiewicz, 1959). However, a change in the numerical values of the parameters (e.g., reduction in the value of $\overline{G}_{Na}$) results in a model that exhibits accommodation (Figure 4.56). Current ramps of high slope give rise to multiple action potentials. As the slope is decreased, the number of action potentials is decreased, and the first action potential occurs later. The later action potentials have smaller peak amplitudes. A further reduction in the cur-

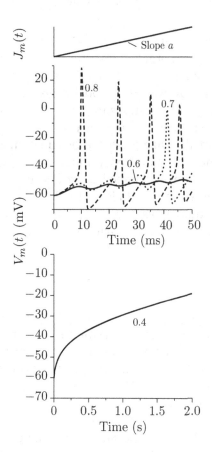

Figure 4.56 Computations of membrane potential responses to membrane current ramps of different slope for the Hodgkin-Huxley model of a space-clamped squid axon with standard parameters except that $\overline{G}_{Na} = 80$ mS/cm^2. The uppermost panel shows the current ramp stimulus; the lower two panels show the membrane potential for different current slopes on two different time scales. The slopes are indicated in $(\mu A/cm^2)/ms$.

rent slope results in no action potential. At a sufficiently low slope, the current can drive the membrane potential through its entire threshold range without producing an action potential (lowest panel in Figure 4.56).

The complexity of the Hodgkin-Huxley model becomes apparent in investigating the property of accommodation. Consider the following argument, which is intuitively appealing, but turns out to be false. Suppose we examine the Hodgkin-Huxley model for arbitrarily slow changes in membrane current density. For these conditions, it seems reasonable to assume that the capacitance current is zero and that the variables m, h, and n take on their steady-state values, m_∞, h_∞, and n_∞, respectively. Therefore, for a space-clamped squid axon, the membrane current density equals the ionic current density, and

$$J_{ion} = \overline{G}_K n_\infty^4(V_m)\,(V_m - V_K)$$
$$+ \overline{G}_{Na} m_\infty^3(V_m) h_\infty(V_m)\,(V_m - V_{Na}) + G_L\,(V_m - V_L)\,. \tag{4.48}$$

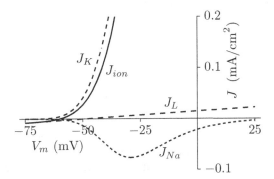

Figure 4.57 Steady-state J_{ion}-V_m characteristic for the Hodgkin-Huxley model of a space-clamped squid axon.

The relation between J_{ion} and V_m is shown in Figure 4.57. Note that J_{ion} has a positive slope for all values of V_m. Thus, for an arbitrarily slowly increasing membrane current, we might expect the membrane potential to increase along this steady-state relation so that no action potential is generated. This argument suggests that the Hodgkin-Huxley model with its standard parameters exhibits accommodation. However, this contention is readily refuted by computing the responses to a sequence of ramp currents for the standard parameters. These computations always give rise to action potentials no matter how small the slope of the ramp current, provided that the computations are carried out for a sufficient time duration. Therefore, the argument based on the steady-state characteristic leads to an incorrect conclusion. This analysis demonstrates that great care must be exercised in examining the stability of a nonlinear dynamic system, such as that governed by the Hodgkin-Huxley equations, with the use of ad hoc approximations. As we shall discuss in Section 4.4.7, for the standard parameters, the Hodgkin-Huxley equations are unstable for a portion of the J_{ion}-V_m characteristic. Therefore, they do not give rise to the accommodation property. This instability can be removed by changing parameters (for example, by reducing $\overline{G}_{Na}$). Under these conditions, the equations show accommodation.

Qualitatively, accommodation occurs because as the membrane potential is slowly increased, both m and n increase, but h decreases. The change in the latter two factors tends to prevent the regenerative action of the sodium conductance, since both act to reduce the membrane potential. Thus, both factors tend to prevent the regenerative increase in both the sodium conductance and the membrane potential that leads to an action potential. Therefore, both refractoriness and accommodation have similar bases. More quantitative explanations of accommodation can be achieved with careful analysis of the equations in the phase plane.

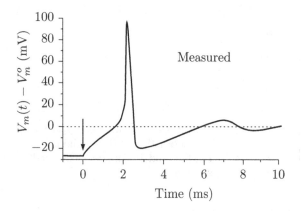

Figure 4.58 Measurement of the phenomenon of anode-break excitation in a space-clamped squid giant axon (adapted from Hodgkin and Huxley, 1952e, Figure 22B). A hyperpolarizing current through the membrane decreased the membrane potential to -26.5 mV below the resting potential. The current was terminated at $t = 0$ (indicated by the arrow). The temperature was 18.5°C.

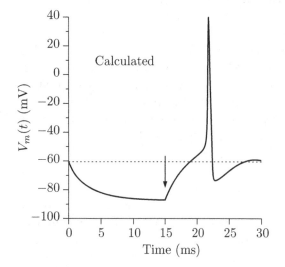

Figure 4.59 Calculation of the phenomenon of anode-break excitation for the Hodgkin-Huxley model of a space-clamped squid giant axon. A hyperpolarizing current pulse was applied at $t = 0$ and terminated 15 ms later (as indicated by the arrow). The current through the membrane decreased the membrane potential to -26.5 mV below the resting potential. The parameters of the model were the standard parameters except that the temperature was set to 18.5°C and $\overline{G}_{Na} = 160$ mS/cm^2. No action potential occurs at this temperature for the standard numerical parameters for which $\overline{G}_{Na} = 120$ mS/cm^2.

4.4.5 Anode-Break Excitation

If an electrically excitable cell is stimulated with a brief pulse of current applied with a pair of extracellular electrodes, the action potential arises with minimum current at the cathode. That is, the action potential arises at the site where the membrane is depolarized. However, in response to a long-duration pulse of current, an action potential arises at the anode when the current is turned off. This phenomenon is therefore called *anode-break excitation*.

As indicated in Figure 4.58, the giant axon of the squid exhibits the phenomenon of anode-break excitation. An action potential occurs at the offset of the hyperpolarizing pulse of current. Figure 4.59 demonstrates that the Hodgkin-Huxley model also exhibits the phenomenon of anode-break excitation. In response to a long-duration hyperpolarizing current—long enough so that all the variables V_m, m, n, and h reach their steady-state values—an action potential occurs at the offset of the current. The hyperpolarizing current leads to an increase in h and a decrease in n, hence in G_K. Both of these results lower the threshold for eliciting an action potential that is produced on depolarizing the membrane from its hyperpolarized value.

4.4.6 Repetitive Activity

Long-duration suprathreshold currents elicit multiple action potentials (Hagiwara and Oomura, 1958), a phenomenon called *repetitive activity*. The pattern of action potentials as a function of the amplitude of the current stimulus is complex. Nevertheless, the Hodgkin-Huxley model accounts for the main features of the response. Computations of the responses to long-duration pulses are shown in Figure 4.60 for a range of current amplitudes. For small-amplitude suprathreshold currents, a single action potential occurs at the current onset. As the current amplitude is increased, the frequency of action potentials increases. In addition, the amplitudes of action potentials that follow the first one are smaller. As the current amplitude increases further, the amplitudes of subsequent action potentials continue to decrease until all subsequent action potentials are blocked, a property called *depolarization block*. This pattern is explained by examining the behavior of the three factors m, n, and h. The depolarization caused by the current tends to reduce h and to increase n, which decreases G_{Na} and increases G_K. These factors reduce the amplitude of the action potential and eventually block it. This has only minimal effect on the first action potential, whose peak occurs before there are appreciable changes in n and h.

4.4.7 Subthreshold Oscillations

In response to a pulse of current density with a small amplitude, the change in the membrane potential shows highly damped oscillations, as shown in Figure 4.61 (Hodgkin and Huxley, 1952e; Mauro et al., 1970). For such small-amplitude currents, the relation between the current and the membrane potential is approximately linear. For example, a reversal of the polarity of the current stimulus approximately results in a reversal of the potential response. The Hodgkin-Huxley equations also exhibit these linear subthreshold oscillations (Hodgkin and Huxley, 1952e; Mauro et al., 1970). While the Hodgkin-Huxley equations are in general nonlinear, they behave linearly for

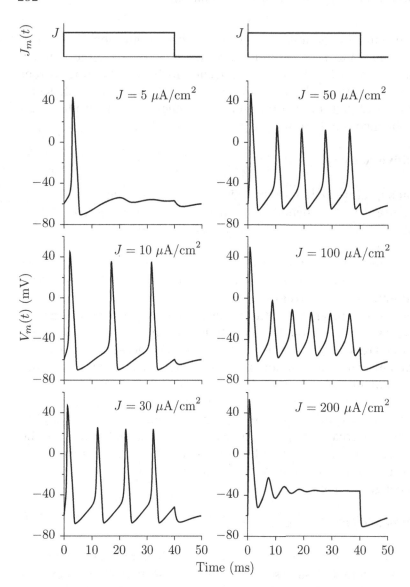

Figure 4.60 Repetitive activity exhibited by the Hodgkin-Huxley model of a space-clamped squid giant axon.

low-amplitude stimuli. Further insight into these subthreshold oscillations can be obtained by linearizing the Hodgkin-Huxley equations about an operating point and examining the linear equations. In a space-clamped axon, the relation between membrane current density and membrane potential is given by Equation 4.3, which depends upon equations that define the Gs. To illustrate the method, we shall derive the linearized relation between the potassium

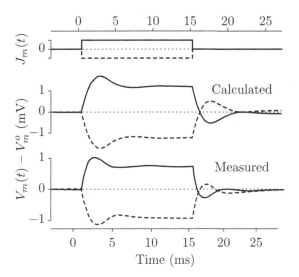

Figure 4.61 Comparison between calculated and measured subthreshold oscillations of the membrane potential in response to both depolarizing (solid line) and hyperpolarizing (dashed line) rectangular pulses of membrane current for a space-clamped squid giant axon (adapted from Hodgkin and Huxley, 1952e, Figure 23). The current density amplitude was $\pm 1.49\ \mu\text{A/cm}^2$; the temperature was 18.5°C.

current density and the membrane potential, and then we shall consider the generalized linear relation.

4.4.7.1   *Linearization of the Relation of the Potassium Current Density to Membrane Potential*

The relation of the potassium current to the membrane potential can be linearized by expanding each of the relevant variables in a Taylor's series and assuming that all nonlinear terms are negligible. Therefore, we can express all the time-varying quantities as sums of quiescent and incremental variables, as follows:

$$V_m(t) = V_m^o + v_m(t),$$

$$J_K(t) = J_K^o + j_K(t),$$

$$G_K(t) = G_K^o + g_K(t),$$

$$\tau_n(t) = \tau_n^o + \delta\tau_n(t),$$

$$n(t) = n_\infty^o + \delta n(t),$$

$$n_\infty(t) = n_\infty^o + \delta n_\infty(t),\tag{4.49}$$

where V_m^o, J_K^o, G_K^o, τ_n^o, and n_∞^o are the quiescent values of the variables and $v_m(t)$, $j_K(t)$, $g_K(t)$, $\delta\tau_n(t)$, $\delta n(t)$, and $\delta n_\infty(t)$ are the incremental variables. Substitution of Equations 4.49 into 4.7 yields

$$J_K(t) = \left(G_K^o + g_K(t)\right)\left(V_m^o + v_m(t) - V_K\right),\tag{4.50}$$

which gives

$$J_K^o = G_K^o (V_m^o - V_K),$$ (4.51)

and, if we assume that the product of incremental quantities $g_K v_m$ is negligibly small,

$$j_K(t) \approx G_K^o v_m(t) + (V_m^o - V_K) g_K(t).$$ (4.52)

Therefore, the relation of $j_K(t)$ to $v_m(t)$ is determined by G_K^o and $g_K(t)$, both of which can be determined from $n(t)$. First, we substitute the quiescent and incremental components of $n(t)$ into Equation 4.9 to obtain

$$G_K(t) = \overline{G}_K \left(n_\infty^o + \delta n(t) \right)^4,$$

which can be expanded in a Taylor's series whose first two terms are

$$G_K(t) \approx \overline{G}_K (n_\infty^o)^4 + 4 \overline{G}_K (n_\infty^o)^3 \delta n(t).$$

Therefore, the quiescent and incremental terms are

$$G_K^o = \overline{G}_K (n_\infty^o)^4$$ (4.53)

and

$$g_K(t) \approx 4 \overline{G}_K (n_\infty^o)^3 \delta n(t).$$ (4.54)

To complete the analysis, we need only linearize the differential equation that determines n. Substitution of Equations 4.49 into Equation 4.12 gives

$$(\tau_n^o + \delta \tau(t)) \frac{d \left(n_\infty^o + \delta n(t) \right)}{dt} + n_\infty^o + \delta n(t) = n_\infty^o + \delta n_\infty(t).$$

Neglecting products of incremental variables, we obtain

$$\tau_n^o \frac{d \delta n(t)}{dt} + \delta n(t) \approx \delta n_\infty(t).$$ (4.55)

Combining Equations 4.54 and 4.55 yields

$$\frac{d g_K(t)}{dt} + \frac{g_K(t)}{\tau_n^o} = \frac{4 \overline{G}_K (n_\infty^o)^3}{\tau_n^o} \delta n_\infty(t).$$ (4.56)

We multiply Equation 4.56 by $(V_m^o - V_K)$, note that from Equation 4.52 $(V_m^o - V_K) g_K(t) = j_K(t) - G_K^o v_m(t)$, and collect terms to obtain

$$\frac{dj_K(t)}{dt} + \frac{j_K(t)}{\tau_n^o} = \overline{G}_K(n_\infty^o)^4 \frac{dv_m(t)}{dt}$$

$$+ \frac{4\overline{G}_K(n_\infty^o)^3(V_m^o - V_K)K + \overline{G}_K(n_\infty^o)^4}{\tau_n^o} v_m(t), \tag{4.57}$$

where $K = \delta n_\infty / v_m = (dn_\infty / dV_m)_{V_m^o}$. K can be computed by obtaining the dependence of n_∞ on V_m from Equations 4.18, 4.19, and 4.20 and computing the derivative at $V_m = V_m^o$. Equation 4.57 relates $j_K(t)$ to $v_m(t)$ so that it can be represented by an incremental equivalent network.

Since we wish to synthesize the incremental model in the form of an electrical network, it is convenient to use impedance methods. To do so, we replace d/dt with the complex frequency s and all variables with their complex amplitudes in Equation 4.57, where $\mathcal{L}\{v_m(t)\}$ and $\mathcal{L}\{j_K(t)\}$ are the complex amplitudes (Laplace transforms) of v_m and j_K, respectively, from which the driving-point admittance is

$$Y_K(s) = \frac{\mathcal{L}\{j_K(t)\}}{\mathcal{L}\{v_m(t)\}} = \frac{\overline{G}_K(n_\infty^o)^4 \left(s + 1/\tau_n^o + 4K(V_m^o - V_K)/n_\infty^o \tau_n^o\right)}{s + 1/\tau_n^o}. \tag{4.58}$$

If we divide the numerator by the denominator and rearrange the remainder term, the admittance can be written in the form

$$Y_K(s) = \frac{1}{R_{Ko}} + \frac{1}{R_{Kn} + L_{Kn}s}, \tag{4.59}$$

where

$$R_{Ko} = \frac{1}{\overline{G}_K(n_\infty^o)^4}, \tag{4.60}$$

$$R_{Kn} = \frac{R_{Ko}}{4\left(\frac{V_m^o - V_K}{n_\infty^o}\right)\left(\frac{dn_\infty}{dV_m}\right)_{V_m^o}}, \tag{4.61}$$

and

$$L_{Kn} = \tau_n^o R_{Kn}. \tag{4.62}$$

Equation 4.59 describes an electrical network comprising a resistance in parallel with the series combination of another resistance and an inductance (Figure 4.62). Note that the network parameters (R_{Ko}, R_{Kn}, and L_{Kn}) depend upon the values of n_∞, dn_∞/dV_m, and τ_n, all evaluated at the operating point. Thus, a change in the operating point results in a change in the resistances and in the inductance.

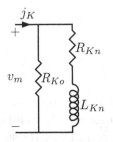

Figure 4.62 Electrical network model that represents the relation between the incremental potassium current density, $j_K(t)$ and the incremental membrane potential, $v_m(t)$.

It is interesting to note that a time-varying chord conductance such as G_K, which is a purely dissipative element as discussed in Section 4.2.3, can have a linearized representation that includes an inductance. The important point is that the chord conductance *is* purely dissipative; that is, it can only dissipate energy and cannot store energy. However, while the total energy must always flow into the element, the element can store incremental energy as does an energy-storage element. This storage of incremental energy is represented in the network by an inductance. Note also that while $R_{Ko} \geq 0$ (Equation 4.60), the signs of both R_{Kn} and L_{Kn} depend upon the product of $V_m^o - V_K$ and dn_∞/dV_m. The latter term is always positive, since n_∞ is a monotonically increasing function of V_m (Figure 4.25). Therefore, for $V_m^o > V_K$, both $R_{Kn} > 0$ and $L_{Kn} > 0$, and for $V_m^o < V_K$, both $R_{Kn} < 0$ and $L_{Kn} < 0$.

4.4.7.2 Linearization of the Relation of the Sodium Current Density to Membrane Potential

The linearization of the relation between sodium current density and membrane potential is similar to the linearization of the potassium current density relation, except that there is both an activation and an inactivation factor to linearize. The result of this linearization can also be represented by an equivalent network, which is shown in Figure 4.63. The network parameters are

$$R_{Nao} = \frac{1}{\overline{G}_{Na}(m_\infty^o)^3 h_\infty^o}, \tag{4.63}$$

$$R_{Nam} = \frac{R_{Nao}}{3\left(\frac{V_m^o - V_{Na}}{m_\infty^o}\right)\left(\frac{dm_\infty}{dV_m}\right)_{V_m^o}} \quad \text{and} \quad R_{Nah} = \frac{R_{Nao}}{\left(\frac{V_m^o - V_{Na}}{h_\infty^o}\right)\left(\frac{dh_\infty}{dV_m}\right)_{V_m^o}}, \tag{4.64}$$

and

$$L_{Nam} = \tau_m^o R_{Nam} \quad \text{and} \quad L_{Nah} = \tau_h^o R_{Nah}. \tag{4.65}$$

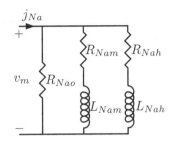

Figure 4.63 Electrical network model that represents the relation between the incremental sodium current density, $j_{Na}(t)$ and the incremental membrane potential, $v_m(t)$.

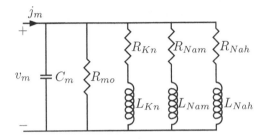

Figure 4.64 Electrical network model that represents the relation between the incremental membrane current density, $j_m(t)$ and the incremental membrane potential, $v_m(t)$. $R_{mo} = R_{Nao} \parallel R_{Ko} \parallel R_L$, where R_L is the specific resistance for leakage and $R_1 \parallel R_2$ denotes the parallel combination of R_1 and R_2.

Each activation factor, which is governed by a first-order differential equation, gives rise to one energy-storage element, which is represented by an inductance in the equivalent network. Just as with the elements of the equivalent network for the potassium branch, the elements in the sodium branch need not be positive quantities. In particular, since m_∞ is a monotonically increasing function of V_m and h_∞ is monotonically decreasing, for $V_m < V_{Na}$, $R_{Nam} < 0$, $L_{Nam} < 0$, $R_{Nah} > 0$, and $L_{Nah} > 0$. The signs of all the elements reverse for $V_m > V_{Na}$.

4.4.7.3 Linearized Hodgkin-Huxley Model

Description
The Hodgkin-Huxley model has four branches for current flow (Figure 4.6), each of which makes a contribution to the linearized equivalent network. In the Hodgkin-Huxley model, both the membrane capacitance and the leakage current have linear current-voltage relations so that they can be combined with the equivalent linearized networks for both the potassium (Figure 4.62) and the sodium (Figure 4.63) currents to yield the linearized equivalent network shown in Figure 4.64. This network has four independent energy-storage elements (C_m, L_{Kn}, L_{Nam}, and L_{Nah}), has four state variables (the voltage across the membrane and the three inductor currents), can be described by a fourth-order differential equation, and contains four natural frequencies when

driven by a current source. All the parameters, with the exception of C_m, depend upon the membrane potential at the operating point.

Natural Frequencies

Because of the presence of the inductances and the capacitance, this network has the potential for oscillations. Since the network elements depend upon the membrane potential at the operating point, the dynamic properties of the network will also depend upon this potential. Furthermore, since the current-voltage characteristic of the membrane exhibits regions of negative slope conductance (e.g., Figure 4.43), it is clear that the network elements need not have positive values. Thus, the oscillations may be damped, sustained, or unstable.

A direct way to examine the network dynamics is to find the four natural frequencies of the network for an applied current at the terminals. These natural frequencies (also called *eigenvalues* or characteristic frequencies), which are poles of the impedance looking into the terminals of the network in Figure 4.64, are in general complex quantities. Thus, they can be displayed in the complex-s plane by placing points at the real and imaginary coordinates $(\mathfrak{R}(s), \mathfrak{I}(s))$ of s that correspond to the natural frequencies. Figure 4.65 shows plots of the natural frequencies of the network for four different values of the membrane potential at the operating point. At a membrane potential of

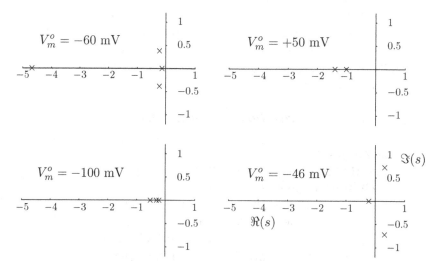

Figure 4.65 Natural frequencies of the linearized Hodgkin-Huxley equations at four different values of the membrane potential at the operating point. There are four natural frequencies at each value of the potential. For all four of these potentials, natural frequencies that occurred outside of the plot range were real and negative. The scale of s is 1/ms.

−100 mV, both the sodium and the potassium activation parameters are near zero and the sodium inactivation parameter is near one. At this potential, all four natural frequencies are real and negative. Thus, any perturbation of the membrane potential around this operating point will result in a transient return of the potential to the operating point with a waveform that is the superposition of four exponentials. Therefore, at this potential there are no oscillations in the subthreshold response. At −60 mV, which is close to the normal resting potential, there are two real natural frequencies with negative real parts and two natural frequencies that are complex with negative real parts. The latter two correspond to a transient response that is a damped oscillation at a radian frequency of near 400 rad/s, which corresponds to a frequency of near 60 Hz. This type of response is illustrated in Figure 4.61. At −46 mV, which corresponds to a depolarization of 14 mV, the potential is in the threshold region for eliciting an action potential with a brief stimulus. At this potential, there are two natural frequencies that are real and negative and two complex natural frequencies with positive real parts. These natural frequencies indicate that the system is unstable at this depolarization. At a potential of +50 mV, both the sodium and the potassium activation factors are near one and the sodium inactivation factor is near zero. At this potential, all four natural frequencies are real and negative. Therefore, for both large hyperpolarizations and large depolarizations, the system is stable with real, negative, natural frequencies. At rest, the system shows an oscillatory but stable response. However, with a depolarization in the range of potential near threshold, the system becomes unstable.

The stability of the system of linearized equations is best seen in a plot of the locus of natural frequencies as the membrane potential at the operating point is varied (Figure 4.66). Loci are shown for the standard parameters for which $\overline{G}_{Na} = 120$ mS/cm^2 and for $\overline{G}_{Na} = 80$ mS/cm^2. In each case, the natural frequencies start as real and negative at hyperpolarized membrane potentials. As the membrane potential increases, two of the natural frequencies approach each other and coincide. After a further increase in potential, the natural frequencies split into a complex conjugate pair whose imaginary and real parts increase. For $\overline{G}_{Na} = 120$ mS/cm^2, the natural frequencies have positive real parts for a range of membrane potential from −54 to −39. For $\overline{G}_{Na} = 80$ mS/cm^2, the real parts are always negative. For a further increase in potential, the real parts decrease and the imaginary parts reach a maximum, and then both the real and the imaginary parts decrease until they coincide and split into a pair of real and negative frequencies. The pattern is qualitatively similar for both $\overline{G}_{Na} = 120$ and 80 mS/cm^2 except for one important difference. The real parts of the natural frequencies become positive for 120,

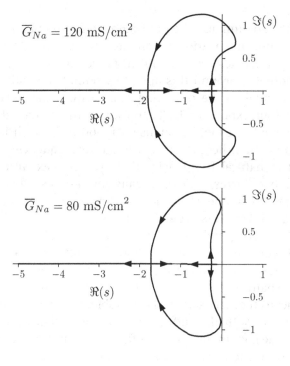

Figure 4.66 Loci of natural frequencies of the linearized Hodgkin-Huxley equations as a function of the membrane potential at the operating point for two different values of $\overline{G}_{Na}$. The natural frequencies were computed over the potential range from -100 to $+50$ mV in increments of 1 mV. The arrows indicate the direction of motion of the natural frequencies as the potential was increased. The scale of s is 1/ms.

but not for 80. Thus, for $\overline{G}_{Na} = 80$ mS/cm^2 the system is stable at all membrane potentials, whereas it is not for 120. This result explains the finding that the Hodgkin-Huxley equations accommodate for 80, but not for 120 mS/cm^2 (Section 4.4.4).

4.4.7.4 Summary

The Hodgkin-Huxley model as well as the squid giant axon show damped oscillations at the resting potential. These results indicate that at the normal resting potential, the membrane of the giant axon of the squid is represented only approximately as a resistance in parallel with a capacitance as the cable model assumes (Chapter 3). Instead, the membrane contains an appreciable equivalent inductive reactance. This inductive reactance has its largest effects at low frequencies and has been observed in measurements of the membrane impedance (Chapter 3). For large hyperpolarizations, the damping increases and the response approaches that of an RC network (Mauro et al., 1970). As the membrane is hyperpolarized, this inductive reactance becomes less prominent. The model shows qualitatively similar behavior for large depolarizations.

4.4.8 Effect of Temperature

Measurements have shown that ionic currents under voltage clamp have a faster time course at higher temperatures with a Q_{10} for the temperature change (Weiss, 1996, Chapter 6, Section 6.2.2) of about three (Hodgkin et al., 1952). This result suggested a way to take the effect of temperature on the action potential into account in an approximate manner (Huxley, 1959). Temperature is assumed to affect the Hodgkin-Huxley model in two ways only. First, the Nernst equilibrium potentials are inversely proportional to absolute temperature. Second, a change in temperature is assumed to scale all the rate constants by a common temperature factor to give a $Q_{10} = 3$. Hence, all the first-order kinetic equations for the activation and inactivation factors can be written in terms of the rate constants, as follows:

$$\frac{dm}{dt} = (\alpha_m - m(\alpha_m + \beta_m)) K_T, \tag{4.66}$$

$$\frac{dh}{dt} = (\alpha_h - h(\alpha_h + \beta_h)) K_T, \tag{4.67}$$

$$\frac{dn}{dt} = (\alpha_n - n(\alpha_n + \beta_n)) K_T, \tag{4.68}$$

where K_T is a temperature factor, which for a $Q_{10} = 3$ is

$$K_T = 3^{(T_c - 6.3)/10}, \tag{4.69}$$

where T_c is the temperature in centigrade. The temperature factor is one when the temperature is 6.3°C.

Measurements of the propagated action potentials recorded from the giant axon of the squid at different temperatures (Figure 4.67, upper panel) show that temperature has a large effect on the action potential. As the temperature is increased, the duration of the action potential is reduced and its amplitude is decreased. The general dependence of action potential waveshape on temperature is reproduced in the computations (Figure 4.67, lower panel). Action potentials apparently occur even at relatively low temperatures approaching the freezing point of seawater. However, action potentials do not occur above some temperature, which varies somewhat from fiber to fiber. This thermal block generally occurs in the range of 35–40°C for a propagated action potential (Hodgkin and Katz, 1949b). Figure 4.68 reveals that the same pattern of the dependence of the response on temperature occurs for computations of the membrane potential by means of the Hodgkin-Huxley model for

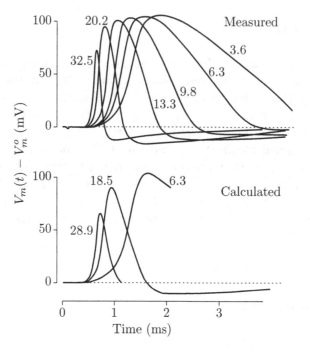

Figure 4.67 Dependence of the action potential on temperature. Measurements obtained from a giant axon of the squid (Hodgkin and Katz, 1949b) are shown in the upper panel. Calculations of the propagated action potential obtained from the Hodgkin-Huxley model (Huxley, 1959) are shown in the lower panel. The temperature in °C is given for each trace.

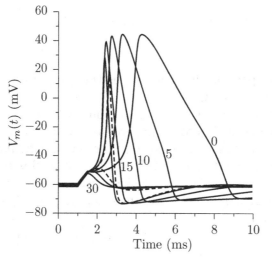

Figure 4.68 Dependence on temperature of the membrane action potential calculated from the Hodgkin-Huxley model. The parameters were the standard parameters of the model. The stimulus was a rectangular pulse of current density of amplitude 20 μA/cm^2 and duration 0.5 ms. The temperatures were 0, 5, 10, 15, 20, 22, 23, 25, and 30°C. Action potentials occur at 22°C (shown dashed) and for lower temperatures. They do not occur at 23°C (shown dashed) and for higher temperatures.

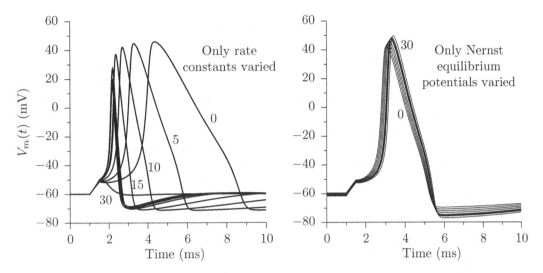

Figure 4.69 Comparison of the relative effects of temperature-induced changes in rate constants (left panel) and in Nernst equilibrium potentials (right panel) on the membrane action potential. The parameters of the Hodgkin-Huxley model are identical to those in Figure 4.68, except that the Nernst equilibrium potentials for sodium and potassium were fixed for the calculations in the left panel and the temperature factor for the rate constants was fixed for the calculations in the right panel. The temperature in °C is indicated on some of the traces.

a space-clamped squid giant axon. Under these conditions and using the standard parameters of the model, thermal block occurs between 22°C and 23°C.

What causes thermal block of the action potential in the Hodgkin-Huxley model? Temperature affects the model in only two ways: through the Nernst equilibrium potentials and through the effect of temperature on the rate constants, as described in Equations 4.66–4.69. The relative importance of these two effects of temperature are compared in Figure 4.69. When only the rate constants are made to depend upon temperature (left panel of Figure 4.69), the results are qualitatively similar, although not identical, to those when both the rate constants and the Nernst equilibrium potentials are made to vary with temperature (Figure 4.68). In general, as the temperature is increased the waveforms become faster, and eventually a thermal block occurs. However, the thermal block does not occur between 22 and 23°C, as in Figure 4.68, but between 25 and 30°C. Thus, the temperature-induced changes in the Nernst equilibrium potentials do affect the temperature at which thermal block occurs. If the rate constants are held fixed as temperature is changed so that only the Nernst equilibrium potentials for sodium and potassium are changed (right panel of Figure 4.69), the temperature-induced changes in the membrane action potential are small and, in the range of 0–30°C, there is no thermal block. Note that an increase in temperature increases the Nernst equilibrium potential for sodium, which results in an increase in the peak of the membrane action potential. An increase in temperature decreases the potassium equilibrium potential, which lowers the resting membrane potential. However, the effects of temperature-induced changes in Nernst equilibrium potentials are relatively small compared to the effects of temperature-induced changes in the rate constants. Therefore, these computations reveal that thermal block is due primarily to the temperature-induced changes in the rate constants.

To investigate why a temperature-induced change in rate constants results in a thermal block of the membrane action potential, we shall examine the underlying variables at two widely separated temperatures, 5°C and 22°C, each below the thermal block temperature. Both the action potentials and the sodium and potassium current densities for these two temperatures are shown in Figure 4.70. Both the membrane action potential and the current densities have more rapid time courses at the higher temperature. We note that the two current densities have amplitudes that are somewhat smaller at the higher temperature. There is also a suggestion that the sodium and potassium current densities are closer to being coincident in time at the higher temperature than they are at the lower temperature.

The timing relation of the current densities and other underlying variables in the Hodgkin-Huxley model are shown in Figure 4.71 on an expanded time scale that allows for examination of the interrelations of these variables at the time of onset of the action potential. The conductances for sodium and potassium have smaller magnitudes and faster time courses at the higher temperature. Furthermore, at the lower temperature the onset of G_{Na} is more rapid than its offset and more rapid than the onset of G_K. This difference in time course is much reduced at the higher temperature. That is, the increase in sodium conductance is reduced and occurs closer in time to the increase in potassium conductance. At the peak of the action potential, G_{Na} = 29.3 and G_K = 2.39 mS/cm^2 at 5°C, and G_{Na} = 13.6 and G_K = 5.94 mS/cm^2 at 22°C. Therefore, at the peak of the action potential G_{Na}/G_K = 12.3 at 5°C and G_{Na}/G_K = 2.29 at 22°C. Therefore, the rapid increase in sodium conductance relative to the potassium conductance is reduced at the higher temperature. Thus, an increase of temperature changes the temporal relation between the sodium and potassium conductances and reduces the amplitude of the sodium conductance relative to the potassium conductance at the peak of the action potential.

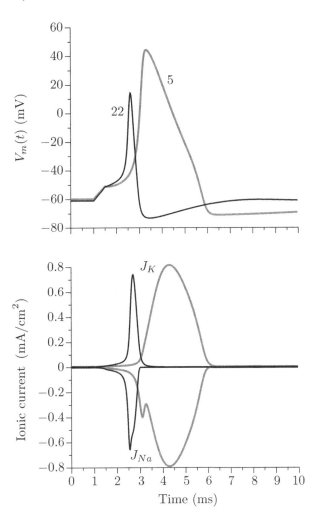

Figure 4.70 Membrane action potentials (upper panel) and sodium and potassium current densities (lower panel) computed for the Hodgkin-Huxley model at temperatures of 5°C (gray line) and 22°C (black line). The parameters were identical to those in Figure 4.68.

Why do these effects occur? The sodium conductance depends upon sodium activation and inactivation factors, i.e., $G_{Na} = \overline{G}_{Na} m^3 h$, and the potassium conductance depends on the potassium activation factor, i.e., $G_K = \overline{G}_K n^4$. The changes in the sodium and potassium conductances must result from changes in the (in)activation factors. The temperature-induced changes in these factors are shown in Figure 4.71 (lower left panel). The important observation is that at the lower temperature, m clearly changes more rapidly than either h or n, but this difference is much reduced at the higher temperature. For example, while neither the peak value of m nor the minimum value of h is changed much, the minimum of h occurs nearer the maximum of m at

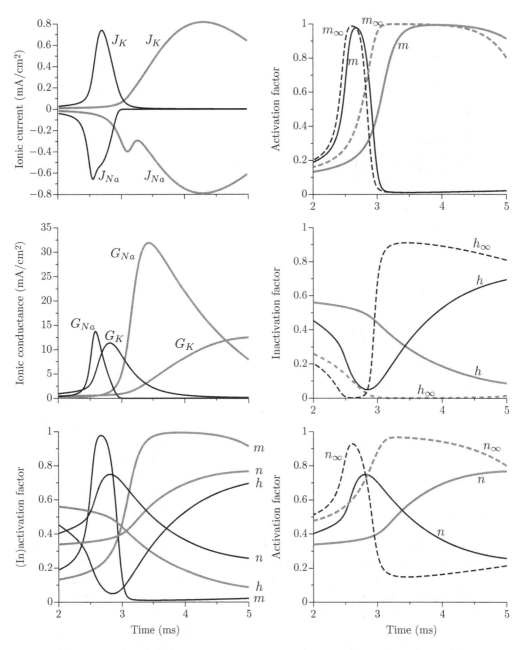

Figure 4.71 Membrane ionic currents, conductances, and (in)activation variables (left panels) and relation of (in)activation variables to their equilibrium values (right panels) computed for the Hodgkin-Huxley model at temperatures of 5°C (gray lines) and 22°C (black lines). The parameters were identical to those in Figure 4.68.

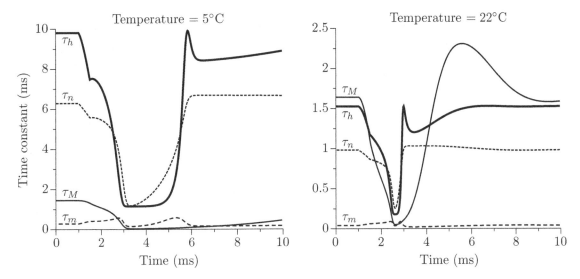

Figure 4.72 Time constants of m, n, and h, as well as the membrane time constant, $\tau_M = C_m/G_m$, as a function of time at 5°C and at 22°C.

the higher temperature than it does at the lower temperature. Therefore, the relative increase in rate of change of h with respect to m that occurs at higher temperature results in a decrease in G_{Na}. Similarly, there are changes in the temperature-induced change in n with respect to m that make G_K occur more synchronously with G_{Na}.

The results shown in Figure 4.71 indicate that the interrelation of the time constants is responsible for thermal block. To see this more directly, consider Figure 4.72, which shows all the relevant time constants at the two temperatures: the membrane time constant, τ_M, and the (in)activation factor time constants, τ_m, τ_h, and τ_n. At the lower temperature, τ_M is less than τ_m near the peak of the action potential, and both τ_M and τ_m are less than both τ_n and τ_h. At the higher temperature, τ_m, τ_n, and τ_h are reduced by about a factor of six. However, τ_M is not changed much. This difference is critical, since the time courses of the factors m, n, and h depend not only on their own kinetics, but on the rate of change of V_m. The effect of temperature on these factors is shown in somewhat more detail in Figure 4.71 (right panels). It can be seen that at the higher temperature, m is closer to m_∞, h is closer to h_∞, and n is closer to n_∞. Thus, as temperature increases, each factor approaches its equilibrium value more closely. At a sufficiently high temperature, the rate of change of each factor is no longer rate limited by its kinetics, i.e., $m \approx m_\infty$,

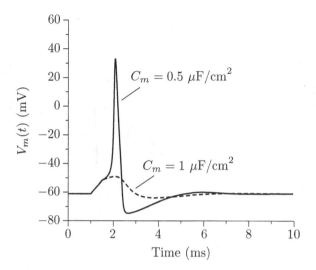

Figure 4.73 Membrane potentials computed for the Hodgkin-Huxley model at a temperatures of 23°C for two values of the membrane capacitance— 1 μF/cm^2 and 0.5 μF/cm^2. The current densities were of duration 0.5 ms for both computations, and the current amplitudes were 20 μA/cm^2 when the capacitance was 1 μF/cm^2 and 10 μA/cm^2 when the capacitance was 0.5 μF/cm^2.

$h \approx h_\infty$, and $n \approx n_\infty$. But since m_∞, h_∞, and n_∞ depend instantaneously on the membrane potential, the rate of change of each factor is limited by the rate at which the membrane potential can change, which is determined by the membrane time constant, τ_M. Thus, m can no longer change more rapidly than h and n, and the difference in timing of the (in)activation factors required to produce an action potential is disrupted.

As a test of the analysis given above, consider a temperature at which the action potential is blocked—23°C. A decrease in the membrane time constant should allow the time courses of the factors to differ in time and should restore the action potential. The membrane time constant can be changed by changing the membrane capacitance. However, a change in the capacitance changes not only the membrane time constant, but also the initial membrane potential caused by the current pulse. To remove the ambiguity of the latter factor, the current amplitude can be reduced to compensate for the reduction in the membrane capacitance as shown in Figure 4.73. The results demonstrate that this procedure results in the same initial value of the membrane potential. The primary result of the computation shows that a reduction in the membrane time constant, obtained by a reduction of the membrane capacitance from 1 to 0.5 μF/cm^2, results in the restoration of the action potential. This analysis of thermal block shows the interrelation of the time courses of the four state variables that define the Hodgkin-Huxley model—V_m, m, h, and n.

4.5 **Summary**

Stimulus-Response Configurations

The phenomena exhibited by the squid giant axon depend upon the configuration of the stimulus and recording arrangement as shown in Figure 4.74. We have described three distinct recording configurations—unclamped, space and current clamped, and space and voltage clamped. An unclamped axon can be represented by a distributed-parameter electrical network consisting of four branches, one for the capacitance current and three ionic branches. The membrane potential is described by a partial differential equation coupled to three first-order differential equations. When stimulated at a point in space with a brief pulse of current of sufficiently low strength, the axon conducts graded potentials decrementally. The Hodgkin-Huxley model for these subthreshold stimuli reduces to the cable model.[7] For suprathreshold current pulses, the axon gives rise to action potentials that propagate at constant velocity.

A space-clamped and current-clamped axon can be represented by a lumped-parameter network driven by a current source. The membrane potential is described by an ordinary differential equation coupled to three first-order differential equations. When stimulated by a step of current of subthreshold amplitude, the membrane potential is graded, and, in the simplest model, the membrane potential increases exponentially to a final value, as does a parallel RC network. When stimulated by a suprathreshold current pulse, the membrane potential is depolarized by the pulse, and, if the threshold potential is exceeded, a membrane action potential occurs.

A space-clamped and step voltage-clamped axon can be represented by a lumped-parameter network driven by a voltage source. No action potential occurs in this configuration. The membrane current is a sum of four current components: a brief surge of capacitance current and three ionic current components. For sufficiently small voltage steps, the ionic current is simply a step.

---

7. The simplest model of the membrane for small perturbations of the membrane potential is that of a parallel resistance and capacitance. This model captures many of the properties of cells. For example, it accounts to first order for the decremental conduction of graded potentials according to the cable model. However, linearization of the time-varying conductances in the Hodgkin-Huxley model leads to a more complex model of the membrane and gives a more accurate representation of measurements of membrane impedance.

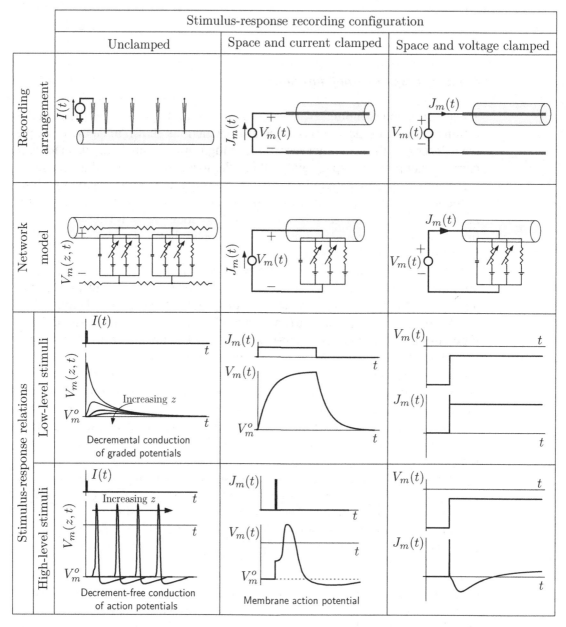

Figure 4.74 Summary of stimulus-response relations for different recording arrangements.

For large voltage steps, the ionic current depends in a nonlinear and time-dependent manner on the membrane potential.

The Hodgkin-Huxley Model

The Hodgkin-Huxley model was formulated on the basis of membrane current measurements made with the space-clamped and step voltage-clamped configuration. However, the model accounts for the electrically excitable properties of the squid giant axon measured when the membrane is not step voltage clamped. The model produces propagated action potentials of approximately the right waveshape and the right conduction velocity in the unclamped configuration and membrane action potentials in the space- and current-clamped configuration. The model also adequately accounts for such diverse phenomena observed in the squid giant axon as the all-or-none property, the existence of a sharp threshold for excitation, refractoriness, the strength-duration relation, accommodation, repetitive activity, anode-break excitation, subthreshold oscillations, and so on. With the incorporation of a single temperature factor, the model also approximately accounts for the effect of temperature on the action potential. Since these types of phenomena are widely seen in electrically excitable cells, the Hodgkin-Huxley model also gives insights into the electrically excitable properties of many cells. While the details may differ from cell to cell, the concepts inherent in the model have had a profound impact on neurobiology.

Appendix 4.1 Properties of Nonlinear, Time-Varying Conductors

Passive, Nonlinear, Time-Invariant Conductors

Definitions

Consider an element that has two terminals. The current through the element is related to the voltage across the element by the constitutive law imposed by the element (Figure 4.75). For a *time-invariant* element, the constitutive law that relates the current to the voltage is not a function of time, even though both the current and voltage may depend upon time. That is, for a time-invariant element, if $V(t)$ is the response to the stimulus $I(t)$, then $V(t - \tau)$ is the response to $I(t - \tau)$. For a time-invariant conductor, the constitutive relation between $V(t)$ and $I(t)$ can be represented as a time-invariant curve

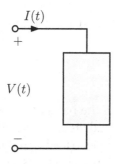

Figure 4.75 Definition of current and voltage for a two-terminal device.

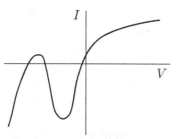

Figure 4.76 Current-voltage relation for a generalized resistive device.

in the current-voltage plane (Figure 4.76); that is, the instantaneous value of the current depends only upon the instantaneous value of the voltage. The relation between current and voltage has no memory for a nonlinear, time-invariant conductor. A *passive conductor* is a conductor that delivers no energy; it only dissipates energy. If the energy flowing into the terminals of a passive conductor, $E(t)$, never decreases as a function of time (Figure 4.77), the power flowing into the terminal, $P(t)$, is a nonnegative function, i.e.,

$$P(t) \equiv \frac{dE(t)}{dt} \geq 0. \tag{4.70}$$

The power flowing into the terminals of any element is $P(t) = V(t)I(t)$, and Equation 4.70 implies that

$$V(t)I(t) \geq 0. \tag{4.71}$$

Equation 4.71 implies that if $V(t) > 0$ then $I(t) > 0$ and if $V(t) < 0$ then $I(t) < 0$. If $V(t) = 0$ then $I(t)$ can have any value. Thus, the current-voltage characteristic of a nonlinear, passive conductor must lie in the first and third quadrants of the current-voltage plane only (Figure 4.78).

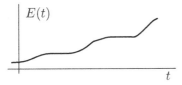

Figure 4.77 Time dependence of energy flow into the terminals of a passive resistive device.

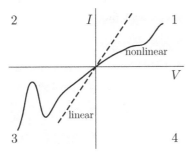

Figure 4.78 Current-voltage relations for linear and nonlinear passive conductors. The characteristic must lie in the first and third quadrants only. These quadrants are shaded.

Chord and Slope Conductance

For a linear conductor, the conductance is defined uniquely by Ohm's law,

$$I(t) = GV(t). \tag{4.72}$$

Note that for a linear conductor, $G = I/V = dI/dV$. However, for a nonlinear conductor, the ratio of the current to the voltage need not equal the slope of the current-voltage characteristic. Hence, we can define two conductances: the *chord conductance* and the *slope conductance* (Figure 4.46). The chord conductance is

$$G(V) \equiv \frac{I}{V}, \tag{4.73}$$

and the slope conductance is

$$g(V) = \frac{dI}{dV}, \tag{4.74}$$

where we have assumed that $I(V)$ is a single-valued function. For a nonlinear conductor, both the chord conductance and the slope conductance depend upon the potential V. However, the two conductances are related. Since $I(V) = G(V)V$,

$$\frac{dI(V)}{dV} = g(V) = \frac{dG(V)}{dV}V + G(V). \tag{4.75}$$

From Equation 4.71 we see that the passivity condition (Equation 4.71) implies that

$$G(V)V^2(t) \geq 0, \tag{4.76}$$

so that for a passive conductor the chord conductance is a nonnegative quantity. However, the slope conductance can be either positive or negative.

Significance of a Negative Slope Conductance

In order to explore the consequences of a negative slope conductance, consider a passive, time-invariant chord conductance. Let the voltage and current each consist of a small perturbation about some operating point, i.e.,

$$V(t) = V_o + v(t) \quad \text{and} \quad I(t) = I_o + i(t), \tag{4.77}$$

where V_o and $I_o = I(V_o)$ are the voltage and the current, respectively, at the operating point. We assume that the perturbations in voltage and current, $v(t)$ and $i(t)$, respectively, are small. With these assumptions, we expand $I(V_o + v)$ in a Taylor's series about V_o and assume that terms of order v^2 and higher have negligible magnitude. Therefore,

$$I(V) = I(V_o + v) = I(V_o) + \left(\frac{dI}{dV}\right)_{V_o} v = I_o + g(V_o)v. \tag{4.78}$$

Hence,

$$i = g(V_o)v. \tag{4.79}$$

Let us define the incremental power delivered to the conductor, $p(t)$, as

$$p(t) \equiv v(t)i(t) = g(V_o)v^2(t). \tag{4.80}$$

Hence, a passive ($G(V) > 0$), time-invariant conductor with a negative slope conductance at some operating point, V_o, can deliver incremental power (Equation 4.80) even though the conductance delivers no total power (Equation 4.76). Note that a passive conductor cannot have a negative slope conductance at an operating point for which $V_o = 0$, since the current-voltage characteristic must lie in the first and third quadrants (Figure 4.78). Therefore, in order for a passive conductor to deliver incremental power, the operating point must be set to some value $V_o \neq 0$ at which the slope conductance is negative. This operating point can be set by using an external source. The incremental power supplied by the conductor must be delivered by this source.

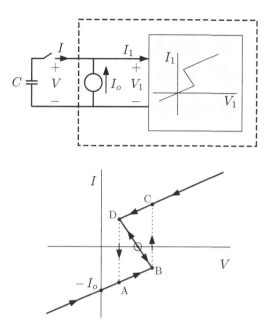

Figure 4.79 Example of an astable network. The astable network has one unstable equilibrium point, which is designated with a single circle on the *I-V* characteristic.

Thresholds and Oscillation in Networks Containing Energy Sources and Conductors with Negative Slope Conductances

A network that contains a passive conductor with a negative slope conductance region, a source to set the operating point in the negative slope conductance region, and a passive energy-storage device can oscillate. To illustrate this, consider the piecewise linear passive conductor shown in Figure 4.79. A constant current source, I_o, is used to set the operating point in the negative slope conductance region of the conductor. The parallel combination of the source and the conductor has terminal variables V and I and is connected to a capacitance through a switch. With the switch closed, the variables V and I must simultaneously satisfy two constraints: they must satisfy the current-voltage characteristic of the parallel combination of the source and conductor, i.e., they must lie on the *I-V* characteristic; and they must satisfy the constitutive relation for the capacitance

$$I = -C\frac{dV}{dt}. \tag{4.81}$$

Equation 4.81 implies that if at any time, t, $I(t) > 0$, then $dV/dt < 0$, i.e., $V(t)$ must decrease. Conversely, if $I(t) < 0$, then $dV/dt > 0$, i.e., $V(t)$ must increase. Arrows have been placed on the characteristic in Figure 4.79 to indicate the voltage and current trajectory on each part of the characteristic.

To see what will happen, suppose the capacitance is initially uncharged and the switch is thrown at $t = 0$. Thus, $V(0) = 0$ and $I(0) = -I_o$ to satisfy the characteristic of the source and conductor. Let us assume that I_o is a positive quantity. Then, since $I(0) < 0$, $dV/dt > 0$ for $t = 0$, and both $V(t)$ and $I(t)$ will increase (with exponential time courses) until the trajectory reaches the point B (in Figure 4.79). At point B, there appears to be a dilemma: the fact that $I < 0$ implies that $dV/dt > 0$, which implies that V must increase, but there is no way for V to increase while V and I are continuous functions of t and remain on the contour. The dilemma is resolved by the fact that the current need not be a continuous function of t. If the current changes discontinuously from B to C (indicated by a dotted line) while the voltage remains continuous, both the current and the voltage satisfy the voltage-current characteristic of the source and the conductor (i.e., at each instant in time the voltage and the current lie somewhere on the I-V characteristic) and Equation 4.81 (which guarantees that for a bounded current, the voltage remains continuous). At point C, $I > 0$, which implies that $dV/dt < 0$ and $V(t)$ decreases (exponentially) along the I-V characteristic toward point D. At point D, I changes discontinuously to point A. From here on, the trajectory continues periodically around the ABCD contour, and the voltage and the current are periodic, piecewise-exponential functions of time. The trajectory in the I-V (phase) plane is a closed loop called a *stable limit cycle*. The network is called *astable*.

The network shown in Figure 4.79 has no stable equilibrium points, but does have one unstable equilibrium point that is in the negative slope conductance region and for which $I = 0$. Thus, the network will enter the stable limit cycle for all initial conditions except for the initial state $I = 0$, which is an unstable equilibrium. Any perturbation from this unstable equilibrium will drive the network into the limit cycle. If a network that consists of a conductor with a negative slope conductance region and a capacitance has a stable equilibrium point, it will not necessarily oscillate. Consider the examples shown in Figures 4.80 and 4.81. One network (Figure 4.80) has a single stable equilibrium point and is called a *monostable network*. The other network (Figure 4.81) has two stable equilibrium points and one unstable equilibrium point and is called a *bistable network*. The significance of the unstable equilibrium point as a threshold point is discussed in Section 4.4.

Passive, Nonlinear, Time-Varying Conductors

By definition a passive conductor, whether time invariant or time varying, is an element that can only dissipate energy (Equation 4.70). A time-invariant conductor exhibits no memory; the instantaneous value of the current is de-

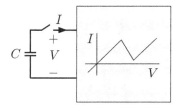

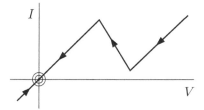

Figure 4.80 Example of a monostable network. The monostable network has one stable equilibrium point indicated with a double circle on the *I-V* characteristic.

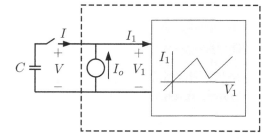

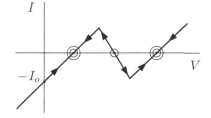

Figure 4.81 Example of a bistable network. The bistable network has two stable equilibrium points (double circles) and one unstable equilibrium point (single circle).

termined by the instantaneous value of the voltage. A time-varying conductor exhibits memory; the present value of the current may depend upon past values of the voltage and the current. Thus, time-varying conductors exhibit memory, which is a property of energy-storage elements. Because of this memory, time-varying conductors can store incremental energy. While total energy is always dissipated in a time-varying conductor, a time-varying conductor can store incremental energy and can have an incremental impedance that has a reactive component.

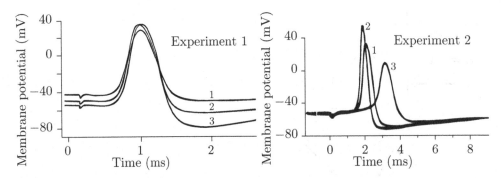

Figure 4.82   Effect of intracellular sodium and extracellular potassium on the action potential (Exercise 4.1).

Exercises

4.1   Figure 4.82 shows results from two separate experiments, Experiments 1 and 2, on the giant axon of a squid. One set of traces shows the effect on the action potential of changing intracellular sodium concentration; the other shows the effect of changing extracellular potassium concentration.

  a. Which experiment shows the results of changing intracellular sodium concentration? Describe the feature or features of the waveforms that indicate changes in intracellular sodium. Which waveform corresponds to the highest concentration of intracellular sodium? Why?

  b. Which experiment shows the results of changing extracellular potassium concentration? Describe the feature or features of the waveforms that indicate changes in extracellular potassium. Which waveform corresponds to the highest concentration of extracellular potassium? Why?

4.2   State whether each of the following is true or false, and give a reason for your answer.

  a. In response to a step of membrane potential in a voltage-clamped axon, the factors $n(t)$, $m(t)$, and $h(t)$ are exponential functions of time.

   b. In response to a step of membrane current in a current-clamped
      axon, the factors $n(t)$, $m(t)$, and $h(t)$ are exponential functions of
      time.

   c. In response to an impulse of membrane current in an unclamped
      axon, the factors $n(t)$, $m(t)$, and $h(t)$ are exponential functions of
      time.

4.3 State whether each of the following is true or false, and give a reason for
     your answer.

   a. An action potential can occur in a voltage-clamped axon.

   b. An action potential can occur in a current-clamped axon.

   c. An action potential can occur in an unclamped axon.

4.4 Both at rest and after a prolonged depolarization, the sodium conduc-
     tance is small compared with the potassium conductance.

   a. Describe the factors that account for this property.

   b. Discuss the consequences of this property for excitation of an action
      potential.

4.5 Consider the Hodgkin-Huxley model for a space- and voltage-clamped
     squid giant axon. You can assume that for $V_m = -50$ mV, $m_\infty = 0.16$,
     $n_\infty = 0.48$, and $h_\infty = 0.26$. Also, for $V_m = -25$ mV, $m_\infty = 0.73$, $n_\infty = 0.77$,
     and $h_\infty = 0.02$.

   a. Sketch the sodium conductance that would result if the applied volt-
      age was -50 mV for a long period of time and then stepped to
      -25 mV at time $t = 0$. No scales are required for conductance, but
      indicate an approximate time scale.

   b. Sketch the sodium conductance that would result if the applied volt-
      age was -25 mV for a long period of time and then stepped to
      -50 mV at time $t = 0$. No scales are required for conductance, but
      indicate an approximate time scale.

   c. Is the value of the sodium conductance at time $t = 0$ in part a greater
      than, less than, or equal to the value of the sodium conductance at
      time $t \to \infty$ in part b? Explain.

   d. Is the maximum value of sodium conductance in part a greater than,
      less than, or equal to the maximum value of sodium conductance in
      part b? Explain.

4.6 The following assertions apply to responses calculated according to the Hodgkin-Huxley model in response to a step of membrane potential applied at $t = 0$. For each assertion, state if it is true or false, and explain your answer.

   a. The leakage conductance is constant.

   b. The sodium conductance is discontinuous at $t = 0$.

   c. The potassium conductance is discontinuous at $t = 0$.

   d. The leakage current is constant.

   e. The sodium current is discontinuous at $t = 0$.

   f. The potassium current is discontinuous at $t = 0$.

   g. The factors $n(t)$, $m(t)$, and $h(t)$ are discontinuous at $t = 0$.

   h. The time constants τ_n, τ_m, and τ_h are discontinuous at $t = 0$.

   i. The steady-state values n_∞, m_∞, and h_∞ are discontinuous at $t = 0$.

4.7 Discuss the following description of the voltage-clamp method (Kandel et al., 1991, page 106): The voltage clamp is a current source connected to two electrodes, one inside and the other outside the cell. By passing current across the membrane, the membrane potential can be stepped rapidly to various predetermined levels of depolarization.

4.8 Discuss the following description of the threshold for eliciting an action potential (Keynes and Aidley, 1991, page 85):

   . . . excitation of a nerve fibre involves the rapid depolarization of the membrane to a critical level normally about 15 mV less negative than the resting potential. The critical level for excitation is the membrane potential at which the net rate of entry of Na^+ ions becomes exactly equal to the net rate of exit of K^+ ions plus the small contribution from an entry of Cl^- ions. Greater depolarization than this tips the balance in favour of Na^+, and the regenerative process . . . takes over and causes a rapidly accelerating inrush of sodium.

4.9 According to the measurements of Curtis and Cole shown in Figure 4.1, the membrane of the squid giant axon can be represented by a parallel network consisting of a constant capacitance in parallel with a circuit consisting of a voltage source $V(t)$ in series with a conductance $G_m(t)$. Both the conductance and the voltage source vary with time. Based on the Hodgkin-Huxley model, explain the basis of the time variation of both $V(t)$ and $G_m(t)$.

4.10 *The Boston Globe* (Monday, August 5, 1985, page 39), in a section called "In the News," contained the following seven-paragraph article:

Squid technology

Live squid have become essential to the study of nerve cells. Since 1936, schools of these curious creatures with a beak like a parrot, tentacles like an octopus and an eye like a human have attracted hundreds of scientists to Cape Cod's shores.

"A whole technology has been developed around the squid," said William J. Adelman Jr., a National Institutes of Health laboratory chief at the Marine Biological Laboratory at Woods Hole. "We now have a fairly good understanding of how (certain parts of the squid nerve cell) work."

Evolution has maintained the chemical formula for these structures since they first evolved. The primary difference is that more complex animals have constructed the nerve cells into more sophisticated structures. The squid's nerve cell, though thousands of times larger, is very similar to the human nerve cell. It has the same basic mechanism.

Electrical signals enter the nerve cell as a chemical reaction through root-like structures called the dendrites. This stimulates tiny gates in the immediate area of the nerve wall to open, letting in positively charged sodium particles and letting out negatively charged potassium particles.

This stimulates a chain reaction that runs the length of the cell. When it reaches the nerve endings the electrical pulse is converted back into a chemical reaction, which either stimulates muscle or the next nerve cell.

As the signal passes along the cell it becomes positively charged. However, it must be made negative again before another signal can pass along it. The cell does this by using sodium-potassium pumps a few molecules in size.

Scientists say they would like to know how sodium-potassium pumps work, and they have begun to study the chemical reactions that take place as the signal is transferred from cell to cell.

Focus on paragraph 6, which contains the author's description of the mechanism of nerve excitation and conduction in squid neurons. Write a concise critique of the mechanism presented in this paragraph. Please stick to the main issues raised in this paragraph; do not give a general discussion of nerve excitation and conduction. Keep the discussion to fifty words or less.

4.11 Figure 4.83 shows the relation between the membrane potential and the membrane current density during a propagated action potential as computed from the Hodgkin-Huxley model. The membrane current density consists of an initial outward current followed by an early inward current whose peak occurs before the peak in the action potential.

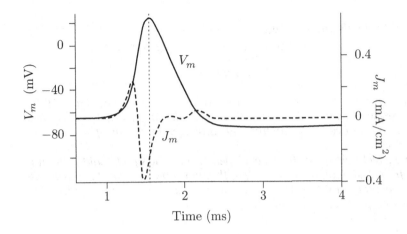

Figure 4.83 Relation of membrane potential and membrane current density during a propagated action potential (Exercise 4.11). The calculation is the same as in Figure 4.32. The dotted vertical line marks the time of occurrence of the peak of the action potential.

a. The initial outward current is due primarily to (choose one of the following)

  i.   an ionic current carried by sodium ions.

  ii. an ionic current carried by potassium ions.

  iii. an ionic current carried by chloride ions.

  iv. an ionic current carried by calcium ions.

  v.   a capacitance current.

b. The early inward current is due primarily to (choose one of the following)

  i.   an ionic current carried by sodium ions.

  ii. an ionic current carried by potassium ions.

  iii. an ionic current carried by chloride ions.

  iv. an ionic current carried by calcium ions.

  v.   a capacitance current.

c. Before the peak of the action potential, the membrane potential increases from its resting value, whereas the membrane current density is first outward (increasing and then decreasing) and then reverses polarity to become inward (decreasing and then increasing again). Discuss this complex relation between membrane potential and current. In particular, explain how the Hodgkin-Huxley model accounts for the fact that the current can be both inward and outward during an interval of time when the membrane potential is depolarizing.

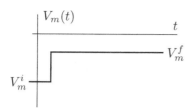

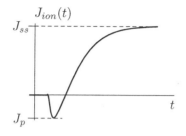

Figure 4.84 Definition of the peak inward current density (J_p) and the steady-state outward current density (J_{ss}) (Exercise 4.12).

4.12 The membrane ionic current density is computed for the Hodgkin-Huxley model of a voltage-clamped squid giant axon. All parameters of the model are at their standard values. The initial voltage V_m^i is set equal to -80 mV, the final voltage V_m^f is -40 mV, and the resulting waveforms are shown in Figure 4.84. In each part of this problem, determine the effect of changing a single parameter of the model on the peak inward current density J_p and on the steady-state outward current density J_{ss}. Indicate whether the change causes the current component to become more positive or more negative or to change only a small amount (less than 10%). For all parts, explain your reasoning.

a. The external concentration of sodium is doubled.

b. The external concentration of potassium is doubled.

c. V_m^f is changed from -40 mV to -30 mV.

d. V_m^i is changed from -80 mV to -70 mV.

Problems

4.1 Certain plant cells are capable of producing action potentials as well as maintaining resting potentials across their membranes. Consider the following results, which are typical of freshwater algae. The approximate compositions of the internal solution, sap, and the external

Table 4.2 Composition of sap and pond water for a freshwater algal plant (Problem 4.1).

	Concentration	
Ion	Pond water (mmol/L)	Sap (mmol/L)
Potassium	0.06	60
Sodium	0.15	60
Chloride	0.05	100

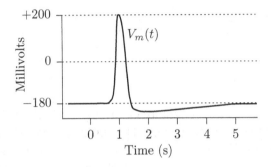

Figure 4.85 Action potential of a freshwater algal plant (Problem 4.1).

solution, pond water, are given in Table 4.2. Other solutes are contained in both solutions in order to maintain osmotic equilibrium. The resting membrane potential is found to be -180 mV, and the peak amplitude of the action potential is found to be 200 mV as shown in Figure 4.85.

a. Propose a model for the membrane of this cell that will account for the size of the resting and action potentials in terms of the selective permeability of the membrane to the three ions Na^+, K^+, and Cl^-.

b. Briefly describe an experiment to test some feature of your model.

4.2   Figures 4.2 and 4.4 show measurements of the effect on the membrane potential of the ion compositions of the extracellular and intracellular solutions, respectively. The results of both experiments show only a small change in the resting potential. How large a change in resting potential would you expect in these measurements? Be as quantitative as possible, and justify your answers.

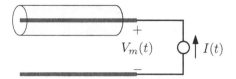

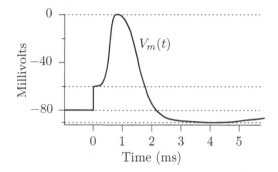

Figure 4.86 Membrane action
potential (Problem 4.3).

4.3 An impulse of current is applied to a space-clamped section of squid
axon. This impulse has an area of $Q = 10^{-8}$ coulombs. The membrane
potential, $V_m(t)$, is shown in Figure 4.86. You may assume that the axon
behaves according to the Hodgkin-Huxley theory. The concentration of
sodium in axoplasm is 10 mmol/L.

   a. Determine the membrane capacitance of this section of squid axon.

   b. Find the minimum sodium concentration in the seawater surrounding
      the axon consistent with the above data.

4.4 Measurements of the capacitance current density of a voltage-clamped
squid giant axon in response to voltage steps is shown in Figure 4.13.

   a. Determine an equivalent network of the axon, and include the resis-
      tance in series with the membrane that results from the seawater,
      cytoplasm, and electrodes.

   b. Determine the membrane capacitance per unit area implied by these
      measurements.

   c. Estimate the value of the series resistance from these measurements.

   d. Determine the response of your network to the voltage steps shown
      in Figure 4.13. Compare your predictions with the measurements and
      the calculations shown in Figure 4.13. Discuss the similarities and
      differences.

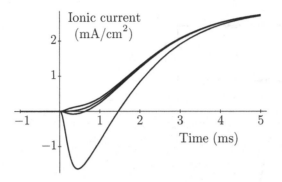

Figure 4.87 Membrane ionic current for four different sodium concentrations measured under voltage clamp from -100 to $+28$ mV (Problem 4.5). The leakage current has been subtracted from the ionic current.

4.5 The ionic component of the membrane current is computed from the Hodgkin-Huxley model of a space-clamped squid giant axon in response to a step of membrane potential from $V_m = -100$ to $V_m = +28$ mV at $t = 0$ in four different external solutions (Figure 4.87). The solutions are identical in potassium concentration, but have the following sodium concentrations: $c_{Na}^0 = 140$, 150, 160, and 450 mmol/L. The membrane ionic currents, with leakage current removed, are designated as J_{140}, J_{150}, J_{160}, and J_{450}. The internal ion concentrations are the same in all these computations.

   a. Clearly indicate by a sketch which ionic current trace corresponds to which external sodium concentration. Give your reasons.

   b. Estimate the Na^+ concentration inside the axon. State your assumptions.

   c. Describe how to separate the currents J_{150} and J_{450} into J_{Na} and J_K. Give explicit formulas for J_{Na} and J_K in terms of J_{150}, J_{450}, V_m, and the ion concentrations only. Sketch J_{Na} and J_K for sodium concentrations of 150 and 450 mmol/L.

   d. Determine and sketch J_{Na} and J_K for $c_{Na}^0 = 50$ mmol/L.

4.6 Figure 4.34 shows a calculation of the propagated action potential and the sodium and potassium conductances as predicted by the Hodgkin-Huxley model.

   a. Indicate how you might calculate the net transfer of sodium and potassium ions associated with the action potential from these curves.

   b. Make some rough approximations to estimate the transfer of Na^+ per action potential in mol/cm^2. Indicate your method.

c. Assume that the initial Na$^+$ concentration in the axoplasm is 50 mol/kg H$_2$O and that the axon diameter is 500 μm. How many action potentials must occur to increase the Na$^+$ concentration by 10%?

4.7 A squid axon is placed in experimental apparatus that ensures that the voltages and currents do not vary in space. The membrane voltage is controlled so that $V_m(t) = 75 - 175u(t)$ mV, where $u(t)$ is the unit step function. Assume that the Hodgkin-Huxley model applies and that the voltage dependence of the parameters τ_m, m_∞, τ_h, h_∞, τ_n, and n_∞ is as shown in Figure 4.25. Determine $G_K(t) = \overline{G}_K n^4$, and sketch the waveform, indicating approximate numerical values of the coordinates at a couple of points on the waveform. $\overline{G}_K = 36$ mS/cm^2.

4.8 Assume that the Hodgkin-Huxley model of the squid axon applies, with $\overline{G}_K = 40, \overline{G}_{Na} = 120, G_L = 0.3$ mS/cm^2, $C_m = 1\mu$F/cm^2, $V_K = -60, V_{Na} = 60$, and $V_L = -40$ mV. The variables m, n, and h are determined by first-order differential equations. That is,

$$\frac{dm}{dt} = \frac{m_\infty - m}{\tau_m}, \tag{4.82}$$

where τ_m, m_∞, τ_h, h_∞, τ_n, and n_∞ are functions of the membrane potential V_m as shown in Figure 4.25. Assume that $\partial V_m/\partial z = 0$ and that $V_m(t)$ is a short depolarization pulse as shown in Figure 4.88.

a. Sketch $J_{Na}(t)$ for $t > 0$. Indicate dimensions on your sketch.

b. The external sodium concentration is reduced to 1/10 of its previous value. For the same $V_m(t)$, sketch $J_{Na}(t)$ for $t > 0$.

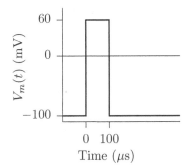

Figure 4.88 Brief voltage pulse (Problem 4.8).

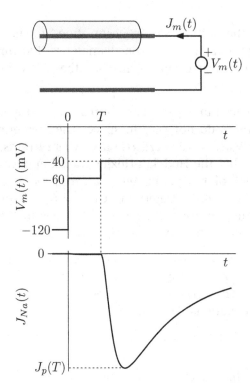

Figure 4.89 Two-step membrane potential profile (Problem 4.9).

4.9 A cell is voltage clamped, and the following membrane potential sequence is applied: the membrane potential is held at −120 mV for a long time, stepped to −60 mV at $t = 0$, and then stepped to −40 mV at $t = T$ (Figure 4.89). The peak sodium current density, $J_p(T)$, in response to the step from −60 to −40 mV is measured as a function of T. Let the normalized current be defined as $y(T) = J_p(T)/J_p(0)$. On the time scale of these measurements, you may assume that the kinetics of the sodium activation factor, m, are sufficiently fast that $m(t)$ reaches its final value instantaneously. The dependences of the parameters m_∞, n_∞, h_∞, τ_m, τ_n, and τ_h on V_m are given in Figure 4.25.

   a. In twenty words or less, describe what this experiment is designed to measure.

   b. Determine an analytic expression for and sketch $y(T)$. Determine *all* of the relevant numerical parameters.

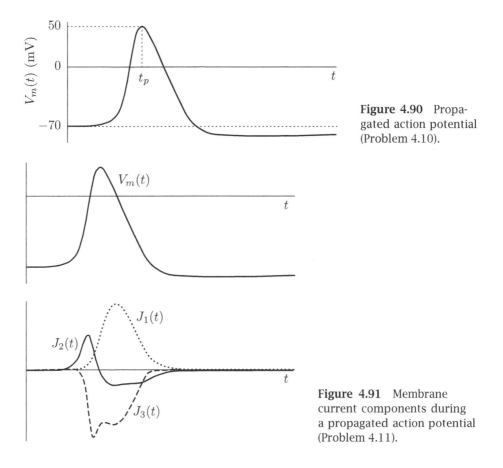

Figure 4.90 Propagated action potential (Problem 4.10).

Figure 4.91 Membrane current components during a propagated action potential (Problem 4.11).

4.10 The propagated action potential shown in Figure 4.90 is recorded intra-cellularly from a squid giant axon. Assume that the only ionic current through the membrane during the interval $(0, t_p)$ is carried by sodium ions and that a total of 1.5 pmol/cm^2 of sodium enter the axon during this time interval. From these data, estimate the value of the membrane capacitance per unit area C_m.

4.11 A propagated action potential of an unmyelinated axon is shown in Figure 4.91, along with three components of the membrane current density, J_1, J_2, and J_3. Outward current density is plotted as positive. Identify which of these currents is the capacity current, J_C, the sodium current, J_{Na}, and the potassium current, J_K. Explain your choices.

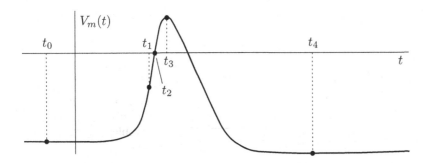

Figure 4.92 Propagated action potential (Problem 4.12).

4.12 A propagated action potential is recorded at one position along an axon. Five points in time during this action potential are identified in Figure 4.92.

- At t_0 the membrane is in its resting state, i.e., $V_m(t_0) = V_m^o$.
- At t_1 the membrane potential has a point of inflection, i.e., $[\partial^2 V_m/\partial t^2]_{t_1} = 0$.
- At t_2 the membrane potential is zero, i.e., $V_m(t_2) = 0$.
- At t_3 the membrane potential has its maximum value.
- At t_4 the membrane potential has its minimum value.

For each of the times t_0, t_1, t_2, t_3, and t_4, determine which, if any, of the following statements apply in a precise manner. If none applies precisely, then indicate "None." Indicate your answers with a check mark in the appropriate places in a table such as Table 4.3.

a. The sodium conductance equals the potassium conductance.

b. The membrane current is zero and changes from inward to outward.

c. The membrane current is zero and changes from outward to inward.

d. The capacitance current is zero.

e. The total ionic current is zero.

f. The sodium conductance has a maximum value.

g. The potassium conductance has a maximum value.

h. The sodium conductance has a minimum value.

i. The potassium conductance has a minimum value.

Table 4.3 Organizer for answers to Problem 4.12.

	t_0	t_1	t_2	t_3	t_4
a	_____	_____	_____	_____	_____
b	_____	_____	_____	_____	_____
c	_____	_____	_____	_____	_____
d	_____	_____	_____	_____	_____
e	_____	_____	_____	_____	_____
f	_____	_____	_____	_____	_____
g	_____	_____	_____	_____	_____
h	_____	_____	_____	_____	_____
i	_____	_____	_____	_____	_____
j	_____	_____	_____	_____	_____
k	_____	_____	_____	_____	_____
None	_____	_____	_____	_____	_____

      j. The magnitude of the potential difference between the membrane potential and the sodium equilibrium potential is a minimum.

      k. The magnitude of the potential difference between the membrane potential and the potassium equilibrium potential is a minimum.

4.13 A set of membrane ionic currents carried entirely by sodium and potassium ions is obtained from a voltage-clamped squid axon (Figure 4.93) in two different solutions. The sodium and potassium concentrations of solution A are c_{Na}^A and c_K^A, respectively, and of solution B are c_{Na}^B and c_K^B, respectively. The potential across the membrane is stepped from its initial value, V_m^i, to its final value, V_m^f.

    a. Find the value of the ratio c_{Na}^A/c_{Na}^B.

    b. What can you say about the value of the ratio c_K^A/c_K^B?

4.14 $V_m(z_o, t)$, shown at the top of Figure 4.94, is the propagated action potential at a point z_o along an axon as computed according to the Hodgkin-Huxley model. The action potential is propagating in the positive z-direction. The other six waveforms represent various components of current associated with this action potential. Outward current through

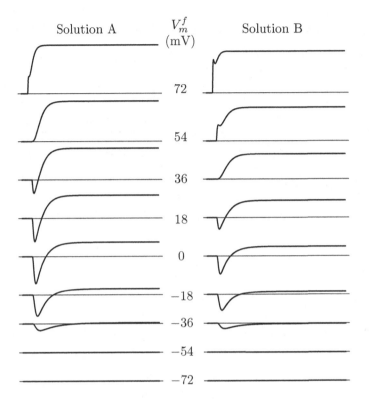

Figure 4.93 Ionic currents as a function of time in solutions A and B (Problem 4.13). The leakage currents have been subtracted, so that the waveforms include the sodium and potassium currents only.

the membrane is defined as positive and is plotted upward, and current in the positive z-direction is defined as positive and is plotted upward. Identify each of the waveforms a through f with one of the choices i through xii listed in Figure 4.94. Briefly state your reason(s) for each choice.

4.15 A length L of squid axon (diameter 500 μm) is bathed in normal seawater and space clamped as indicated in Figure 4.95. An external current $I(t)$ is passed through the membrane, and the membrane action potential $V_m(t)$, shown in Figure 4.95, is obtained. The specific capacitance of the membrane is 1 μF/cm^2.

    a. Determine L.

    b. Half the sodium ions in the seawater that bathes the axon are replaced by ions to which the membrane is impermeable. Estimate the changes in V_1, V_2, V_3, and V_4 caused by this change in seawater composition. Explain your estimates.

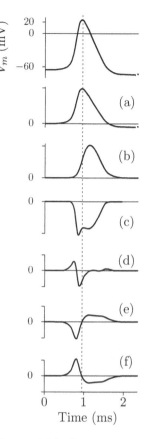

(i) The total ionic current.

(ii) The internal longitudinal current.

(iii) The sodium current.

(iv) The potassium current.

(v) The calcium current.

(vi) The leakage current.

(vii) The capacitance current.

(viii) The inductive current.

(ix) The total membrane current.

(x) The sodium pump current.

(xi) The potassium pump current.

(xii) The external longitudinal current.

Figure 4.94 Current components during a propagated action potential (see Problem 4.14).

4.16 A squid giant axon is immersed in seawater that has a concentration of potassium and sodium of 10 and 400 mmol/L, respectively. Two measurements are made on this axon: the membrane current density in response to a step of membrane potential (Figure 4.96, left panel) and the membrane potential in response to a brief pulse of membrane current density (Figure 4.96, right panel).

a. Find the smallest range of values of internal sodium concentration that is consistent with these measurements. Explain.

b. Find the smallest range of values of internal potassium concentration that is consistent with these measurements. Explain.

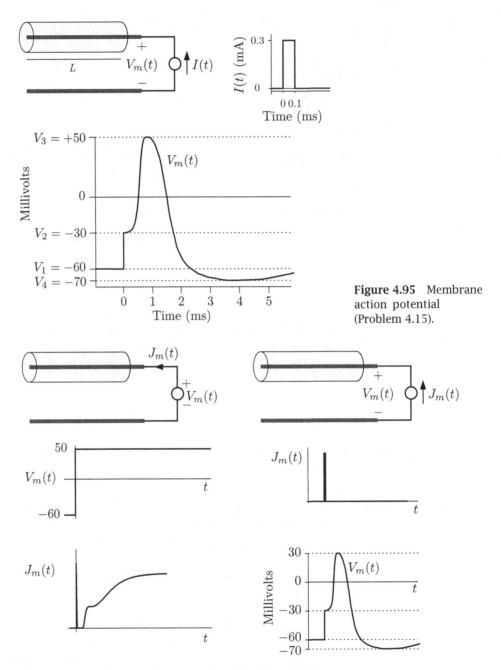

Figure 4.95 Membrane action potential (Problem 4.15).

Figure 4.96 Voltage and current clamp (Problem 4.16).

Table 4.4 Numerical values of the Hodgkin-Huxley parameters of an axon (Problem 4.17). $\overline{G}_{Na}$ and $\overline{G}_K$ are the maximum values of the sodium and potassium conductances; G_L is the leakage conductance; c_{Na}^o, c_{Na}^i, c_K^o, and c_K^i are the concentrations, outside and inside the cell, of sodium and potassium ions; V_L is the equilibrium potential for the leakage current density; and C_m is the membrane capacitance.

Parameter	Value
$\overline{G}_{Na}$ (mS/cm^2)	120
$\overline{G}_K$ (mS/cm^2)	36
G_L (mS/cm^2)	0.3
c_{Na}^o (mmol/L)	491
c_{Na}^i (mmol/L)	50
c_K^o (mmol/L)	20.11
c_K^i (mmol/L)	400
V_L (mV)	−49
temperature (°C)	6.3
C_m (μF/cm^2)	1

4.17 This problem deals with the cable and Hodgkin-Huxley models of an axon. Assume that a giant axon of the squid whose diameter is 500 μm is space clamped and that its behavior is represented accurately by the Hodgkin-Huxley model with parameters given in Table 4.4. In response to a suprathreshold pulse of membrane current density (of amplitude 20 μA/cm^2 and duration 0.5 ms), a membrane action potential occurs; the resting, maximum, and minimum values of selected variables are shown in Table 4.5.

a. Determine the value of the resting potential.

b. In a few sentences, explain why J_C, J_{ion}, and J_m are zero for the resting conditions.

c. Let the maximum value of J_C be $(J_C)_{max}$ and the minimum value of J_{ion} be $(J_{ion})_{min}$. Explain why $(J_C)_{max} = -(J_{ion})_{min}$ for the simulation summarized in Table 4.5.

d. Explain why J_{Na} is negative, J_K is positive, and J_L changes sign for the simulation summarized in Table 4.5.

Now the space-clamp electrodes are removed, and the unclamped axon is stimulated by a current pulse applied at the center of the axon.

Table 4.5 Summary of values of key variables of the membrane action potential computed from the Hodgkin-Huxley model of a space-clamped axon (Problem 4.17). V_m is the membrane potential; G_{Na} and G_K are the sodium and potassium conductances per unit area; J_{Na}, J_K, J_L, and J_C are the sodium, potassium, leakage, and capacitance current densities; J_{ion} is the total ionic current density; and J_m is the total membrane current density.

Variable	Resting	Minimum	Maximum
V_m (mV)		−71.2	44.2
m	0.0536	0.0136	0.994
h	0.593	0.0764	0.593
n	0.319	0.319	0.768
G_{Na} (mS/cm^2)	0.0109	5.65×10^{-5}	31.6
G_K (mS/cm^2)	0.374	0.374	12.5
J_{Na} (μA/cm^2)	−1.26	−788	−0.00712
J_K (μA/cm^2)	4.53	4.36	822
J_L (μA/cm^2)	−3.27	−6.65	28
J_C (μA/cm^2)	0	−68.6	298
J_{ion} (μA/cm^2)	0	−298	68.6
J_m (μA/cm^2)	0	0	20

The current amplitude is so small that the response of the axon remains in its linear range. You may assume that the external longitudinal resistance per unit length is much smaller that the internal longitudinal resistance per unit length ($r_o \ll r_i$) and that the resistivity of axoplasm is $\rho_i = 100 \ \Omega \cdot \text{cm}$.

e. Estimate the space constant, λ_C, of the axon. Explain your reasoning.

f. Estimate the time constant, τ_M, of the axon. Explain your reasoning.

4.18 For each instant of time during a propagated action potential along a normal squid axon, the first derivative of the membrane potential dV_m/dt and the membrane potential V_m are used to determine the coordinates of a point on the phase-plane trajectory shown in Figure 4.97. Certain points in time during the action potential are indicated on the phase-plane trajectory.

a. Sketch the phase-plane trajectory, and place arrows on the trajectory to indicate the direction of increasing time during the action potential.

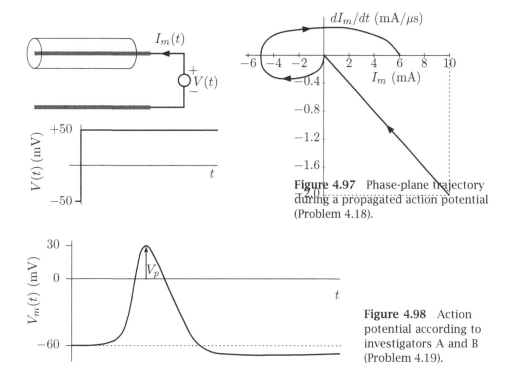

Figure 4.97 Phase-plane trajectory during a propagated action potential (Problem 4.18).

Figure 4.98 Action potential according to investigators A and B (Problem 4.19).

b. For which if any of the times t_1, t_2, t_3, t_4, or t_5 is the capacitance current density through the membrane zero? Explain.

c. For which if any of the times t_1, t_2, t_3, t_4, or t_5 is the membrane current density zero? Explain.

4.19 A giant crustacean muscle fiber is known to produce action potentials as shown in Figure 4.98 with a peak value of $+30$ mV and a resting potential of -60 mV. The concentrations of key ions in the internal and external solutions are given in Table 4.6. Two investigators, A and B, have recently published additional measurements on this preparation. They both measured the peak amplitude of the action potential as a function of the extracellular calcium concentration. Although the conditions of the experiment were reported to be identical, the results obtained were

Table 4.6 Ion concentrations (Problem 4.19). Other substances are present to make the solutions isosmotic.

Ion	c_n^i (mmol/L)	c_n^o (mmol/L)
K^+	160	16
Na^+	40	400
Cl^-	100	100
Ca^{++}	2	20

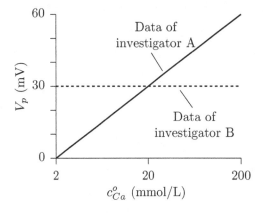

Figure 4.99 Dependence of the peak value of the action potential on calcium concentration (Problem 4.19).

dramatically different (Figure 4.99). Both investigators have presented simple models that are consistent with their data as well as with the original data, but not with the other investigator's data. Describe a model that might have been proposed by A *and* one that might have been proposed by B.

4.20 During the propagation of an action potential along an axon, energy flows into and out of the membrane. This problem concerns the power and energy associated with a propagating action potential as recorded at some position z_0 along an axon. With the usual definitions of the reference directions for membrane potential and membrane current, e.g., as shown in Figure 4.100, the instantaneous power density flowing into the membrane at position z_0 at time t is defined as

$$\mathcal{P}_m(z_0, t) = V_m(z_0, t) J_m(z_0, t),$$

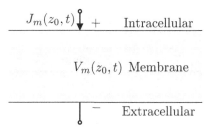

Figure 4.100 Reference directions for membrane potential and membrane current in an axon (Problem 4.20).

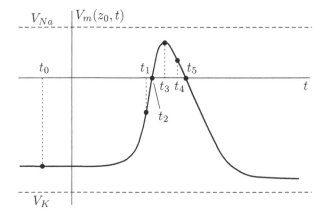

Figure 4.101 Propagated action potential with several points in time indicated (Problem 4.20). V_{Na} and V_K are Nernst equilibrium potentials for sodium and potassium, respectively.

and the energy density flowing into the membrane at position z_0 in the time interval (t_a, t_b) is defined as

$$\mathcal{E}_m(z_0) = \int_{t_a}^{t_b} V_m(z_0, t) J_m(z_0, t) \, dt.$$

Figure 4.101 shows a propagated action potential recorded at position z_0 as a function of time. Points t_0, t_1, t_2, t_3, t_4, and t_5 are indicated on the waveform, and these define the time intervals τ_0, τ_1, τ_2, τ_3, and τ_4. Definitions of all these quantities are given in Table 4.7.

a. Assume that the core conductor model applies, and list those intervals of time, τ_0, τ_1, τ_2, τ_3, or τ_4, during which there is a net flow of energy out of the membrane.

   In the remaining parts of the problem (b through d), assume that the relation between the membrane potential and the membrane current density can be represented by the Hodgkin-Huxley model (Figure 4.6). We

Table 4.7 Definitions of times and intervals during the propagated action potential shown in Figure 4.101 (Problem 4.20).

Definitions of t_0, t_1, t_2, t_3, t_4, and t_5		Definitions of τ_0, τ_1, τ_2, τ_3, and τ_4
$V_m(z_0, t) = V_m^o$	at $t = t_0$	$\tau_0 = t_1 - t_2$
$\frac{\partial^2 V_m(z_0,t)}{\partial t^2} = 0$	at $t = t_1$	$\tau_1 = t_2 - t_1$
$V_m(z_0, t) = 0$	at $t = t_2$	$\tau_2 = t_3 - t_2$
$\frac{\partial V_m(z_0,t)}{\partial t} = 0$	at $t = t_3$	$\tau_3 = t_4 - t_3$
$\frac{\partial^2 V_m(z_0,t)}{\partial t^2} = 0$	at $t = t_4$	$\tau_4 = t_5 - t_4$
$V_m(z_0, t) = 0$	at $t = t_5$	

use the term *sodium branch* to refer to particular branches in the model of Figure 4.6 in which J_{Na} flows. Similarly, the *potassium branch* is the branch in which J_K flows.

b. List those intervals of time during which there is a net flow of energy out of the potassium branch.

c. List those intervals of time during which there is a net flow of energy out of the sodium branch.

d. Which of the phrases i through viii makes the following statement valid during the interval τ_2?

According to the model, during this interval energy is supplied:

i.   by electrostatic stored energy in the membrane capacitance.

ii.   by the ionic conductances.

iii. by metabolic storehouses.

iv.   by the batteries.

v.   at the expense of the potential energy stored in ionic concentration differences.

vi.   by an external source.

vii. only by the sodium branch.

viii. only by the potassium branch.

4.21 A squid giant axon is placed in an external solution whose sodium concentration is unknown and step voltage clamped from -50 mV to $+50$ mV. The total membrane current, $I_m(t)$, and its first derivative, $dI_m(t)/dt$, are plotted at each instant in time for $t > 0$ as shown in Fig-

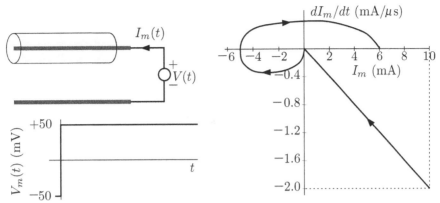

Figure 4.102 Phase-plane trajectory for membrane current during a voltage clamp from -50 to $+50$ mV (Problem 4.21). The left panel shows the recording arrangement and the voltage step. The right panel shows the phase-plane trajectory for the current.

ure 4.102. The arrows indicate the direction of increasing time along the phase-plane trajectory.

a. Components of the membrane current can be ascribed to current flow through the membrane capacitance, current carried by sodium ions, and current carried by potassium ions. Determine the approximate values of the following:

   i.   The maximum membrane capacitance current.

   ii. The maximum sodium current.

   iii. The maximum potassium current.

b. Is the sodium equilibrium potential $V_{Na} > 50$ mV, or is $V_{Na} < 50$ mV? Explain.

c. The sodium concentration of the external solution is now changed to make the sodium equilibrium potential $V_{Na} = 50$ mV. Was the sodium concentration increased or decreased? Explain.

d. Determine a simple electrical network that will account for the capacitance current *only*. Determine the numerical value of each element in your electrical network model, and discuss the physical basis for the element.

4.22 A squid giant axon is voltage clamped at a constant membrane potential of $V_m = 0$ mV. The total membrane current density is found to be $J_m =$

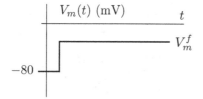

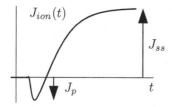

Figure 4.103 Ionic current under voltage clamp (Problem 4.23). The definitions of the peak ionic current density, J_p, and the steady-state ionic current density, J_{ss}, are shown.

1.5 mA/cm^2. The ratio of extracellular to intracellular sodium concentration is $c^o_{Na}/c^i_{Na} = 1$, and the sodium conductance is $G_{Na} = 5$ mS/cm^2. If the extracellular sodium concentration is increased by a factor of ten without changing the membrane potential or the potassium concentration, estimate the new steady-state value of the membrane current density. You may assume that the Hodgkin-Huxley model applies and that for this axon the leakage current is negligible, so that only sodium and potassium flow through the membrane. State any other assumptions that you make.

4.23 The membrane ionic current density under voltage clamp is measured for a squid giant axon as shown in Figure 4.103. The peak inward current density, J_p, and the steady-state outward current density, J_{ss}, are measured as a function of the final value of the membrane potential, V^f_m. The results are shown for two different extracellular solutions in Figure 4.104. Solution A is a seawater solution diluted with sucrose; Solution B is another seawater solution diluted with sucrose (not necessarily in the same proportion as Solution A) that contains the new neural toxin XXX.

a. Are the sodium concentrations in Solutions A and B the same or different? Explain.

b. The axon is placed in Solution A, and the membrane potential is clamped at $t = 0$ to -30 mV and changed to V^f_m at the peak of the early current component whose amplitude is J_p, as shown in Figure 4.105.

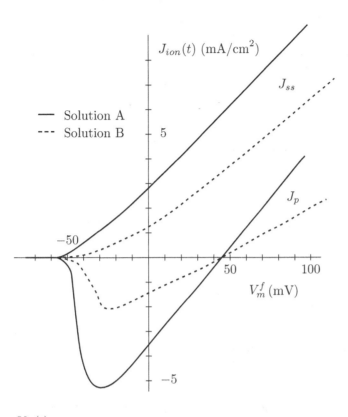

Figure 4.104 Peak and steady-state ionic current densities under voltage clamp (Problem 4.23).

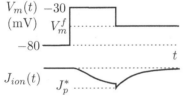

Figure 4.105 Peak ionic current under a two-step voltage clamp (Problem 4.23).

Plot the locus of instantaneous values of the membrane ionic current, J_p^*, as a function of V_m^f. Explain your method.

4.24 A network consists of a voltage-dependent conductance in series with a battery whose voltage is V_0, as shown in Figure 4.106. The objective of this problem is to examine the relation between the current-voltage relation and the conductance-voltage relation for this network.

a. For each of the three voltage-dependent conductances, sketch on one axis the relation of J to V for $V_0 = -60$, 0, and $+60$ mV.

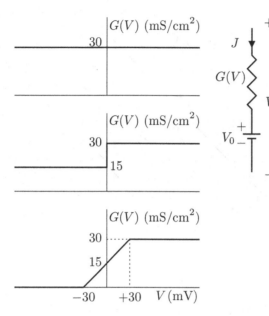

Figure 4.106 Voltage-dependent conductances (Problem 4.24). The left panels show three different voltage-dependent conductances. The right panel shows an electrical network that contains a conductance.

b. With the results of part a in mind, discuss the relation between the results shown in Figures 4.16 and 4.27.

4.25 Figure 4.107 shows ionic currents measured from a giant axon of a squid. Both capacitance and leakage currents have been subtracted from the measured membrane currents to yield the ionic currents shown.

a. Which ion carries the transient inward current in these measurements? Explain.

b. Why is there virtually no outward current in these measurements? Explain.

4.26 This problem concerns the Hodgkin-Huxley model of a space-clamped, current-clamped giant axon of the squid. The voltage dependence of the equilibrium values and time constants of the activation and inactivation factors of the model are shown in Figure 4.25. Additional parameters of the Hodgkin-Huxley model are $\overline{G}_{Na} = 120$ mS/cm^2, $\overline{G}_K = 36$ mS/cm^2, $G_L = 0.3$ mS/cm^2, $V_{Na} = +55$ mV, $V_K = -72$ mV, $V_L = -49$ mV, and $C_m = 1$ μF/cm^2. It is found that the resting potential of the model is about -60 mV, and the threshold amplitude for eliciting an action potential for a 0.5 ms current pulse is about 13 μA/cm^2.

a. Suppose a constant outward membrane current of 4 mA/cm^2 is applied to the membrane for a long time. Estimate the steady-state

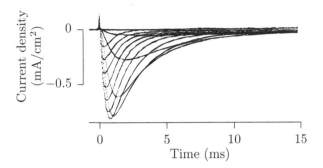

Figure 4.107 Superimposed traces of ionic currents measured under voltage clamp from a giant axon of a squid (adapted from Keynes and Aidley, 1981, Figure 4.12) (Problem 4.25). Currents were obtained from an axon bathed in K-free artificial seawater that contained 103 mmol/L NaCl and 344 mmol/L Tris buffer and dialyzed intracellularly with 350 mmol/L CsF for voltage steps from a holding potential of -70 mV to potentials that varied from -50 to $+60$ mV in 10 mV steps. The temperature was 5°C.

value of the membrane potential. Explain any approximations you make.

b. Suppose the membrane is depolarized to a potential of -10 mV by means of a constant applied membrane current whose amplitude is unknown. An outward pulse of current of amplitude 100 μA/cm^2 and duration 0.5 ms is superimposed on the constant current. Will an action potential occur in response to the current pulse? Explain briefly.

4.27 This problem deals with the Hodgkin-Huxley model under space-clamp conditions in response to a pulse of membrane potential (upper left panel in Figure 4.108). The equations and the voltage-dependent parameters of the Hodgkin-Huxley model are summarized in Section 4.2.3. Figure 4.108 shows a collection of waveforms purported to be responses of variables in the Hodgkin-Huxley model to the voltage pulse. The variable is not identified, but the numerical values have the units mS/cm^2 for conductance, mA/cm^2 for current, mV for potential, and ms for time constants. The (in)activation factors are dimensionless. For each of the following variables, choose the appropriate waveform that represents its response to the voltage pulse, and briefly justify your choice.

a. $m(V_m, t)$

b. $m_\infty(V_m)$

c. $\tau_m(V_m)$

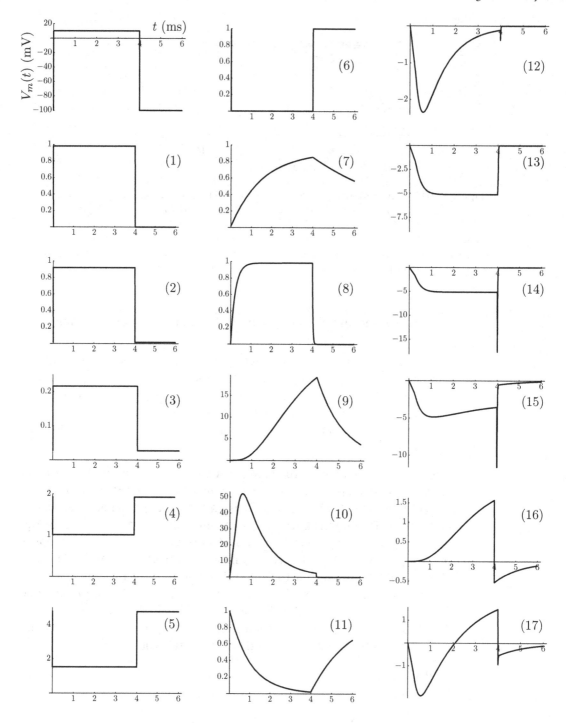

d. $G_K(V_m, t)$

e. $G_{Na}(V_m, t)$

f. $J_{Na}(V_m, t)$

g. $J_K(V_m, t)$.

4.28 Each of the panels in Figure 4.109 shows an action potential computed from the Hodgkin-Huxley model of a space-clamped axon in response to a current pulse of duration 0.5 ms and of amplitude 20 μA/cm^2. In each panel, the dashed curve shows the response for the standard parameters of the Hodgkin-Huxley model (Section 4.2.3). Each of the solid curves was obtained by computing the response of the Hodgkin-Huxley model with parameters identical to the dashed curve except that one parameter of the model was changed. For each of the waveforms a through d, determine which one of the following statements is consistent with the computation:

1. The leakage conductance was reduced from $G_L = 0.3$ to 0.01 mS/cm^2.

2. The temperature was increased from 6.3 to 10°C.

3. The membrane capacitance was increased from $C_m = 1$ to 1.1 μF/cm^2.

4. The intracellular sodium concentration was reduced from 50 to 25 mmol/L.

4.29 This problem concerns the sodium and potassium currents predicted by the Hodgkin-Huxley model for a squid giant axon that is internally perfused with a solution that contains pronase (which blocks inactivation, i.e., sets $h = 1$), a high concentration of sodium ions, and a low concentration of potassium. The axon is bathed in a solution that contains a high concentration of potassium and a low concentration of sodium. The membrane potential is held for a long time at -100 mV, then stepped to 0 mV at $t = 0$ and held at that level. The voltage dependences of rate factors of the Hodgkin-Huxley model are given in Figure 4.25. The temperature is such that a perfect sodium electrode has a slope of 59 mV/decade when the potential is plotted versus the logarithm of the concentration. The other parameters are as follows: $C_m = 1$ μF/cm^2, $\overline{G}_{Na} = 120$ mS/cm^2, $\overline{G}_K = 36$ mS/cm^2, $G_L = 0.3$ mS/cm^2, $c_{Na}^o = 40$ mmol/L, $c_{Na}^i = 400$ mmol/L, $c_K^o = 400$ mmol/L, and $c_K^i = 40$ mmol/L.

Figure 4.108 Waveforms purported to be responses to a pulse of membrane potential (upper left panel) (Problem 4.27). The pulse consists of a depolarization of 110 mV lasting for 4 ms superimposed on a DC potential of -100 mV.

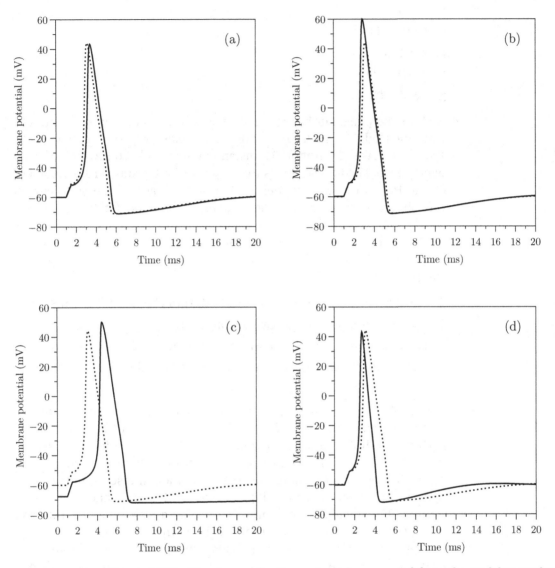

Figure 4.109 Waveforms of action potentials computed from the Hodgkin-Huxley model (Problem 4.28).

    a. Determine and sketch the sodium current density, $J_{Na}(t)$, as a function of time. Indicate numerical values for amplitudes and time constants.

    b. Determine and sketch the potassium current density, $J_K(t)$, as a function of time. Indicate numerical values for amplitudes and time constants.

    c. Briefly compare the sodium and potassium current densities found in parts a and b to the sodium and potassium current densities predicted by the Hodgkin-Huxley model for normal concentrations and normal sodium inactivation in response to the same membrane potential step.

References

Books and Reviews

Aidley, D. J. (1989). *The Physiology of Excitable Cells*. Cambridge University Press, Cambridge, England.

Cole, K. S. (1968). *Membranes, Ions, and Impulses*. University of California Press, Berkeley, CA.

Eccles, J. C. (1973). *The Understanding of the Brain*. McGraw-Hill, New York.

Hille, B. (1977). Ionic basis of resting and action potentials. In Brookhart, J. M. and Mountcastle, V. B., eds., *Handbook of Physiology*, sec. 1, *The Nervous System*, vol. 1, *Cellular Biology of Neurons*, pt. 1, 99–136. American Physiological Society, Bethesda, MD.

Hille, B. (1992). *Ionic Channels of Excitable Membranes*. Sinauer, Sunderland, MA.

Hodgkin, A. L. (1964). *The Conduction of the Nervous Impulse*. Charles C. Thomas, Springfield, IL.

Hodgkin, A. L. (1977). Chance and design in electrophysiology: An informal account of certain experiments on nerve carried out between 1934 and 1952. In *The Pursuit of Nature*, 1–21. Cambridge University Press, Cambridge, England.

Huxley, A. F. (1964). Excitation and conduction in nerve: Quantitative analysis. *Science*, 145:1154–1159.

Junge, D. (1992). *Nerve and Muscle Excitation*. Sinauer, Sunderland, MA.

Kandel, E. R., Schwartz, J. H., and Jessell, T. M. (1991). *Principles of Neural Science*. Elsevier, New York.

Katz, B. (1939). *Electric Excitation of Nerve*. Oxford University Press, London.

Katz, B. (1966). *Nerve, Muscle and Synapse*. McGraw-Hill, New York.

Keynes, R. D. and Aidley, D. J. (1981). *Nerve and Muscle*. Cambridge University Press, Cambridge, England.

Keynes, R. D. and Aidley, D. J. (1991). *Nerve and Muscle*. Cambridge University Press, Cambridge, England.

Weiss, T. F. (1996). *Cellular Biophysics*, vol. 1, *Transport*. MIT Press, Cambridge, MA.

Original Articles

Armstrong, C. M. (1969). Inactivation of the potassium conductance and related phenomena caused by quaternary ammonium ion injection in squid axons. *J. Gen. Physiol.*, 54:553–575.

Atwater, I., Bezanilla, F., and Rojas, E. (1969). Sodium influxes in internally perfused squid giant axon during voltage clamp. *J. Physiol.*, 201:657–664.

Baker, P. F., Hodgkin, A. L., and Shaw, T. I. (1961). Replacement of the protoplasm of a giant nerve fibre with artificial solutions. *Nature*, 190:885–887.

Bezanilla, F., Rojas, E., and Taylor, R. E. (1970). Sodium and potassium conductance changes during a membrane action potential. *J. Physiol.*, 211:729–751.

Cole, K. S. (1949). Dynamic electrical characteristics of the squid axon membrane. *Arch. Sci. Physiol.*, 22:253–258.

Cole, K. S. (1955). Ions, potentials, and the nerve impulse. In Shedlovsky, T., ed., *Electrochemistry in Biology and Medicine*, 121–140. John Wiley & Sons, New York.

Cole, K. S., Antosiewicz, H. A., and Rabinowitz, P. (1955). Automatic computation of nerve excitation. *J. Soc. Ind. Appl. Math.*, 3:153–172.

Cole, K. S. and Curtis, H. J. (1939). Electric impedance of the squid giant axon during activity. *J. Gen. Physiol.*, 22:649–670.

Cole, K. S. and Moore, J. W. (1960). Ionic current measurements in the squid giant axon membrane. *J. Gen. Physiol.*, 44:123–167.

Cooley, J. W. and Dodge, F. A. (1965). Digital computer solutions for excitation and propagation of the nerve impulse. Technical report RC 1496, IBM Research.

Cooley, J. W. and Dodge, F. A. (1966). Digital computer solutions for excitation and propagation of the nerve impulse. *Biophys. J.*, 6:583–599.

Cooley, J. W., Dodge, F., and Cohen, H. (1965). Digital computer solutions for excitable membrane models. *J. Cell. Comp. Physiol.*, 66:99–110.

Curtis, H. J. and Cole, K. S. (1940). Membrane action potentials from the squid giant axon. *J. Cell. Comp. Physiol.*, 15:147–157.

Curtis, H. J. and Cole, K. S. (1942). Membrane resting and action potentials from the squid giant axon. *J. Cell. Comp. Physiol.*, 19:135–144.

Fitzhugh, R. (1955). Mathematical models of threshold phenomena in the nerve membrane. *Bull. Math. Biophys.*, 17:257–278.

Fitzhugh, R. (1960). Thresholds and plateaus in the Hodgkin-Huxley nerve equations. *J. Gen. Physiol.*, 43:867–896.

Fitzhugh, R. (1961). Impulses and physiological states in theoretical models of nerve membrane. *Biophys. J.*, 1:445–466.

Fitzhugh, R. (1969). Mathematical models of excitation and propagation in nerve. In Schwann, H., ed., *Biological Engineering*, 1–85. McGraw-Hill, New York.

Fitzhugh, R. and Antosiewicz, H. A. (1959). Automatic computation of nerve excitation: Detailed corrections and additions. *J. Soc. Indust. Appl. Math.*, 7:447–458.

Foster, W. R., Unger, L. H., and Schwaber, J. S. (1993). Significance of conductances in Hodgkin-Huxley models. *J. Neurophys.*, 70:2502–2518.

Frankenhaeser, B. and Vallbo, A. B. (1965). Accomodation in myelinated nerve fibres of *Xenopus laevis* as computed on the basis of voltage clamp data. *Acta Physiol. Scand.*, 63:1–20.

Grundfest, H. (1965). Julius Bernstein, Ludimar Hermann, and the discovery of the overshoot of the axon spike. *Arch. Ital. Biol.*, 103:483–490.

Guttman, R. and Barnhill, R. (1966). Temperature characteristics of excitation in space-clamped axons. *J. Gen. Physiol.*, 49:1007–1018.

Guttman, R. and Barnhill, R. (1968). Temperature dependence of accomodation and excitation in space-clamped axons. *J. Gen. Physiol.*, 51:759–769.

Guttman, R. and Barnhill, R. (1970). Oscillations and repetitive firing in squid axons: Comparison of experiments with computations. *J. Gen. Physiol.*, 55:104–118.

Hagiwara, S. and Oomura, Y. (1958). The critical depolarization for the spike in the squid giant axon. *Jap. J. Physiol.*, 8:234–245.

Hodgkin, A. L. (1938). The subthreshold potentials in a crustacean nerve fibre. *Proc. R. Soc. London, Ser. B*, 126:87–121.

Hodgkin, A. L. and Huxley, A. F. (1939). Action potentials recorded from inside a nerve fibre. *Nature*, 144:710.

Hodgkin, A. L. and Huxley, A. F. (1952a). The components of membrane conductance in the giant axon of *Loligo*. *J. Physiol.*, 116:473–496.

Hodgkin, A. L. and Huxley, A. F. (1952b). Currents carried by sodium and potassium ions through the membrane of the giant axon of *Loligo*. *J. Physiol.*, 116:449–472.

Hodgkin, A. L. and Huxley, A. F. (1952c). The dual effect of membrane potential on sodium conductance in the giant axon of *Loligo*. *J. Physiol.*, 116:497–506.

Hodgkin, A. L. and Huxley, A. F. (1952d). Movement of sodium and potassium ions during nervous activity. In *Cold Spring Harbor Symposia on Quantitative Biology, Vol. 17*, pages 43–52. Long Island Biological Society, Cold Spring Harbor, NY.

Hodgkin, A. L. and Huxley, A. F. (1952e). A quantitative description of membrane current and its application to conduction and excitation in nerve. *J. Physiol.*, 117:500–544.

Hodgkin, A. L. and Huxley, A. F. (1953). Movement of radioactive potassium and membrane current in a giant axon. *J. Physiol.*, 121:403–414.

Hodgkin, A. L., Huxley, A. F., and Katz, B. (1949). Ionic currents underlying activity in the giant axon of the squid. *Arch. Sci. Physiol.*, 3:129–150.

Hodgkin, A. L., Huxley, A. F., and Katz, B. (1952). Measurement of current-voltage relations in the membrane of the giant axon of *Loligo*. *J. Physiol.*, 116:424–448.

Hodgkin, A. L. and Katz, B. (1949a). The effect of sodium ions on the electrical activity of the giant axon of the squid. *J. Physiol.*, 108:37–77.

Hodgkin, A. L. and Katz, B. (1949b). The effect of temperature on the electrical activity of the giant axon of the squid. *J. Physiol.*, 109:240–249.

Huxley, A. F. (1959). Ion movements during nerve activity. *Ann. N.Y. Acad. Sci.*, 81:221–246.

Marmont, G. (1949). Studies on the axon membrane. *J. Cell. Comp. Physiol.*, 34:351–382.

Mauro, A. (1961). Anomalous impedance: A phenomenological property of time-variant resistance. *Biophys. J.*, 1:353–372.

Mauro, A., Conti, F., Dodge, F., and Schor, R. (1970). Subthreshold behavior and phenomenological impedance of the squid giant axon. *J. Gen. Physiol.*, 55:497–523.

Takashima, S. (1979). Admittance change of squid axon during action potentials. *Biophys. J.*, 26:133–142.

Vallbo, A. B. (1964). Accommodation related to inactivation of the sodium permeability in single myelinated nerve fibres. *Acta Physiol. Scand.*, 61:429–444.

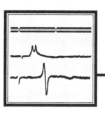

5

Saltatory Conduction
in Myelinated Nerve Fibers

In 1935, I was convinced that narcotics act with extreme rapidity upon the nodes of Ranvier but not on the myelin-covered portion of the nerve. At the same time, I was very much puzzled by the fact that the rate of transmission could be reduced promptly down to 50 per cent of the normal value or still less by an application of a narcotizing solution of adequate concentration upon a single nerve fiber. Why can such a pronounced slowing of transmission velocity occur if the narcotic acts only at the nodes of Ranvier which occupy a length of about 0.02 per cent of the total length of the nerve fiber? This question remained unanswered until the end of 1938. Nevertheless, this series of work, done with an old, rusty Helmholtz pendulum combined with a pair of fine dissecting needles, and published only in the Japanese language, has given me the title of Doctor of Medicine.

Thus, all the experimental results described in this book have been obtained in Japan, an island next to the one where Robinson Crusoe had been secluded. The unfortunate warfare had made our cultural isolation from the rest of the world virtually complete until the summer of 1948, when Dr. Davis sent me a set of the Annual Review of Physiology *and his reprints, for which my friends in Japan and I are very grateful. The fact that the manuscript of this book was written in a place where no foreign physiological journals were yet available can probably be a partial excuse if I have omitted some of the important literature in this field of physiology.*
—Tasaki, 1953

5.1   Structure of Myelinated Nerve Fibers

5.1.1   Gross Morphology

Neurons are found in close association with large numbers of nonneuronal cells called *supporting cells*. Supporting cells are found both in the central nervous system or CNS (the brain and spinal cord) and in the peripheral nervous system or PNS (the portion of the nervous system that is outside the central

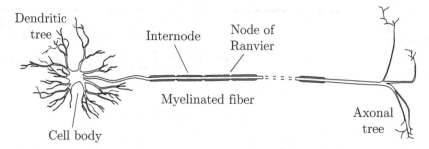

Figure 5.1 Schematic diagram of a neuron with a myelinated fiber. The myelinated fiber consists of an axon plus a sequence of myelin internodes separated by gaps in the myelin called nodes of Ranvier. Each internode is formed by a supporting cell.

nervous system). In the CNS, a large fraction of the cells (estimated to exceed 90% of the total number of cells) are not neurons, but supporting cells called *glial cells*. There are several types of glial cells, including astrocytes, oligodendrocytes, and microglia. In the PNS, the supporting cell associated with neurons is the *Schwann cell*. A variety of functions have been ascribed to supporting cells, including providing mechanical support for neurons, buffering extracellular solution composition, guiding the migration of neurons during development of the nervous system, nourishing neurons, participating in the formation of the blood-brain barrier, disposing of cellular waste, and so on. In this chapter, we are concerned with one important function of supporting cells: they give rise to myelin to form myelinated nerve fibers. Specifically, myelin is formed by Schwann cells in the PNS and by oligodendrocytes in the CNS.

Figure 5.1 shows a schematic diagram of a neuron with a myelinated nerve fiber. A myelinated fiber consists of an axon ensheathed by myelin. The myelin sheath is a segmented cylindrical sleeve around the axon. The segments of myelin are called *internodal segments* or *internodes,* and these are separated by narrow gaps called *nodes of Ranvier* (Figure 5.2). The internodes are not part of the neuron, but are formed by supporting cells. In the PNS, a single Schwann cell supplies the myelin for a single internode. Thus, a peripheral myelinated nerve fiber consists of an axon and a sequence of Schwann cells, one for each internode. In the CNS, a single oligodendrocyte may provide the myelin for the internodes of many (as many as fifty) different axons, as well as the myelin for several internodes in the same fiber.

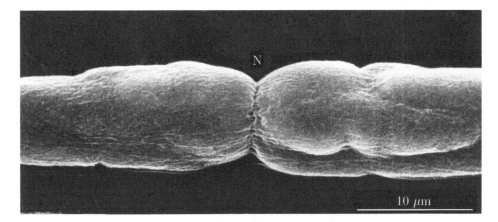

Figure 5.2 Scanning electron micrograph of the node of Ranvier of a myelinated fiber of 10 μm diameter from the sciatic nerve of a rat. The node (N) separates the two internodal segments on either side (adapted from Landon and Hall, 1976, Figure 1.3). Dr. Landon kindly supplied the photograph.

5.1.2 Relation of Supporting Cells to Axons

Nerve fibers are categorized as either myelinated or unmyelinated according to the presence or absence of myelin. Both types of fibers are ensheathed by supporting cells. A defining distinction between the two types of fibers is the number of layers of supporting cell membrane that ensheaths the fiber. As shown schematically in Figure 5.3 and in a transmission electron micrograph in Figure 5.4, unmyelinated fibers in vertebrates are typically found ensheathed by supporting cell. A supporting cell may ensheath one or many (ten is not unusual) unmyelinated fibers. The same Schwann cell does not ensheath the unmyelinated fibers over their entire lengths, but rather a sequence of Schwann cells ensheaths different longitudinal segments (about 20–500 μm long) of the unmyelinated fibers. An invertebrate giant axon, such as the giant axon of the squid, is surrounded with a number of supporting cells that envelope the axon circumferentially.

In contrast to the relation between supporting cells and unmyelinated fibers, myelinated fibers are surrounded by multiple layers of membrane of supporting cells. Figure 5.5 shows a schematic diagram illustrating the relation between a Schwann cell and an axon in a myelinated fiber. During development, the Schwann cell wraps itself spirally around the axon

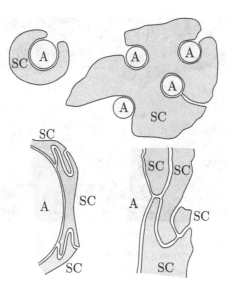

Figure 5.3 Relations of Schwann cells (SC) and axons (A) in different types of unmyelinated peripheral nerve fibers (adapted from Hodgkin, 1964, Figure 9). An individual unmyelinated axon may be surrounded by a single Schwann cell (upper left panel), or several such fibers may share a Schwann cell (upper right panel). An invertebrate unmyelinated giant axon may be surrounded by a number of Schwann cells (lower panels).

in jelly roll fashion. The relation between the Schwann cell and the axon is shown in Figure 5.5 with the Schwann cell unwrapped. The unwrapped Schwann cell is flat and has a trapezoidal shape. Portions of the Schwann cell have rather little cytoplasm, so that the membranes on the two cytoplasmic faces of the cell have their cytoplasmic surfaces fused to make *compacted membrane* (which is not shown in the cross sections in Figure 5.5, but is shown clearly in the transmission electron micrographs in Figures 5.4 and 5.6). Ridges containing cytoplasm are found around the perimeter of the

Figure 5.4 Electron micrograph of a transverse section of a myelinated nerve fiber (Ax_1) sectioned through an internode and several unmyelinated fibers from the sciatic nerve of an adult rat (adapted from Peters et al., 1991, Figure 6-4). This section contains many unmyelinated fibers ensheathed by Schwann cells. For example, the Schwann cell labeled SC ensheaths a number of unmyelinated fibers (e.g., Ax_2 and Ax_3).

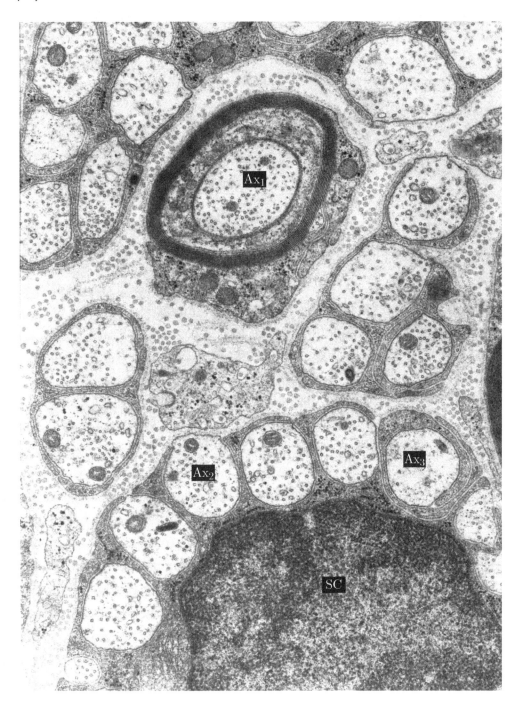

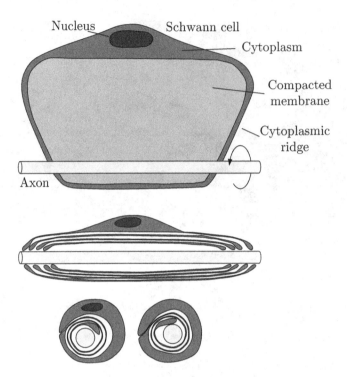

Figure 5.5 Schematic diagram of the relation between a Schwann cell and its axon (adapted from Raine, 1977, Figure 7). The upper panel shows a Schwann cell unwrapped from its axon, and the lower panels show longitudinal and cross-sectional views of the Schwann cell wrapped around the axon. One cross section is taken through the cell nucleus; the other is not. For clarity, this diagram shows only a few layers of myelin membrane, and the spacing between layers has been exaggerated.

Schwann cell. The ridges of cytoplasm appear as pockets of cytoplasm in longitudinal and cross sections, and these ridges pile up in the region of the internode on both sides of the node of Ranvier called the *paranodal region.*

In a fully developed fiber, the membrane of the supporting cell is so tightly wrapped around the axon that the cytoplasm is squeezed out and the myelin

Figure 5.6 Electron micrograph of a transverse section of a myelinated nerve fiber (Ax) sectioned through an internode from the sciatic nerve of an adult rat (adapted from Peters et al., 1991, Figure 6-5). The abbreviations are the same as in Figure 5.4. The myelin appears as an alternating sequence of dense lines (DL) separated by intraperiod lines (IL).

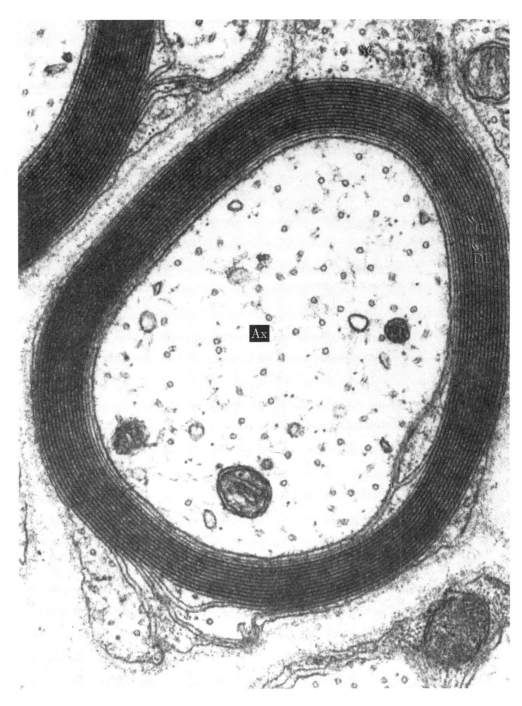

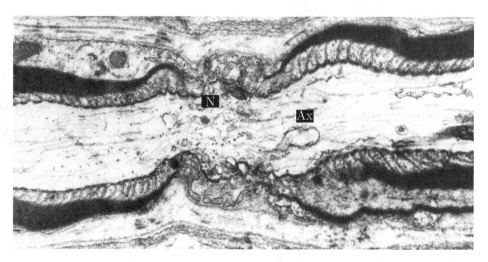

Figure 5.7 Electron micrograph of a longitudinal section through a node of Ranvier (N) of a myelinated nerve fiber (Ax) from the sciatic nerve of an adult rat (adapted from Peters et al., 1991, Figure 6-15).

consists almost entirely of concentric layers of membrane with rather little intervening cytoplasm (Figure 5.6).[1] In high-power electron micrographs (Figure 5.6), the myelin appears as a sequence of alternating dense lines separated by intraperiod lines. The dense lines are formed by the apposed cytoplasmic faces of the Schwann cell membrane, whereas the intraperiod lines are formed by the apposed external faces of the Schwann cell membranes. Measurement of the period between dense line gives values that are about two plasma membranes thick if allowances are made for tissue shrinkage incurred during histological processing. The myelin from adjacent internodes stops at the node of Ranvier, so the axonal membrane communicates with interstitial space at the node of Ranvier (Figure 5.7).

---

1. Thus, myelin has been a useful material for the study of the molecular structure of membranes. It has been studied with optical methods and X-ray diffraction, methods that require crystalline arrays of membranes in order to yield information on the molecular structure of membranes. Early estimates of the thickness of cellular membranes and indications of the transmembrane arrangements of lipids were obtained from studies of myelin.

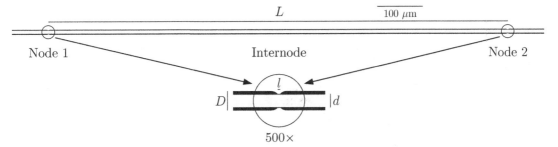

Figure 5.8 Schematic diagram of a myelinated fiber showing definitions of key dimensions. These dimensions have been drawn to scale for a myelinated fiber for which $L = 1$ mm, $D = 10$ μm, $d = 7$ μm, and $l = 1$ μm. The inset shows the region around a node scale by 500×.

5.1.3   Dimensions of Myelinated Fibers

Myelinated fibers are found predominantly in vertebrates, where they range in diameter from 1 to 20 μm, although myelinlike sheaths are also found in some invertebrates (crustaceans and annelids). Unmyelinated fibers in vertebrates have diameters that are generally in the range of 0.1-1 μm. Invertebrate nerve fibers are predominantly unmyelinated and span the range from under 1 μm to as large as 1 mm.

The length of a node (typically a few μm) is about 10^{-3} times the length of the internode (typically about 1 mm in large fibers), as indicated in Figure 5.8. Thus, the nodes are difficult to visualize on a scale in which an entire internode is observable. The definitions of the internodal length, L, the fiber diameter, D, the axis cylinder diameter, d, and the length of the node of Ranvier, l, are shown to scale in a schematic diagram of a myelinated fiber shown in Figure 5.8. In many fibers, the spacing of nodes are periodic, although there are clearly exceptions to this rule.

Initially, quantitative data on the dimensions of myelinated fibers were obtained in the PNS, where fibers are technically simpler to study, but such information has become available for central fibers as well. In the PNS, it appears that d and L are roughly proportional to D (Figures 5.9 and 5.10), at least for a functionally homogeneous population of fibers, e.g., the population of α-motoneurons that innervate skeletal muscles. However, when heterogeneous populations of fibers are pooled, the proportionality does not necessarily hold. Axons with larger diameters also have a proportionally larger number of myelin layers or lamellae (Figure 5.11). Hence, when d increases,

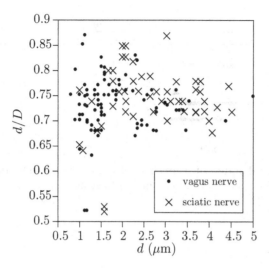

Figure 5.9 Ratio of axon diameter to total fiber diameter plotted against axon diameter for all fibers in one vagus and one sciatic nerve of a mouse (adapted from Friede and Samorajski, 1967, Figure 4). The thickness of the sheath was calculated from the measured number of myelin lamellae and the axon diameters from the measured circumference, assuming a circular cross section. The average value of d/D was 0.74 for these fibers. The average $\pm$ SD was: 0.731 ± 0.056 for the vagus nerve; 0.742 ± 0.065 for the sciatic nerve.

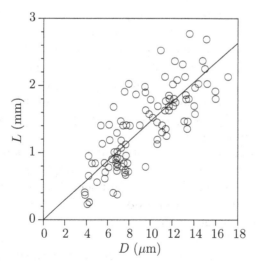

Figure 5.10 Internodal distance, L, plotted against fiber diameter, D, for 100 fibers of the bullfrog (adapted from Tasaki, 1953, Figure 67). The line that goes through the origin and best fits the data has the equation $L = 0.146D$, where L is in mm and D in μm. The correlation coefficient is 0.80.

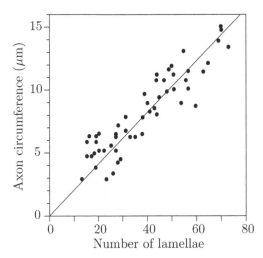

Figure 5.11 Relation between fiber diameter and the number of myelin lamellae (adapted from Friede and Samorajski, 1967, Figure 3F). The line that goes through the origin and best fits the data has the equation $C = 0.206n$, where C is the circumference in μm and n is the number of lamellae. The correlation coefficient is 0.89. The shaded band indicates the range of axon circumferences for a population of unmyelinated nerve fibers in the same nerve.

D increases, because the number of lamellae increases. A large fiber, with a diameter of 20 μm, may contain 200–300 lamellae of myelin.

We note that a peripheral fiber that has a length of 100 cm and is 10 μm in diameter will contain about 600 internodes, each due to a single Schwann cell. Thus, such a myelinated fiber consists of one nerve process plus 600 Schwann cells. The internodal length of this fiber is approximately 1 mm, and the length of the node is on the order of 1 μm.

5.2 Physiological Evidence for Saltatory Conduction

In the 1930s and 1940s, methods were developed to dissect single myelinated fibers from peripheral nerves and to explore the excitable properties of these isolated fibers (Stämpfli and Hille, 1976; Rogart and Ritchie, 1977). These studies showed that the physiological properties of the internodes differed from those of the nodes of Ranvier and that action potentials were initiated at the nodes and not in the internodes. Thus, the activity associated with the conduction of the action potential hops from node to node, a conduction mechanism called *saltatory conduction*. Later, methods were developed to explore conduc-

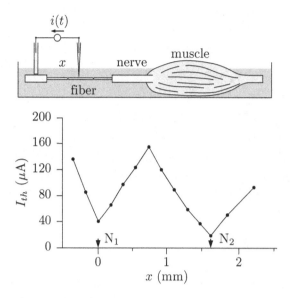

Figure 5.12 Threshold for eliciting a twitch from a gastrocnemius muscle of a toad versus location of electrical stimulus to the motonerve fiber (adapted from Tasaki, 1953, Figure 3). N_1 and N_2 indicate the locations of adjacent nodes of Ranvier. The inset shows a schematic diagram of the nerve and muscle preparation with a stimulating current applied to the nerve.

tion in myelinated fibers that had not been dissected, thus providing evidence for saltatory conduction in undissected fibers.

5.2.1 Generation of Action Potentials at Nodes

5.2.1.1 Low Threshold for Electrical Stimulation at Nodes of Ranvier

If a stimulating electrode is placed on the surface of a myelinated fiber that is attached via its nerve to a muscle, the threshold current required to elicit a just-noticeable twitch of the muscle can be measured as a function of the position of the stimulating electrode along the nerve (Figure 5.12). Current minima occur at the locations of the nodes of Ranvier, demonstrating the higher sensitivities of these nodes to electric currents. Furthermore, if the threshold is measured at various locations in a plane around a myelinated nerve fiber, it is found that the contours of constant threshold are circular and concentric with the node (Figure 5.13). Thus, the threshold depends only on distance from the node and not on direction.

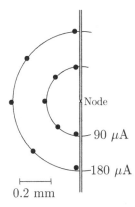

Figure 5.13 Threshold for eliciting a twitch from a muscle versus the distance of the stimulating electrode from a node of Ranvier (adapted from Tasaki, 1968, Figure 5). The contours of constant threshold current are circular and centered on a node of Ranvier.

5.2.1.2 Location of Maximum Extracellular Action Potential

Figure 5.14 shows the results of an experiment in which a myelinated nerve fiber was stimulated to conduct an action potential. The extracellular potential was measured at various locations around the fiber. The extracellular action potential is a maximum near a node of Ranvier and decreases in magnitude as an electrode is moved in any direction away from the node. Thus, the node of Ranvier appears to be the source of current that gives rise to the extracellular potential.

5.2.1.3 Blockage of the Action Potential at a Node

If anesthetics or nerve blockers, such as cocaine, are used to block the action potential, it is found that a lower concentration is required at the node than at the internode. Furthermore, application of a hyperpolarizing current to a nerve fiber can also block the action potential. Lower current strengths are required to block the action potential when applied at a node than at an internode. These results suggest that it is easier to interfere with the conduction of action potentials in the fiber by manipulating the electrical and pharmacological environment of the node than by manipulating those of the internode.

5.2.2 Extracellular Current Flow between Adjacent Nodes

External currents flowing in the internodal regions are critical for the propagation of the action potential along a myelinated nerve fiber. This property was demonstrated in experiments such as the one illustrated in Figure 5.15.

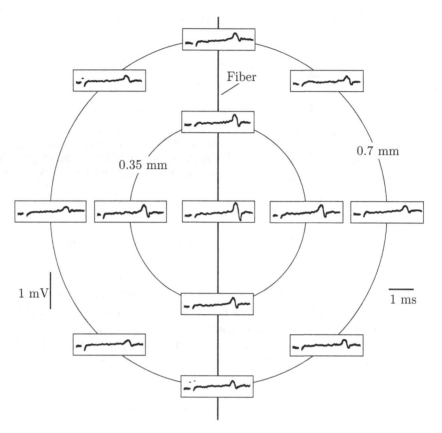

Figure 5.14 Extracellular potential as a function of the position of a recording electrode around a node of Ranvier (adapted from Tasaki, 1953, Figure 30). The vertical line indicates the position of a nerve fiber that was placed in a thin layer of Ringer's solution. A node of Ranvier is in the center of the field. Superimposed on this schematic diagram are shown the extracellular action potentials recorded at the indicated locations as an action potential traveled downward along the nerve fiber. The extracellular potential is largest near the node of Ranvier and decreases in magnitude at locations that are farther from the node of Ranvier.

Here, a nerve that innervates a muscle is dissected so that a single fiber connects the central branch of the nerve with the muscle. An internodal region spans an insulated region between two microscope slides. With the two microscope slides insulated from each other, electrical stimulation of the nerve causes no muscle contraction. If a saline-soaked thread is used to electrically connect the two microscope slides, nerve stimulation causes muscle contraction. Thus, nerve conduction from the node to the left of the insulated gap

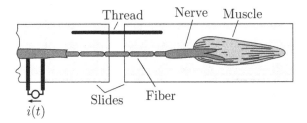

Figure 5.15 Schematic diagram of the method used to determine the role of external current in the propagation of the action potential of a myelinated fiber (Huxley and Stämpfli, 1949a). The two microscope slides contain pools of Ringer's solution that are electrically insulated from each other by the air gap between the slides. A frog sciatic nerve with its attached gastrocnemius muscle is placed on the two slides. The nerve has been dissected so that the internode of a single fiber connects the pools of Ringer's solution on the two slides. A moist thread is used to connect the two pools electrically. The nerve is stimulated electrically with the current $i(t)$.

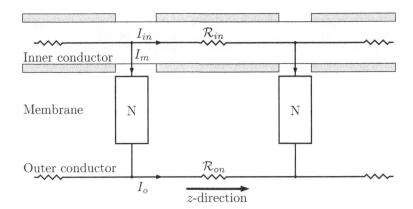

Figure 5.16 Simple model of a myelinated nerve fiber in which no current flows through the internode.

causes excitation of the node to the right only if an extracellular path for flow of electric current is present.

5.2.3 The Saltatory Conduction Hypothesis

The observations described earlier in this section led to the hypothesis that action potentials are generated only at nodes of Ranvier. This hypothesis needs to be made more precise so that it can be tested. Suppose that the internodes are assumed to be perfectly insulating so that no current flows through the membrane at an internode. According to this hypothesis, an equivalent network model of myelinated fibers would have the form shown in Figure 5.16. In

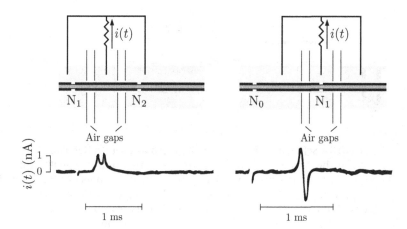

Figure 5.17 Comparison of membrane currents from a region of a toad motor nerve fiber that contains a node of Ranvier with one that does not (adapted from Tasaki, 1959, Figure 11). The upper panels show the arrangement of electrodes for recording the membrane current from a region of a fiber that is insulated from regions to the left and right by air gaps. Thus, all the membrane current, $i(t)$, from the insulated region is collected by the electrode. The lower panels show the membrane current recorded when the insulated region contains no node of Ranvier (left) and when it contains a node of Ranvier (right).

this model, membrane current, I_m, flows only at the nodes. Since there is no membrane current in the internode, the longitudinal current is independent of z in an internode, i.e., $\partial I_o / \partial z = 0$ for z in an internode. These ideas were well formulated by the 1930s and tested by two groups of research workers in the 1940s.

5.2.4 Experimental Test of Saltatory Conduction

5.2.4.1 *Difference of Currents through the Nodal and Internodal Membranes*

Membrane currents were measured from short segments of a myelinated fiber that contained either a single node of Ranvier or only an internode (Figure 5.17). It was found that membrane currents could be recorded through the internodal membrane as well as through the nodal membrane. This observation conflicts with the simple model shown in Figure 5.16. However, the membrane current from a segment of a fiber that contains only an internode differs from that when the segment includes a node of Ranvier. The current through the internode is outward only, whereas the current through the nodal membrane has a large inward component. This outcome fits with the idea that the nodes are electrically excitable and generate action potentials, whereas the internodes are electrically inexcitable and do not generate action potentials. Recall from Chapter 4 that in unmyelinated fibers the occurrence of an action potential is accompanied by an inward current carried by sodium.

In the record shown in Figure 5.17, the membrane current through the internode is outward and shows two peaks. This result also fits with the observation that an inward current occurs only at the nodes. To interpret this two-peaked waveform, consider the two nodes on either side of the internode. The first peak results when the action potential reaches one of these nodes, and the second peak results when it reaches the other. An action potential at a node is accompanied by an inward current in that node that flows outward through the neighboring internode. Thus, the two peaks result from the inward current through the nodes on the two ends of the internode.

The result shown in Figure 5.17 clearly demonstrates that there is a difference in the membrane current through nodes and internodes. However, further progress in understanding conduction in myelinated fibers required measurements in which the nodal and internodal membrane currents could be separated more fully and measured with a higher degree of spatial resolution.

5.2.4.2 Estimates of the Membrane Current in the Node and the Internode

The first definitive test of the basis of conduction in myelinated fibers came in the late 1940s (Huxley and Stämpfli, 1949a). In these experiments, the extracellular potential was measured around a single myelinated fiber passed through an insulating barrier. The fiber was held between two forceps attached to a micromanipulator so that the fiber could be positioned in the partition. The insulating partition was made either of paraffin oil held between cover slips or of a glass capillary tube embedded in plastic. The relation between the fiber and the insulating partition is shown schematically in Figure 5.18. The core conductor equation, $\partial V_o / \partial z = -r_o I_o$ (Equation 2.21), shows that the difference of potential across the distance Δz is approximately $\Delta V_o \approx -r_o \Delta z I_o$, so that the external voltage across the partition is, to a first approximation,

$$\Delta V_o = -\mathcal{R}_o I_o(z, t), \tag{5.1}$$

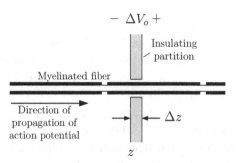

$-\,\Delta V_o\,+$

Insulating
partition

Myelinated fiber

Direction of
propagation of
action potential

Δz

z

Figure 5.18 Schematic diagram of the arrangement for recording the extracellular potential of a myelinated nerve fiber. The myelinated fiber is threaded through an insulating partition that separates two pools of saline. The potential across the partition, which has width Δz, is ΔV_0.

where $\mathcal{R}_o$ is the resistance to longitudinal current flow through the partition, $\mathcal{R}_o = r_o \Delta z$, and where it is assumed that the partition is sufficiently thin that $I_o(z) \approx I_o(z + \Delta z)$. With the partitions used in the experiments, it was found that $\mathcal{R}_o = 0.5$ MΩ, which was sufficient to make ΔV_o large enough to be measured.

With this technique, it was possible to measure $\Delta V_o(z, t)$ and, hence, $I_o(z, t)$ versus z as an action potential propagated along the fiber. The results are shown in Figure 5.19. $I_o(z, t)$ is shown at three positions in each internode. These results demonstrate that $I_o(z, t)$ is not independent of z in an internode and, hence, that the membrane current through the internode is not zero as is required by the simple model of saltatory conduction (Figure 5.16). The amplitude of $I_o(z, t)$ decreases systematically with distance in an internode, although the peak current occurs at almost the same time at all positions in an internode. There is a delay in the time of the peak current on the two sides of a node. Both the attenuation of the amplitude in an internode and the delay of the peak of the longitudinal current at each node are shown more clearly in Figure 5.20. These measurements demonstrate that I_o is dependent on z in an internode.

To estimate the membrane current per unit length, the core conductor equation (Equation 2.19) is used to give

$$K_m(z, t) \approx \frac{I_o(z + \Delta z, t) - I_o(z, t)}{\Delta z}. \tag{5.2}$$

The first difference in $I_o(z, t)$ is proportional to $K_m(z, t)$. Thus, estimates of $K_m(z, t)$ can be obtained by subtracting records of $I_o(z, t)$ from adjacent segments (Figure 5.21). These records show that the membrane current through the internodal region has a relatively small magnitude and is outward. Current through the membrane of the node is larger and is initially outward, then inward as in unmyelinated axons conducting an action potential (Figure 4.32).

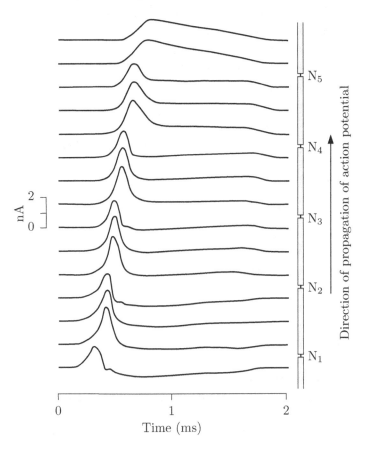

Figure 5.19 Measurements of longitudinal current taken at a number of locations along a myelinated nerve fiber (adapted from Huxley and Stämpfli, 1949a, Figure 6). The right-hand side shows a schematic diagram of a myelinated nerve fiber with the nodes of Ranvier indicated by the N_js. The longitudinal currents are arranged to indicate the position along the fiber at which each record was obtained. The external longitudinal current is plotted in a direction opposite to the direction of propagation.

5.2.4.3 *Estimates of the Membrane Potential*

We have already seen that $K_m(z, t)$ can be estimated from first differences of measurements of $I_o(z, t)$ as shown in Figure 5.19. $V_m(z, t)$ can also be estimated from measurements of $I_o(z, t)$. From the core conductor model, we have

$$V_o(z, t) - V_o(-\infty, t) = -\int_{-\infty}^{z} r_o I_o(z', t) \, dz' \tag{5.3}$$

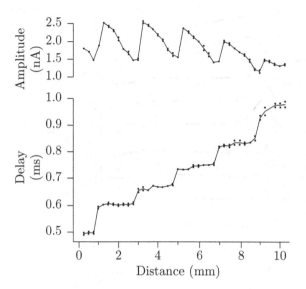

Figure 5.20 Amplitude (upper panel) and time of peak (lower panel) of the longitudinal current plotted versus position along a myelinated nerve fiber (adapted from Huxley and Stämpfli, 1949a, Figure 7) for the measurements shown in Figure 5.19.

and

$$V_m(z, t) - V_m(-\infty, t) = (r_o + r_i) \int_{-\infty}^{z} I_o(z', t) \, dz'. \tag{5.4}$$

Thus, numerical integration of records of $I_o(z, t)$ versus z give estimates of the waveform of $V_m(z, t)$. Such a numerical integration can be achieved from records such as those shown in Figure 5.19 by summing the currents up to some position $z = z_1$ at a fixed time interval. That summation determines $V_m(z, t)$ at the position z_1. The integration can be carried out for different values of z. The results are shown in Figure 5.22 as a function of z. The abscissa shows the distance scale in terms of the number of nodes. While the membrane potential is a continuous function of z, its slope is discontinuous. The discontinuities occur at the nodes. This property is anticipated, since the slope of the membrane potential is proportional to the longitudinal current (Equation 5.4), which is discontinuous at the nodes. Since the membrane current is outward in the internodes and since the membrane current is proportional to the second derivative of the potential, the membrane potential in the internodes must have a positive second derivative, i.e., the membrane potential is a concave function of position in the internodes. The computation also shows that the change in membrane potential at any time is appreciably different from zero for many nodes, i.e., the spatial extent of the propagated action potential is greater than ten nodes for this case.

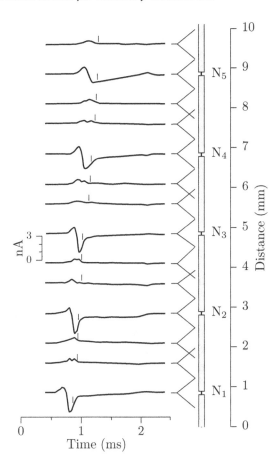

Figure 5.21 Measurements of membrane current along a myelinated nerve fiber (adapted from Huxley and Stämpfli, 1949a, Figure 12). Differences of longitudinal currents at two points (indicated on the diagram of the fiber on the right by the —≺ symbol) separated by 0.75 mm along the fiber were used to compute the membrane current. The vertical line along each trace shows the time of the peak of the membrane potential at that position along the fiber.

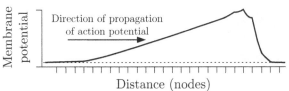

Figure 5.22 A diagram of the dependence of the membrane potential on distance along a myelinated nerve fiber at one instant in time (adapted from Huxley and Stämpfli, 1949a, Figure 13). The shaded portion of the waveform was computed by numerical integration of measurements of longitudinal current as a function of time at different locations along the nerve as shown in Figure 5.19. The tick marks indicate the locations of nodes of Ranvier.

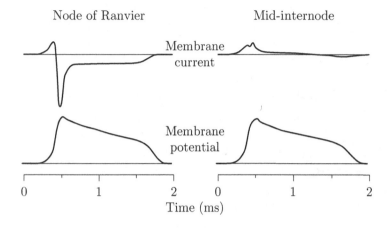

Figure 5.23 Relation between membrane potential and membrane current in an internode and at a node of Ranvier (adapted from Huxley and Stämpfli, 1949b, Figure 7).

The numerical integration of $I_o(z, t)$ can be done at different times so that $V_m(z, t)$ can be estimated as a function of time. Thus, both membrane potential and current can be obtained as a function of time. From the determinations of $K_m(z, t)$ and $V_m(z, t)$, it is possible to examine whether the relation of K_m to V_m is consistent with that for a linear resistance-capacitance cable (as described in Chapter 3). This consistency can be checked, for example, by computing K_m from V_m assuming (Equation 3.6) as follows:

$$K_m = c_m \frac{\partial V_m}{\partial t} + g_m(V_m - V_m^o). \tag{5.5}$$

Equation 5.5 shows that for time intervals during the rising portion of the membrane potential, when both the potential change and its slope are positive, the membrane current predicted by the cable model will be positive, i.e., outward through the membrane. Measurements (Figure 5.23) show that this relation between V_m to K_m occurs at an internode, but not at a node. At the node, there is a large inward current during a time interval when the membrane potential and its slope are positive (see both Figure 5.21 and Figure 5.23). Thus, the relation between current and voltage is qualitatively consistent with the predictions of the cable model in the internode, but not at the node of Ranvier.

5.2.4.4 Evidence from Undissected Fibers

Most of the evidence of saltatory conduction presented thus far is based on dissected peripheral nerve fibers. Thus, it is possible that saltatory con-

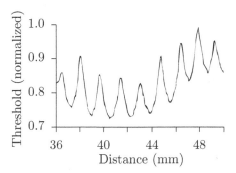

Figure 5.24 Threshold for eliciting an action potential from a single myelinated nerve fiber in the sciatic nerve of the frog (*Rana pipiens*) as a function of position along the fiber (adapted from Carley and Raymond, 1987, Figure 2). The threshold was measured automatically as the stimulating electrode was driven from 50 mm to 36 mm from the recording electrode at 0.5 mm/min. Test currents were presented every 2 s. The threshold current is normalized to its maximum value during the excursion of the stimulus electrode. The nerve sheath had been removed for this measurement. With the nerve sheath intact, the pattern of threshold maxima and minima is similar, but shows less spatial definition.

duction is a property of dissected fibers only and does not occur in situ. However, a variety of results obtained with less invasive dissection procedures essentially rule out this possibility. For example, the results shown in Figure 5.24 were obtained by dissecting the sciatic nerve from a frog. A single fiber was isolated in one end of the nerve, and an electrode was used to record the potential from this fiber. The bulk of the fiber remained in the nerve. The whole nerve was stimulated with current pulses delivered by an electrode mounted on a manipulator that was driven along the nerve by a motor. The amplitude of the current was adjusted automatically under computer control to determine the threshold current at which half the current pulses gave rise to action potentials. In this way, the threshold of a single fiber was measured continuously as a function of location along the whole nerve. The maxima and minima occur at spacings consistent with the internodal distance. Thus, the threshold pattern seen for dissected fibers (Figure 5.12) occurs in fibers undissected at the site of stimulation.

Results of another approach to the examination of saltatory conduction in single fibers using an even less invasive method are shown in Figure 5.25. In this case, single motonerve fibers were stimulated in a muscle in the tail

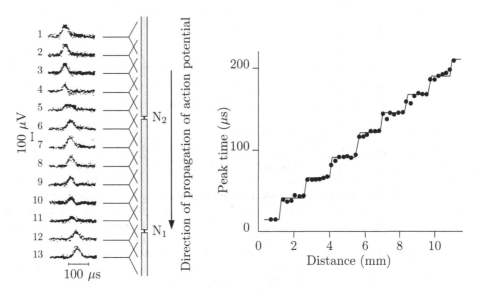

Figure 5.25 Measurement of conduction time versus distance along a nerve fiber (adapted from Rasminsky and Sears, 1972, Figure 2). The technique was to record the extracellular potential gradient every 0.2 mm along an intact ventral root of the spinal cord that contained a single myelinated nerve fiber that had been stimulated electrically. Each trace represents the potential difference between the two points indicated along the fiber by the $-\!\!\!<$ symbol, with the potential defined as the potential of the upper electrode minus that of the lower electrode. The sites of the nodes of Ranvier of the fiber are inferred from the potential records and are shown schematically to the right of the potentials. The time of occurrence of the peak of the potential is plotted versus distance along the nerve root on the right. The temperature was 30°C.

of a rat. The motonerve was cut at its entrance to the spinal cord, and the otherwise intact nerve was placed on a pair of closely spaced (400–600 μm) recording electrodes. The difference of potential between these electrodes, the potential gradient, was measured as a function of position along the nerve. This potential gradient would be proportional to the external longitudinal current if all the assumptions of the core conductor model applied to the recording situation. The potential gradient shows a spatial pattern that in many ways resembles the spatial pattern of external longitudinal current recorded from isolated single myelinated nerve fibers (Figures 5.19 and 5.20) as an action potential travels up the fiber. The time of occurrence of the peak of the potential gradient as a function of distance along the fiber shows the periodic delay (of about 20 μs) that occurs at each node and results in an average conduction

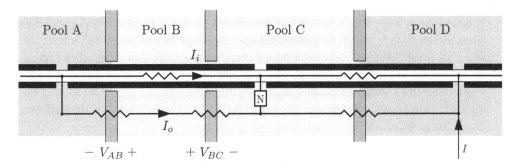

Figure 5.26 Schematic diagram of gap techniques for recording from a node of Ranvier. The voltage-current characteristics of the nodal membrane in Pool C are represented by a two-terminal element labeled N.

velocity, in this case, of about 50 m/s or 50 mm/ms. Thus, the evidence available from undissected myelinated fibers, while less direct than that obtained from dissected fibers, also supports the notion that conduction in myelinated fibers is normally saltatory.

5.3   Electrical Properties of Myelinated Nerve Fibers

5.3.1   Electrical Properties of Nodes of Ranvier

The results obtained by Tasaki and by Huxley and Stämpfli clearly indicated that the process of generation of action potentials normally takes place at the nodes of Ranvier. Hence, it became important to determine the electrical properties of the nodal membrane. A variety of techniques were developed in the 1950s and 1960s to make possible studies of nodal membrane characteristics without requiring the introduction of an intracellular micropipette that might injure such small-diameter cells. These techniques, called *gap techniques,* are indicated schematically in Figure 5.26.

In these gap techniques, three insulating partitions are used to divide the myelinated fiber into four parts, each in a different pool of extracellular fluid (pools A through D). The insulating partitions consist of air (air gap technique), oil (oil gap technique), or deionized sucrose (sucrose gap technique). Pools A and D are isotonic, but have a high concentration of KCl, which depolarizes the nodes in A and D and decreases the membrane resistance at these two nodes (represented by a short circuit in the figure). By electronic means (using a feedback circuit), the voltage V_{AB} is set to zero. This condition guarantees that I_o and I_i are zero and that V_{BC} equals the potential across the node in pool C (which contains an extracellular solution of normal composition). Hence, the current through the membrane of the node in pool C can be controlled via I and the potential across this membrane can be measured via V_{BC}, or, conversely, the membrane potential of the node can be controlled and the current measured.

Measurements were initially obtained on frog and toad nerve fibers. Complete characterizations of sodium and potassium currents were obtained that were similar to those obtained by Hodgkin and Huxley for squid giant axon. The characterization showed many similarities. For example, in the toad node (Dodge and Frankenhaeuser, 1958, 1959), it was found that the ionic current included three time-varying components:

$$I_{Na}(V_m, t) = \bar{I}_{Na}(V_m)m^2(V_m, t)h(V_m, t),$$

$$I_K(V_m, t) = \bar{I}_K(V_m)n^2(V_m, t), \tag{5.6}$$

$$I_p(V_m, t) = \bar{I}_p(V_m)p^2(V_m, t),$$

where I_p is a sodium current with different kinetic properties than I_{Na}, and m, n, h, and p satisfy first-order kinetic equations. An additional difference of the results obtained in squid axons was that the quantities $\bar{I}_{Na}$, $\bar{I}_K$, and $\bar{I}_p$ were not constants, but showed some nonlinear dependence on V_m.

More recently, techniques have been developed to record membrane currents under voltage clamp from nodes of Ranvier of mammalian peripheral myelinated nerve fibers. Comparison of these with frog and rabbit voltage-clamp currents shows many qualitative similarities, but also some differences. For example, rabbit nodes have very small potassium currents, and the kinetics of sodium currents are faster in rabbit nodes (Figure 5.27). It was found that the measured ionic current of rabbit nodes could be fit only with a sodium current whose form was

$$I_{Na}(V_m, t) = \bar{I}_{Na}(V_m)m^2(V_m, t)h(V_m, t) \tag{5.7}$$

where $\bar{I}_{Na}(V_m)$ is a nonlinear function of V_m, and m and h are governed by first-order kinetics. When these data were used to construct a model of rabbit nerve fiber nodal membrane, the membrane action potential computed from the model resembled that of the measurements (Figure 5.28). The process of generation of membrane action potentials at nodes of Ranvier is qualitatively similar to that at a patch of squid giant axon membrane. This similarity can

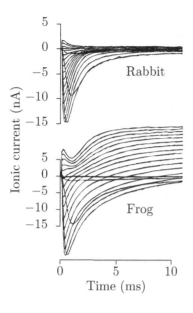

Figure 5.27 Ionic currents in nodes of Ranvier of rabbit and frog myelinated nerve fibers (adapted from Chiu et al., 1979, Figure 4). The measurements were all obtained at 14°C, and the leakage currents have been subtracted. The ionic currents were obtained under voltage clamp for membrane potentials from −80 to +55 mV in equal voltage steps.

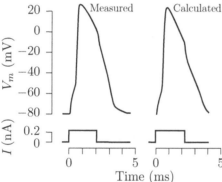

Figure 5.28 Comparison of measured and calculated membrane action potentials of a rabbit node of Ranvier (adapted from Chiu et al., 1979, Figure 11).

be seen in the reconstruction of the events that underlie a membrane action potential at frog and rabbit nodes as shown in Figure 5.29.

5.3.2 Electrical Properties of Internodes

The measurements of Tasaki, Huxley and Stämpfli, and subsequent workers have resulted in estimates of the electrical characteristics of the internodes of myelinated fibers. Typical values of electrical and geometric properties of frog myelinated fibers are given in Table 5.1.

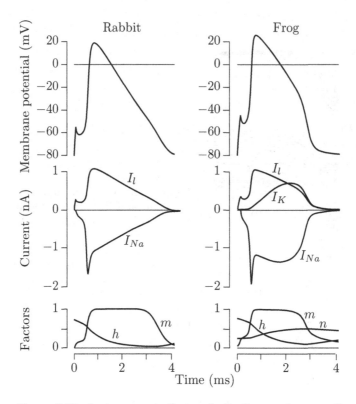

Figure 5.29 Ionic currents that underlie the membrane action potential of frog and rabbit nodes at 14°C (adapted from Chiu et al., 1979, Figure 12). Note that the outward ionic current includes both a leakage and a potassium current in frog nodes, but only a leakage current in rabbit nodes.

5.4 Model of Saltatory Conduction in Myelinated Nerve Fibers

The electrical properties of internodes and nodes have suggested that a myelinated fiber can be represented as a periodic network where each internode is represented by a distributed-parameter model consisting of an RC cable and each node is represented by a lumped-parameter network that has characteristics similar to those of the Hodgkin-Huxley model of a membrane patch (Figure 5.30). With numerical values for the internodal cable parameters and with fits of the membrane nodal current with a model like the Hodgkin-Huxley model, the equations for a myelinated fiber can be solved numerically (Fig-

Table 5.1 Summary of electrical and geometric characteristics of frog myelinated fibers (Hodgkin, 1964) obtained from a number of sources (Huxley and Stämpfli, 1949a; Huxley and Stämpfli, 1951a; Tasaki, 1955).

Fiber geometry	
Fiber diameter (D)	14 μm
Thickness of myelin ($(D - d)/2$)	2 μm
Distance between nodes (L)	2 mm
Estimated area of nodal membrane (πdl)	22 μm^2
Internodal characteristics	
Resistance per unit length of axis cylinder (r_i)	140 MΩ/cm
Resistivity of axoplasm (ρ_i)	110 $\Omega \cdot$cm
Capacity per unit length of myelin sheath (c_m)	10–16 pF/cm
Capacity per unit area of myelin sheath (C_m)	0.0025–0.005 μF/cm^2
Dielectric constant of myelin sheath (κ_m)	5–10
Resistance of a unit area of myelin sheath (R_m)	0.1–0.16 M$\Omega \cdot$cm^2
Resistivity of myelin sheath (ρ_m)	500–800 M$\Omega \cdot$cm
Space constant of internode (λ_C)	4.4–5.6 mm
Time constant of internode (τ_M)	250–800 μs
Nodal characteristics	
Capacity of node ($\pi dl C_{mn}$)	0.6–1.5 pF
Capacity per unit area of nodal membrane (C_{mn})	3–7 μF/cm^2
Resistance of resting node ($R_{mn}/\pi dl$)	40–80 MΩ
Resistance of a unit area of nodal membrane (R_{mn})	10–20 $\Omega \cdot$cm^2
Space constant of node (λ_{mn})	0.2 mm
Time constant of node (τ_{mn})	30–140 μs
Resting and action potential characteristics	
Action potential	116 mV
Resting potential	−71 mV
Peak inward current density	20 mA/cm^2
Conduction velocity	23 m/s

ure 5.31). Note that the membrane potential is within 50% of its peak value for about ten nodes. Hence, the notion that the action potential hops from node to node is a bit misleading. Computation of the membrane currents reveals that an inward current occurs only at the nodes and that the current through the internodes is much smaller than that through a node of Ranvier (Figure 5.32). Thus, the membrane current appears more discontinuous than

Schematic diagram of a myelinated nerve fiber

Block diagram model of a myelinated nerve fiber

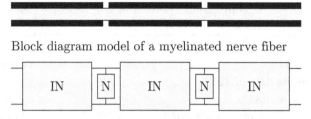

Distributed-parameter (cable) model of an internode

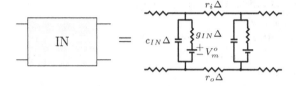

Lumped-parameter model of a node

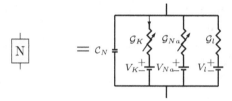

Figure 5.30 Model of a myelinated nerve fiber. The upper panel shows a schematic diagram of a myelinated nerve fiber below which is shown a block diagram in which a myelinated fiber is represented by alternate blocks that represent the node of Ranvier (N) and the internode (IN). Electrical network models of the node and internode are shown in the lower panels.

the membrane potential—i.e., the membrane current is more nearly *saltatory* than is the membrane potential. The computations are at a greater spatial resolution than was possible for the measurements, but are qualitatively similar to the measurements. Note that the membrane current in the internode has two peaks both for the calculations (Figure 5.32) and for the measurements (Figure 5.21). According to the cable model, these two peaks result from the spread of action currents from the two nodes terminating the internode. For each position in the internode, the larger peak in the membrane current corresponds to the nearer node. Taken together, all these results show that the representation of a myelinated fiber as a periodic cable structure with electri-

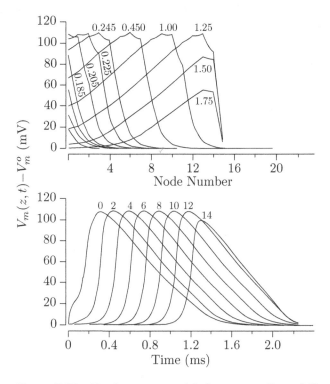

Figure 5.31 Membrane potential along a myelinated fiber computed according to a model of the electrical characteristics of the node and the internode (adapted from Rogart and Ritchie, 1977, Figure 13). The upper panel shows the membrane potential along the fiber at several times indicated on the traces. The positions of the nodes of Ranvier are indicated on the abscissa. The lower panel shows the membrane potential as a function of time at the nodes indicated on each trace. In all traces, the membrane potential deviation from the resting potential is shown.

cally excitable nodes is adequate to explain the normal conduction of action potentials along these fibers.

5.5 Conduction Velocity of Myelinated Nerve Fibers

In contrast to results obtained for unmyelinated fibers that indicate conduction velocity is proportional to the square root of the fiber diameter (see Section 2.4.3), the conduction velocity of myelinated nerve fibers appears to be proportional to fiber diameter. The evidence is based on direct measurements of action potentials obtained from single fibers and on histological

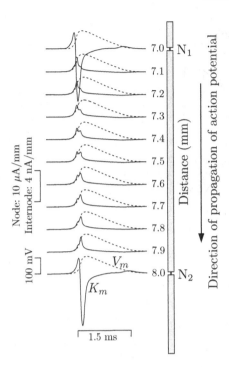

Figure 5.32 Membrane potential (dashed line) and membrane current per unit length (solid line) computed from a model of a myelinated fiber (adapted from Rogart and Ritchie, 1977, Figure 12). Note that the scale for current in the node is 2,500 times that for current in the internode.

determinations of the diameters of the fibers as in Figure 5.33, as well as on measurements of the compound action potential of a population of fibers and histological determinations of the diameters of the largest fibers in the nerve as in Figure 5.34.

The proportionality between conduction velocity and fiber diameter is a simple result, but the explanation is a bit involved because of the complexity of the conduction mechanism in myelinated fibers. The following argument, which is similar to ones given elsewhere (Rushton, 1951; Goldman and Albus, 1968; Fitzhugh, 1973), shows that the proportionality between conduction velocity and fiber diameter follows from the model of a myelinated fiber presented in Section 5.4 provided that the dimensions of fibers of different diameter scale in a particular manner. According to the model, a myelinated fiber consists of periodic internodes represented by continuous cables separated by nodes that are represented by lumped-parameter elements as shown in Figure 5.30.

Assume that the longitudinal direction is the z-direction and that nodes are located at locations $z = nL$, where n ranges over the positive and negative integers. The membrane potential in the internodes is $V_m(z, t)$, and the mem-

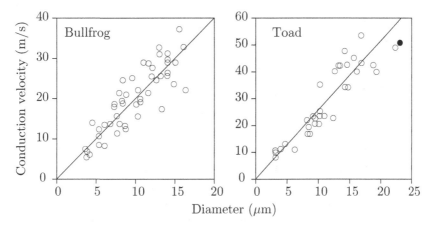

Figure 5.33 Conduction velocity versus fiber diameter for single myelinated nerve fibers in amphibians: left panel for bullfrog at 24°C (adapted from Tasaki, 1953, Figure 65); right panel for the African clawed toad *Xenopus laevis* at 23°C (adapted from Hutchinson et al., 1970, Figure 1). The filled circle (in the right panel) represents the maximum conduction velocity in a whole nerve trunk plotted against the diameter of the largest nerve fiber as determined by histological section. Lines that passed through the origin and best fit the data had the equations $v = 2.0D$ with a correlation coefficient of 0.67 (left panel) and $v = 2.6D$ with a correlation coefficient of 0.88 (right panel), where v is the conduction velocity in m/s and D is the fiber diameter in μm.

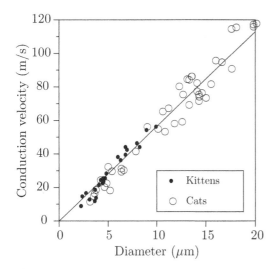

Figure 5.34 Conduction velocity as a function of fiber diameter (adapted from Hursh, 1939, Figure 2). The conduction velocities of nerves in kittens and adult cats were measured for the compound action potential that represents the response of the population of nerve fibers in the nerve. The earliest peak in this potential was assumed to reflect the responses of the fastest-conducting fibers, which were assumed to be the largest fibers in the nerve. Fiber diameter was based on histological determination of the diameter of the largest fiber in each nerve. The line that passed through the origin and best fit the data had the equation $v = 5.7D$ with a correlation coefficient of 0.97, where v is the conduction velocity in m/s and D is the fiber diameter in μm.

brane potential at node n is $V_{mn}(t)$. The equations for the voltage and current must satisfy the following conditions:

■ The membrane potential in the internode must satisfy the cable equation (Equation 3.15). Thus,

$$\lambda_C^2 \frac{\partial^2 V_m(z,t)}{\partial z^2} = V_m(z,t) + \tau_M \frac{\partial V_m(z,t)}{\partial t}, \tag{5.8}$$

where λ_C is the cell space constant and τ_M is the membrane time constant. As we have seen in Chapter 3, for an unmyelinated fiber τ_M is independent of the dimensions of the cable. A similar argument shows that τ_M is independent of the dimensions of a myelinated fiber. If we approximate the myelin in an internode as composed entirely of membrane that has resistivity ρ_m and permittivity ϵ_m, the resistance and capacitance of a unit length of membrane are $r_m = (\rho_m/2\pi) \ln(D/d)$ and $c_m = 2\pi\epsilon_m/\ln(D/d)$, respectively. Thus, $\tau_M = r_m c_m = \rho_m \epsilon_m$, which does not depend upon cable dimensions. However, λ_C does depend upon cable dimensions. From Equation 3.13, $\lambda_C^2 = 1/(g_m(r_i + r_o))$. We shall assume that $r_i \gg r_o$, so that the dependence of the space constant on dimensions is as follows:

$$\lambda_C^2 = \frac{(\rho_m/2\pi)\ln(D/d)}{\rho_i/(\pi(d/2)^2)} = \frac{\rho_m d^2}{8\rho_i} \ln\left(\frac{D}{d}\right). \tag{5.9}$$

We can express Equation 5.8 in terms of a distance scale normalized to the internodal length L. Define $\lambda = z/L$ so that Equation 5.8 becomes

$$\alpha \frac{\partial^2 V(\lambda,t)}{\partial \lambda^2} = V(\lambda,t) + \tau_M \frac{\partial V(\lambda,t)}{\partial t}, \tag{5.10}$$

where $V(\lambda,t) = V_m(z/L,t)$ and

$$\alpha = \left(\frac{\rho_m}{8\rho_i}\right)\left(\frac{d}{L}\right)^2 \ln\left(\frac{D}{d}\right). \tag{5.11}$$

Therefore, the membrane potential expressed in a distance scale normalized to the internodal length has two parameters: α, which depends on the dimensions of the myelinated fiber, and τ_M, which does not.

■ The current-voltage relation for the node is given by

$$I_n = \pi dl \left(C_{mn} \frac{dV_{mn}(t)}{dt} + (J_{ion})_n \right), \tag{5.12}$$

where I_n is the total current through the membrane of node n, $(J_{ion})_n$ is the ionic current density through node n, and C_{mn} is the specific capacitance of the nodal membrane (in $\mu F/cm^2$). The ionic current density through the node depends in a complex fashion on the membrane potential and on time through intermediate activation and inactivation variables (see Section 5.3), whose details are not important for this development. The important property is that the relation between the ionic current density through the nodal membrane and the membrane potential does not depend upon the dimensions of the myelinated fiber; it depends only on the electrical characteristics of nodal membrane.

▪ The membrane potential and membrane currents obey boundary conditions at the nodes. The membrane potential is continuous at the node, so that

$$V_m(nL, t) = V_{mn}(t) \quad \text{or} \quad V(n, t) = V_{mn}(t).\tag{5.13}$$

In addition, Kirchhoff's current law must be satisfied at the node n. Thus, $I_n(t) = I_i(z-, t) - I_i(z+, t)$, where $I_i(z-, t)$ and $I_i(z+, t)$ are the internal longitudinal currents in the internodes to the left and right of node n, respectively. For any z in an internode, $I_i(z, t) = -I_o(z, t)$. This relation and Equation 2.23 give

$$I_n(t) = \frac{1}{(\rho_i/(\pi(d/2)^2))} \left(\left(\frac{\partial V_m(z, t)}{\partial z}\right)_{z=nL+} - \left(\frac{\partial V_m(z, t)}{\partial z}\right)_{z=nL-} \right).\tag{5.14}$$

Equation 5.14 can be written in normalized coordinates as follows:

$$I_n(t) = \frac{\pi d^2}{4\rho_i L} \left(\left(\frac{\partial V(\lambda, t)}{\partial \lambda}\right)_{\lambda=n+} - \left(\frac{\partial V(\lambda, t)}{\partial \lambda}\right)_{\lambda=n-} \right).\tag{5.15}$$

We can combine Equations 5.12 with 5.15 to yield

$$C_m \frac{dV_{mn}(t)}{dt} + (J_{ion})_n = \beta \left(\left(\frac{\partial V(\lambda, t)}{\partial \lambda}\right)_{\lambda=n+} - \left(\frac{\partial V(\lambda, t)}{\partial \lambda}\right)_{\lambda=n-} \right),\tag{5.16}$$

where

$$\beta = \frac{d}{4\rho_i Ll}.\tag{5.17}$$

The boundary condition for the current, when expressed in a distance scale normalized to the internodal length, depends upon two parameters: β, which depends upon fiber dimensions, and C_m, which does not.

Thus, we see that the membrane potential along the model of a myelinated fiber, when expressed in a distance scale normalized to the internodal length, must satisfy Equation 5.10 subject to the boundary conditions given in Equations 5.13 and 5.16. The solution depends upon two parameters, α and β, that depend upon the dimensions of the myelinated fiber as well as several parameters that depend upon the specific properties of the membrane and cytoplasm. Now consider fibers that differ only in dimensions D, d, L, and l, but have identical specific properties; that is, the cytoplasmic resistivity, ρ_i, the membrane time constant of the internode, τ_M, the membrane capacitance of the node, C_{mn}, the membrane resistivity, and the relation between ionic current density in a node and the membrane potential are all identical. That is, we consider fibers of different dimensions that have been made from the same axon membrane and myelin membrane material. Thus, the solutions to the equations for the myelinated fiber expressed in normalized distance scale will be identical if α and β are constants. A sufficient set of conditions to guarantee the constancy of α and β is that d/L, D/d, and l are constants, i.e., if the nodal gap l is constant and if the axon diameter d and the internodal length L are proportional to the fiber diameter D. If these conditions are met, the solutions to the equations for fibers of different dimensions are identical. Therefore, the internodal conduction time will be the same for all these fibers, i.e., the time taken to go a distance z/L is a constant. This result implies that the velocity in z is proportional to L, which is proportional to D. This argument shows that the proportionality of the conduction velocity to the fiber diameter predicted by the model is a consequence of the scaling of certain dimensions of myelinated fibers.

Measurements of the dimensions of myelinated fibers show a great deal of scatter, but are generally consistent with the assumptions that d is proportional to D (Figure 5.9) and that L is proportional to D (Figure 5.10). What happens if these assumptions do not hold? The results of computations on a model of a myelinated fiber are shown in Figure 5.35 for three different assumptions. As expected, the computation of conduction velocity when $L \propto D$ shows a proportionality between conduction velocity and fiber diameter. This result is not new, since, by construction, the model obeys the scaling property given above. Note, however, that if either the internodal length is assumed to be constant or the myelin thickness is assumed to be constant for fibers of different diameter, the proportionality between conduction velocity and fiber diameter is destroyed.

The physical basis for the proportionality between conduction velocity and fiber diameter involves a number of interacting factors. Which factors are critical? To investigate this issue, the conduction velocity was computed for different values of the underlying parameters of a myelinated fiber (Brill et al.,

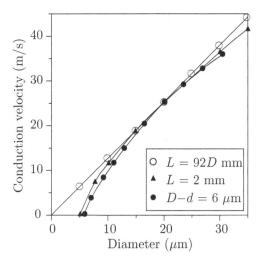

Figure 5.35 Dependence of conduction velocity on fiber diameter computed from a model of conduction in a myelinated nerve fiber (adapted from Goldman and Albus, 1968, Figure 3). Computations are shown for three conditions: proportionality of internodal length to fiber diameter ($L = 92D$ mm with D in μm), constant internodal length ($L = 2$ mm), and constant myelin thickness ($D - d = 6$ μm).

1977; Moore et al., 1978). The conduction velocity was relatively insensitive to changes in a number of nodal parameters including nodal length, nodal capacitance, and nodal ionic conductances. The conduction velocity was also relatively insensitive to some internodal parameters. For example, changes in internodal length around 1–5 mm changed the velocity very little (Figure 5.36). Changes in myelin conductance also had a small effect on conduction velocity. However, the conduction velocity was very sensitive both to the myelin capacitance and to the axoplasm resistance. Both of these factors are important for determining the time it takes to charge the nodal membrane.

This discussion suggests a simple heuristic model of the proportionality of conduction velocity to fiber diameter. Measurements indicate that the time delay of the longitudinal current occurs primarily at the nodes (Figure 5.20). The computations indicate that the time it takes for the action current at one node to charge the membrane potential at the next node apparently is limited by the time it takes to charge the intervening internode and not by the time it takes to charge the node alone. Since the time to charge an internode, T, is determined by the time constant of the internode, which is independent of fiber diameter, T is also independent of fiber diameter provided that the fibers have the same geometry. Therefore, the action potential moves an internodal distance L in time T. Therefore, the conduction velocity of the action potential is $v = L/T$. But since $L \propto D$, the conduction velocity is proportional to the diameter.

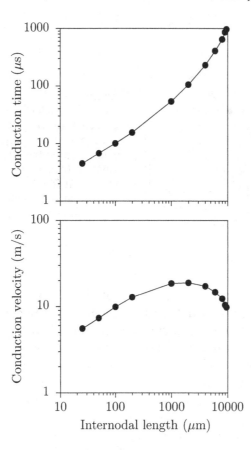

Figure 5.36 Dependence of internodal conduction time and conduction velocity on internodal length computed from a model of conduction in a myelinated nerve fiber (adapted from Brill et al., 1977, Figure 1).

5.6 Causes of Saltatory Conduction

Is saltatory conduction "caused" by the presence of the myelin, or does the myelin simply enhance a mechanism for saltatory conduction that is inherent in differences in the electrical characteristics of the axon membrane between the node and the internode? Several different methods have shown that the axon membrane at the node differs from that at the internode (Waxman and Foster, 1980; Waxman and Ritchie, 1985). These include the differential binding of radioactive TTX and STX, the differential density of particles seen in freeze-fracture studies of membranes, as well as other cytochemical methods. Although estimates based on these studies vary widely, they suggest that the density of sodium channels exceeds 1,000 sites/μm^2 at the node and is less than 25 sites/μm^2 at the internode. Thus, the electrical characteristics of

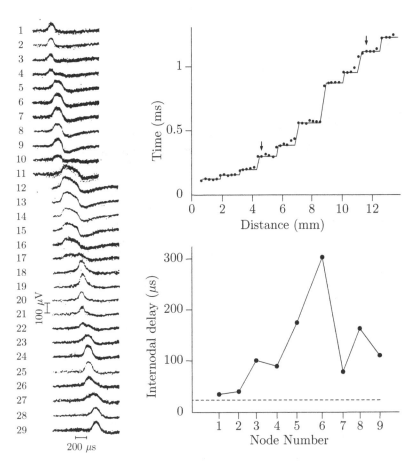

Figure 5.37 Measurement of conduction time versus distance along a partially demyelinated nerve fiber (adapted from Rasminsky and Sears, 1972, Figure 4). The records were obtained in a manner similar to those in Figure 5.25. The arrows in the upper right panel delineate the first and last record shown on the left. The lower right panel shows the conduction time versus distance (same scale as above); successive nodes are indicated. The dashed line shows the normal internodal conduction time.

the node and internode do differ. But can this difference alone account for saltatory conduction? One physiological approach to answering this question has been to attempt to prepare myelinated fibers with the myelin removed. If conduction remained saltatory, this would prove that the myelin may be important, but not necessary to saltatory conduction.

A number of diseases affect myelin in humans (e.g., the demyelinating disease, multiple sclerosis). Chronically demyelinated nerve fibers have been prepared in animal models (e.g., by injecting diptheria toxin systemically), and conduction in such fibers has been investigated and compared with conduction in normal fibers. Figure 5.37 shows the results obtained from chronically, but only partially, demyelinated fibers using recording techniques described in connection with Figure 5.25. In this study, conduction is still saltatory, but the internodal conduction time is greatly increased above that seen in

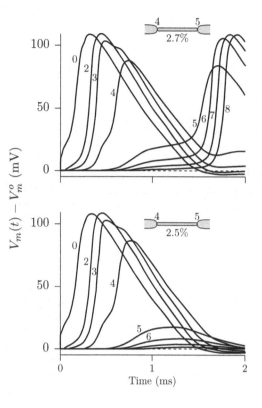

Figure 5.38 Computer simulation of the effects of demyelination on conduction in myelinated nerve fibers (adapted from Koles and Rasminsky, 1972, Figure 3). Computed action potentials are shown at nodes 0, 2-8 as a function of time. In the upper panel, internode 4-5 was assumed to be composed of myelin of 2.7% of normal thickness; in the lower panel, this internode was composed of myelin of 2.5% of the normal thickness. These computations should be compared with those for a normal fiber, as shown in the lower panel of Figure 5.31.

normally myelinated fibers (Figure 5.25). Hence, the conduction velocity is reduced. The effect of reducing or eliminating myelination on conduction of the action potential can be investigated theoretically by increasing the internodal membrane capacitance and decreasing the internodal membrane resistance in models of myelinated fibers and computing the action potential (Figure 5.38). Experimental results such as those shown in Figure 5.37 can be accounted for by such calculations. Both the experimental studies and the theoretical studies show that a reduction in the myelin thickness results in a reduction in the conduction velocity.

These experiments have significance both for the effect of the myelin on conduction velocity and for understanding the pathophysiological basis of demyelinating diseases. However, they have not unambiguously determined whether conduction in demyelinated fibers is saltatory or continuous. A difficulty with experiments on chronically demyelinated nerve fibers is that it takes several days to obtain appreciable demyelination. While continuous conduction has been observed in chronically demyelinated fibers (Bostock and Sears, 1976), it may have resulted from a redistribution of ion channels normally segregated at the nodes (Foster et al., 1980).

Some of the technical problems with chronically demyelinated fibers are circumvented by applying lysolecithin, a detergent that produces demyelination, to produce acutely demyelinated nerve fibers in about 45 minutes. With this technique, it has been possible to produce different demyelinated nerve fiber preparations, including preparations consisting of a single demyelinated internode with the nodes at either end removed. With these methods, the differences in composition of ion channels at the node and internode have been demonstrated more directly (Chiu and Ritchie, 1981, 1982; Grissmer, 1986; Chiu, 1987a; Shrager, 1987; Jonas et al., 1989; Röper and Schwarz, 1989). The conclusion of these studies is that the axonal membrane at the node of Ranvier is richly populated with sodium channels and only sparsely populated with potassium channels, whereas the membrane at the internode has a complementary composition of ion channels. Present evidence suggests that both the inhomogeneity in the composition of ion channels at the node of Ranvier and in the internode and the effect of the myelin on the cable properties of the internode are important in producing saltatory conduction.

5.7 Summary

Figure 5.39 is a summary of the key features that are responsible for saltatory conduction. The density of voltage-gated sodium channels is much higher at nodes of Ranvier than in the internodes. Therefore, a large inward sodium current occurs only at the nodes. This inward sodium current is regenerative and is the basis of the action potential in myelinated axons. The myelin sheath increases the space constant of the internode so that the membrane potential at one node is not appreciably reduced at the next node. In fact, current spread from one node is sufficient to excite nodes that are several internodes away. This *safety factor* implies that action potential propagation can proceed past a succession of even five or six inexcitable nodes. The safety factor gives

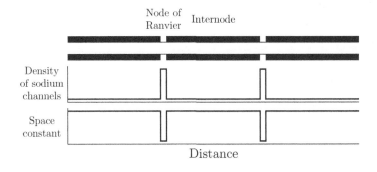

Figure 5.39 Schematic summary of two factors known to be important in saltatory conduction.

propagation of action potential in myelinated fibers a degree of robustness against local fiber pathologies.

Saltatory conduction in myelinated fibers achieves a much higher conduction velocity for a given fiber diameter than occurs in unmyelinated fibers. Thus, vertebrates have evolved fibers that can transmit action potentials at a velocity of over 100 m/s in a fiber that is only 20 μm in diameter. In contrast, the giant axon of the squid, with a diameter of 0.5 mm, transmits action potentials at only about 20 m/s. Thus, myelinated fibers use less space for a given conduction velocity.

For the same conduction velocity, myelinated fibers have a much smaller diameter than do unmyelinated fibers. Hence, these myelinated fibers have a smaller reserve of internal ions. If action potentials were to occur along the entire length of such a fiber, the metabolic load required to maintain concentrations of ions would be large. However, the sodium inward current is restricted predominantly to the nodes, which constitute only a small fraction of the surface area of the fiber. Hence, conduction in myelinated fibers is energetically efficient. The metabolic energy required to maintain concentrations, which are unbalanced by the passage of an action potential, is less than that which would be required if conduction were continuous rather than saltatory.

Exercises

5.1 The membrane current through a myelinated fiber at the internode differs from that at the node of Ranvier, while the membrane potential is quite similar at the two locations. For both the internode and the node of Ranvier, determine whether the relation between membrane current and membrane potential given in Figure 5.23 is qualitatively consistent with the cable model or not. Briefly discuss the significance of your finding for the mode of conduction of action potentials in myelinated fibers.

5.2 Explain why the membrane current through the internode of a myelinated nerve fiber contains two peaks of outward current as an action potential propagates down the fiber (Figure 5.17).

5.3 Estimate the conduction velocity of the action potential from the measurements on a single myelinated nerve fiber shown in Figures 5.19–5.21.

5.4 State whether each of the following is true or false, and give a reason for your answer.

a. Myelinated fibers conduct action potentials more rapidly than do un-myelinated fibers.

b. The action potential of a myelinated nerve fiber hops from node to node.

c. The action current of a myelinated nerve fiber hops from node to node.

d. Saltatory conduction results because the internodes are covered by the insulating myelin, while the nodes are not.

5.5 Define the safety factor.

5.6 What is the range of diameters of vertebrate myelinated nerve fibers? What is the range of diameters of vertebrate unmyelinated nerve fibers?

5.7 Why are there cusps (points of discontinuity in the derivative) in the spatial dependence of the membrane potential, but not in the temporal dependence of the membrane potential shown in Figure 5.31?

5.8 In the first paragraph of the quotation at the beginning of this chapter, Tasaki poses a question about the significance of the effects of narcotics on the conduction velocity of a nerve fiber. What is the answer to the question posed?

5.9 Suppose the density of sodium channels in a myelinated nerve fiber of the bullfrog sciatic nerve is 1,000 channels/μm^2 at the node of Ranvier and 25 channels/μm^2 in the axonal membrane in the internode. For bullfrog sciatic nerve fiber with a fiber diameter (D) of 15 μm:

a. Estimate the number of sodium channels in a single node of Ranvier.

b. Estimate the number of sodium channels in a single internode.

c. Briefly discuss the significance of your findings for saltatory conduction in this myelinated nerve fiber.

Problems

5.1 A myelinated axon, shown schematically in Figure 5.8, has the dimensions $L = 2$ mm, $D = 14$ μm, $d = 10$ μm, and $l = 0.7$ μm. The resistivity of the cytoplasm $\rho_i = 110$ $\Omega \cdot$cm, and the resistance of the extracellular space can be assumed to be negligible. The membrane potential, V_m, and the current per unit length, K_m, are shown in Figure 5.40 for a location

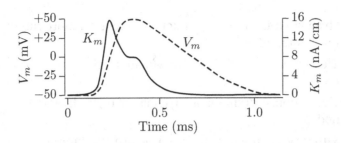

Figure 5.40 Membrane potential and current at an internode (Problem 5.1).

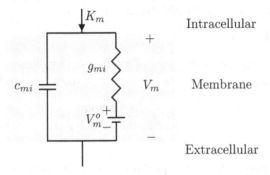

Figure 5.41 Model of voltage-current relation at an internode (Problem 5.1).

at an internode as an action potential propagates down the fiber. Assume that the internode can be represented by a linear cable with the equivalent membrane model shown in Figure 5.41.

a. Estimate the conductance, g_{mi}, and capacitance, c_{mi}, per unit length of internode from the data given. (Hint: the capacitance current is zero when $dV_m/dt = 0$ and largest when dV_m/dt is large.)

b. Find the values of the membrane time constant, τ_{Mi}, and the axon space constant, λ_{Ci}, of the internode.

c. Find the specific conductance, G_{mi}, and the capacitance, C_{mi}, per unit area of internodal myelin.

d. Given that the myelin is composed of 150 lamellae, find the specific conductance, G_m, and the capacitance, C_m, per unit area for a single layer of myelin membrane. How do these values compare to those of unmyelinated fibers?

e. Now consider an unmyelinated fiber whose diameter is 10 μm and whose membrane has a specific capacitance, C_m, and a conductance, G_m, per unit area equal to that found in part d. What is the time constant, τ_M, and the space constant, λ_C, of this fiber? Compare these

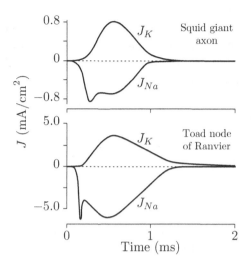

Figure 5.42 Comparison of ionic currents during an action potential for an unmyelinated squid giant axon and a myelinated toad node of Ranvier (Problem 5.2). These ionic currents are based on calculations of models of the squid giant axon (adapted from Cooley and Dodge, 1965, Figure 2.4) and toad node of Ranvier (adapted from Frankenhaeuser and Huxley, 1964, Figure 6).

results to those found in part b. What is the physiological significance of the difference?

5.2 A squid giant axon (which is an unmyelinated axon) has a diameter of 500 μm. The ionic currents during the passage of one action potential are shown in Figure 5.42. The normal internal concentration of sodium is 40 mmol/L. In contrast, consider a frog myelinated fiber for which the axon diameter (not including the myelin) is 10 μm, the fiber diameter (including the myelin) is 14 μm, the internodal length is 2 mm, the nodal length is 0.7 μm, and the nodal area is 22 μm^2. We shall assume that action potentials occur only at the nodes. The ionic currents at the node of Ranvier during the passage of an action potential are also shown in Figure 5.42. You may assume that the sodium current is negligible in the internodes. The normal internal concentration of sodium is 10 mmol/L in the frog fiber. Both the squid unmyelinated fiber and the frog myelinated fiber conduct action potentials with about the same conduction velocity. This problem concerns the energetic efficiency of these two fibers.

a. Compute the number of moles of sodium entering each fiber per action potential per unit length of fiber.

b. Assume that the energy expended to pump the accumulated sodium out of the cell can be measured in terms of the number of ATP molecules hydrolyzed to ADP, and assume that 3 moles of Na$^+$ are transported out of the axon for every mole of ATP hydrolyzed to ADP inside the axon by the $(Na^+-K^+)ATPase$ pump. Find the ratio of energy expended per unit length per action potential in order to pump out the accumulated sodium for the squid unmyelinated fiber to that for the frog myelinated fiber.

c. Describe the advantages of the frog myelinated fiber over the squid unmyelinated fiber.

5.3 Figure 5.31 shows the membrane potential, $V_m(z, t)$, and Figure 5.32 shows the membrane current per unit length, $K_m(z, t)$, associated with a propagating action potential in a model of a myelinated nerve fiber. In Figure 5.32, the membrane potential and the current per unit length are shown at different points along a single myelinated nerve fiber during a propagated action potential. The scales for currents and voltages are shown at the left. Note that the nodal current scale is much greater than the internodal current scale, so nodal current density is enormous in comparison to internodal current density. Upward deflection represents outward current. In Figure 5.31 (upper panel), the membrane potential along a myelinated nerve fiber is plotted versus position along the fiber at a variety of times as indicated. Distance along the fiber is indicated by node number. In Figure 5.31 (lower panel), the membrane potential at alternate nodes along a myelinated nerve fiber is plotted as a function of time. Use the plots and your knowledge of nerve fiber models to answer the questions below. Consider the locations between nodes 2 and 12. Assume that the nodes are uniformly spaced 1 mm apart.

a. Estimate the conduction velocity of the action potential.

b. Is $V_m(z, t) = f(t - z/v)$? Explain your answer briefly.

c. Explain the feature of the electrical behavior of the node (as compared to the internode) that causes the apparent discontinuity in $\partial V_m(z, t)/\partial z$ that occurs at the nodes for some values of V_m.

d. Suppose that two adjacent nodes lost their ability to change conductance in response to a potential change. Describe qualitatively what might happen when an action potential arrived at these nodes. Would an action potential occur at locations beyond the altered nodes? Explain your thinking as precisely as possible. Be brief and to the point.

5.4 This problem involves the analysis of a simple model of the dependence of the conduction velocity, v, of a myelinated fiber on the length of the internodes, L, with all other parameters of the myelinated fiber held

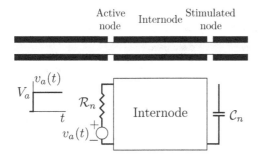

Figure 5.43 Electrical network model of a myelinated nerve fiber (Problem 5.4).

fixed. In this model, the time it takes to generate an action potential at a node is assumed to result from the time it takes for one node to charge the membrane capacitance of a contiguous node until some threshold potential is reached. A schematic diagram of a single internode separating two nodes is shown in Figure 5.43. To simplify the analysis, assume that (1) the myelin is a perfect insulator, i.e., no current flows through the membrane in the internode; (2) the active node is represented by a network with a nodal resistance, R_n ohms, in series with a step change in the membrane potential from its resting value of amplitude V_a volts; (3) the stimulated node is represented by the nodal capacitance, C_n farads, and the electrical load of the internode to the right of the stimulated node is assumed to be negligible; (4) the internal longitudinal resistance per unit length of axoplasm in the internode, r_i Ω/cm, is much larger than the external resistance per unit length, r_o Ω/cm; and (5) the stimulated node produces an action potential when the membrane potential change equals V_{th}.

a. Draw an equivalent electrical network that represents the active and stimulated nodes as well as the intervening internode.

b. Find the time of occurrence of the action potential at the stimulated node in terms of the internodal length and the other parameters specified above.

c. Find and sketch the dependence of the conduction velocity on the internodal length, and explain the physical significance of your answer.

d. Which of assumptions 1 to 5 would you expect to represent myelinated fibers most poorly for large values of L? Explain your choice, and briefly suggest how you would examine the effects of removing this assumption. Clearly indicate (on the sketch in part c) how removal of this assumption would affect the dependence of conduction velocity on L.

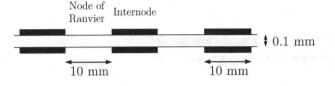

Figure 5.44 Schematic diagram of a myelinated nerve fiber (Problem 5.5).

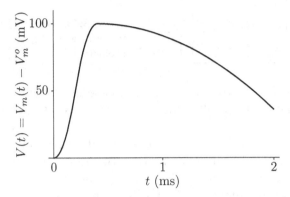

Figure 5.45 Membrane potential waveform (Problem 5.5).

5.5   Nobel P. Seeker has discovered that the species *Saltatoris giganticus* contains giant myelinated nerve fibers with unusually large nodes of Ranvier (see Figure 5.44). Nobel is intent on testing the hypothesis that the membrane potential in this myelinated fiber is conducted decrement free (by propagated action potentials) in the nodal region and decrementally (according to the cable model) in the internodal region. You are a member of Nobel's crack staff of scientists and are asked to determine the membrane current per unit length in the nodal and internodal regions that would be predicted by this hypothesis. To simplify the calculations, the membrane potential waveshapes are assumed to be identical (except for a translation of the time origin) in the nodal and internodal regions, and this waveshape is represented by the piecewise polynomial approximation shown in Figure 5.45, where $V(t)$ is the difference of the membrane potential from its resting value. $V(t)$ is defined as

$$V(t) = \begin{cases} 1250t^2, & 0 \le t < 0.2, \\ 50 + 500(t - 0.2) - 1250(t - 0.2)^2, & 0.2 \le t < 0.4, \\ 100 - 25(t - 0.4)^2, & 0.4 \le t < 2.0, \end{cases}$$

where $V(t)$ is expressed in mV and t in ms. The internodal membrane (including the myelin) has a capacitance and conductance per unit length of 100 pF/cm and 100 nS/cm, respectively. The conduction velocity of

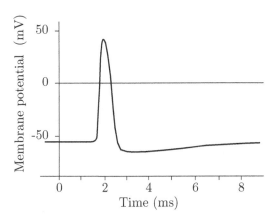

Figure 5.46 Membrane potential of an unmyelinated nerve fiber at a fixed position along the fiber plotted versus time (Problem 5.6).

the action potential in the nodal region is 100 m/s. The internal resistance per unit length of axoplasm is 2.5 MΩ/cm and greatly exceeds the external resistance per unit length of axon.

a. Determine and sketch the membrane current per unit length of axon in the internode. Carefully state your assumptions and approximations (if any).

b. Determine and sketch the membrane current per unit length of axon in the node of Ranvier. Carefully state your assumptions and approximations (if any).

c. Is it reasonable to assume that the membrane potential waveshapes are similar in the nodal and internodal regions even though the mechanism of conduction is different in these two regions? Explain your answer in a quantitative manner.

5.6 This problem deals with comparisons between the propagation of action potentials in unmyelinated (part a) and myelinated (part b) nerve fibers.

a. The membrane potential, $V_m(z_0, t)$, of a propagated action potential at position z_0 along an invertebrate unmyelinated axon of diameter 10 μm is shown as a function of time in Figure 5.46. The action potential propagates in the $+z$-direction with velocity 2 m/s.

   i. Sketch the membrane potential $V_m(z, t_0)$ as a function of position z at some time t_0.

   ii. Determine the distance along the axon over which the membrane potential is positive.

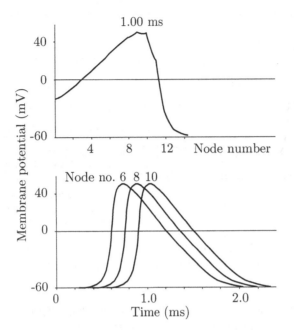

Figure 5.47 Membrane potential of a myelinated nerve fiber (Problem 5.6). The membrane potential is plotted as a function of position (upper panel) at a time that is 1 ms after initiation of an action potential at node 1. The distance scale is in units of internodal distance, and the node numbers are indicated on the abscissa. Note that there are cusps in the membrane potential as a function of position and that these cusps occur at the nodes of Ranvier. The membrane potential is plotted versus time (lower panel) for nodes 6, 8, and 10.

b. Figure 5.47 shows the membrane potential, $V_m(z, t)$, for a propagating action potential of a myelinated nerve fiber whose (axon) diameter is 10 μm. Assume that the nodes of Ranvier are uniformly spaced $L = 1.38$ mm apart.

    i.   Estimate the conduction velocity of the action potential. Clearly indicate your method of estimation with a sketch.

    ii.   Does $V_m(z, t)$ satisfy the wave equation? Explain briefly in terms of the results shown in Figure 5.47.

    iii. For which values of z, if any, does the relation $V_m(z, t) = V_m(t - z/v)$ apply? Explain your answer briefly.

    iv. Determine the distance along the myelinated fiber over which the membrane potential is positive.

5.7   The membrane potential and membrane current per unit length at three sites in an internode of a myelinated fiber are shown in Figure 5.48. The relations between the traces and the locations are unknown a priori.

a. If Trace 1 was recorded at location A, what is the direction of propagation of the action potential — i.e., in the positive z-direction or the negative z-direction?

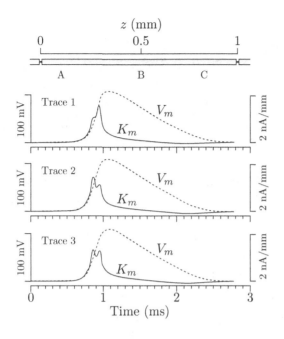

Figure 5.48 Membrane potential and membrane current in an internode (Problem 5.7). The measurement locations are A, B, and C. Traces 1, 2, and 3 are three pairs of traces obtained at different locations.

b. If the action potential were propagated in the positive z-direction, determine the correspondence between the traces and the recording locations. Explain your choice.

c. Estimate the conduction velocity of the action potential of this fiber. Explain your method.

5.8 Assume that myelin can be represented electrically by a homogeneous medium with resistivity ρ_m $\Omega \cdot$cm and permittivity ϵ F/cm. Assume that the myelin is a cylindrical sleeve on the unmyelinated axon with inner diameter d and outer diameter D. Show that the resistance and the capacitance of a unit length of myelinated fiber are $r_m = (\rho_m/2\pi)\ln(D/d)$ and $c_m = 2\pi\epsilon/\ln(D/d)$.

5.9 In this problem, you will explore the internodal cable properties of myelinated fibers of a fixed outer diameter D but with different myelin thicknesses. You may assume that the longitudinal resistance per unit length of inner conductor greatly exceeds that of the external conductor, i.e., $r_i \gg r_o$.

a. Determine the value of d/D that maximizes the internodal space constant λ_C.

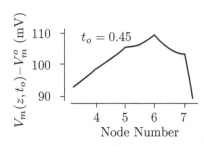

Figure 5.49 Membrane potential along a myelinated fiber computed from a model of electrical characteristics of the node and internode (Figure 5.31) (Problem 5.10). The membrane potential is plotted as a function of distance along the fiber (expressed in units of internodal length, where $L = 1.38$ mm).

b. Give a physical, not a mathematical, explanation of why there is a maximum in the space constant as a function of d/D for fixed values of D.

c. Explain the significance of this maximum for the conduction velocity of a myelinated nerve fiber.

d. Compare the value of d/D you computed in part a with the results shown in Figure 5.9. Briefly discuss any differences.

5.10 Figure 5.49 shows a detail of a propagating action potential calculated using a model of a myelinated nerve fiber (Figure 5.31).

a. Describe a method by which the data in Figure 5.49 could be analyzed to estimate the current, I_m, flowing out of a node. Apply your method to calculate the current flowing out of node 6 at $t = 0.45$ ms. Assume that $r_i = 140 \text{M}\Omega/\text{cm} \gg r_o$.

b. Describe a method by which the data in Figure 5.49 could be analyzed to estimate the current density, K_m, flowing out of an internode. Apply your method to determine whether current is flowing into or out of the internode between nodes 5 and 6 at $t_0 = 0.45$ ms.

References

Books and Reviews

Eccles, J. C. (1973). *The Understanding of the Brain.* McGraw-Hill, New York.

Hodgkin, A. L. (1964). *The Conduction of the Nervous Impulse.* Charles C. Thomas, Springfield, MA.

Keynes, R. D. and Aidley, D. J. (1991). *Nerve and Muscle.* Cambridge University Press, Cambridge, England.

Landon, D. N. (1976). *The Peripheral Nerve.* Chapman and Hall, London.

Landon, D. N. and Hall, S. (1976). The myelinated nerve fibre. In Landon, D. N., ed., *The Peripheral Nerve*, 1–105. Chapman and Hall, London.

Morell, P. (1977). *Myelin*. Plenum, New York.

Peters, A., Palay, S. L., and Webster, H. F. (1991). *The Fine Structure of the Nervous System*. Oxford University Press, New York.

Porter, K. R. and Bonneville, M. A. (1964). *An Introduction to the Fine Structure of Cells and Tissues*. Lea & Febiger, New York.

Raine, C. S. (1977). Morphological aspects of myelin and myelination. In Morell, P., ed., *Myelin*, 1–49. Plenum, New York.

Schadé, J. P. and Ford, D. H. (1965). *Basic Neurology*. Elsevier, New York.

Stämpfli, R. (1954). Saltatory conduction in nerve. *Physiol. Rev.*, 34:101–112.

Stämpfli, R. and Hille, B. (1976). Electrophysiology of the peripheral myelinated nerve. In Llinás, R. and Precht, W., eds., *Frog Neurobiology*, 1–32. Springer-Verlag, New York.

Tasaki, I. (1953). *Nervous Transmission*. C. C. Thomas, Springfield, Il.

Tasaki, I. (1959). Conduction of the nerve impulse. In Field, J., ed., *Handbook of Physiology*, sec. 1, *Neurophysiology*, vol. 1, 75–121. American Physiological Society, Washington, DC.

Tasaki, I. (1968). *Nerve Excitation*. C. C. Thomas, Springfield, Il.

Waxman, S. G. (1978). *Physiology and Pathobiology of Axons*. Raven, New York.

Waxman, S. G. (1983). Action potential propagation and conduction velocity: New perspectives and questions. *Trends Neurosci.*, 6:157–161.

Zagoren, J. C. and Fedoroff, S. (1984). *The Node of Ranvier*. Academic Press, New York.

Original Articles

Adrian, E. D. and Bronk, D. W. (1928). The discharge of impulses in motor nerve fibres: I. Impulses in single fibres of the phrenic nerve. *J. Physiol.*, 66:81–101.

Arbuthnott, E. R., Ballard, K. J., Boyd, I. A., and Kalu, K. U. (1980). Quantitative study of the non-circularity of myelinated peripheral nerve fibres in the cat. *J. Physiol.*, 308:99–123.

Arbuthnott, E. R., Boyd, I. A., and Kalu, K. U. (1980). Ultrastructural dimensions of myelinated peripheral nerve fibres in the cat and their relation to conduction velocity. *J. Physiol.*, 308:125–157.

Bostock, H. and Sears, T. A. (1976). Continuous conduction in demyelinated mammalian nerve fibers. *Nature*, 263:786–787.

Boyd, I. A. and Kalu, K. U. (1979). Scaling factor relating conduction velocity and diameter for myelinated afferent nerve fibres in the cat hind limb. *J. Physiol.*, 289:277–297.

Brill, M. H., Waxman, S. G., Moore, J. W., and Joyner, R. W. (1977). Conduction velocity and spike configuration in myelinated fibres: Computed dependence on internode distance. *J. Neurol. Neurosurg. Psychiatry*, 40:769–744.

Brismar, T. (1973). Effects of ionic concentration on permeability properties of nodal membrane in myelinated nerve fibres on *Xenopus laevis*. Potential clamp experiments. *Acta Physiol. Scand.*, 87:474–484.

Brismar, T. (1980). Potential clamp analysis of membrane currents in rat myelinated nerve fibres. *J. Physiol.*, 298:171–184.

Brismar, T. and Frankenhaeuser, B. (1981). Potential clamp analysis of mammalian myelinated fibres. *Trends Neurosci.*, 4:68–70.

Carley, L. R. and Raymond, S. A. (1987). Comparison of the after-effects of impulse conduction on threshold at nodes of Ranvier along single frog sciatic axons. *J. Physiol.*, 386:503–527.

Chiu, S. Y. (1987a). Sodium and potassium currents in acutely demyelinated internodes of rabbit sciatic nerves. *J. Physiol.*, 391:631–649.

Chiu, S. Y. (1987b). Sodium currents in axon-associated Schwann cells from adult rabbits. *J. Physiol.*, 386:181–203.

Chiu, S. Y. (1993). Differential expression of sodium channels in acutely isolated myelinating and non-myelinating Schwann cells of rabbits. *J. Physiol.*, 470:485–499.

Chiu, S. Y. and Ritchie, J. M. (1980). Potassium channels in nodal and internodal axonal membrane of mammalian myelinated fibres. *Nature*, 284:170–171.

Chiu, S. Y. and Ritchie, J. M. (1981). Evidence for the presence of potassium channels in the paranodal region of acutely demyelinated mammalian single nerve fibres. *J. Physiol.*, 313:415–437.

Chiu, S. Y. and Ritchie, J. M. (1982). Evidence for the presence of potassium channels in the internode of frog myelinated nerve fibres. *J. Physiol.*, 322:485–501.

Chiu, S. Y. and Ritchie, J. M. (1984). On the physiological role of internodal potassium channels and the security of conduction in myelinated nerve fibres. *Proc. R. Soc. London, Ser. B*, 220:415–422.

Chiu, S. Y., Ritchie, J. M., Rogart, R. B., and Stagg, D. (1979). A quantitative description of membrane currents in rabbit myelinated nerve. *J. Physiol.*, 292:149–166.

Cooley, J. W. and Dodge, F. A. (1965). Digital computer solutions for excitation and propagation of the nerve impulse. Technical report RC 1496, IBM Research.

Dodge, F. A. and Frankenhaeuser, B. (1958). Membrane currents in isolated frog nerve fibre under voltage clamp conditions. *J. Physiol.*, 143:76–90.

Dodge, F. A. and Frankenhaeuser, B. (1959). Sodium currents in the myelinated nerve fibre of *Xenopus laevis* investigated with the voltage clamp technique. *J. Physiol.*, 148:188–200.

England, J. D., Gamboni, F., Levinson, S. R., and Finger, T. E. (1990). Changed distribution of sodium channels along demyelinated axons. *Proc. Natl. Acad. Sci. U.S.A.*, 87:6777–6780.

Fitzhugh, R. (1962). Computation of impulse initiation and saltatory conduction in a myelinated nerve fiber. *Biophys. J.*, 2:11–21.

Fitzhugh, R. (1973). Dimensional analysis of nerve models. *J. Theor. Biol.*, 40:517–541.

Foster, R. E., Whalen, C. C., and Waxman, S. G. (1980). Reorganization of the axon membrane in demyelinated peripheral nerve fibers: Morphological evidence. *Science*, 210:661–663.

Fraher, J. P. (1972). A quantitative study of anterior root fibres during early myelination. *J. Anat.*, 112:99–124.

Fraher, J. P. (1973). A quantitative study of anterior root fibres during early myelination: II. Longitudinal variation in sheath thickness and axon circumference. *J. Anat.*, 115:421–444.

Fraher, J. P. (1976). The growth and myelination of central and peripheral segments of ventral motoneurone axons: A quantitative ultrastructural study. *Brain Res.*, 105:193–211.

Fraher, J. P. (1978a). Quantitative studies on the maturation of central and peripheral parts of individual ventral motoneuron axons: I. Myelin sheath and axon calibre. *J. Anat.*, 126:509–533.

Fraher, J. P. (1978b). Quantitative studies on the maturation of central and peripheral parts of individual ventral motoneuron axons: II. Internodal length. *J. Anat.*, 127:1–15.

Fraher, J. P., Kaar, G. F., Bristol, D. C., and Rossiter, J. P. (1988). Development of ventral spinal motoneurone fibres: A correlative study of the growth and maturation of central and peripheral segments of large and small fibre classes. *Prog. Neurobiol.*, 31:199–239.

Frankenhaeuser, B. (1957a). The effect of calcium on the myelinated nerve fibre. *J. Physiol.*, 137:245–260.

Frankenhaeuser, B. (1957b). A method for recording resting and action potentials in the isolated myelinated nerve fibre of the frog. *J. Physiol.*, 135:550–559.

Frankenhaeuser, B. (1959). Steady state inactivation of sodium permeability in myelinated nerve fibres of *Xenopus laevis*. *J. Physiol.*, 148:671–676.

Frankenhaeuser, B. (1960a). Quantitative description of sodium currents in myelinated nerve fibres of *Xenopus laevis*. *J. Physiol.*, 151:491–501.

Frankenhaeuser, B. (1960b). Sodium permeability in toad nerve and in squid nerve. *J. Physiol.*, 152:159–166.

Frankenhaeuser, B. (1962a). Delayed currents in myelinated nerve fibres of *Xenopus laevis* investigated with voltage clamp technique. *J. Physiol.*, 160:40–45.

Frankenhaeuser, B. (1962b). Instantaneous potassium currents in myelinated nerve fibres of *Xenopus laevis*. *J. Physiol.*, 160:46–53.

Frankenhaeuser, B. (1962c). Potassium permeability in myelinated nerve fibres of *Xenopus laevis*. *J. Physiol.*, 160:54–61.

Frankenhaeuser, B. (1963a). Inactivation of the sodium-carrying mechanism in myelinated nerve fibres of *Xenopus laevis*. *J. Physiol.*, 169:445–451.

Frankenhaeuser, B. (1963b). A quantitative description of potassium currents in myelinated nerve fibres of *Xenopus laevis*. *J. Physiol.*, 169:424–430.

Frankenhaeuser, B. (1965). Computed action potential in nerve from *Xenopus laevis*. *J. Physiol.*, 180:780–787.

Frankenhaeuser, B. and Huxley, A. F. (1964). The action potential in the myelinated nerve fibre of *Xenopus laevis* as computed on the basis of voltage clamp data. *J. Physiol.*, 171:302–325.

Frankenhaeuser, B. and Moore, L. E. (1963a). The effect of temperature on the sodium and potassium permeability changes in myelinated nerve fibres of *Xenopus laevis*. *J. Physiol.*, 169:431–437.

Frankenhaeuser, B. and Moore, L. E. (1963b). The specificity of the initial current in myelinated nerve fibres of *Xenopus laevis*: Voltage clamp experiments. *J. Physiol.*, 169:438–444.

Friede, R. L. and Samorajski, T. (1967). Relation between the number of myelin lamellae and axon circumference in fibers of vagus and sciatic nerves of mice. *J. Comp. Neur.*, 130:223–232.

Gasser, H. S. and Grundfest, H. (1939). Axon diameters in relation to the spike dimensions and the conduction velocity in mammalian A fibers. *Am. J. Physiol.*, 127:393–414.

Goldman, L. and Albus, J. S. (1968). Computation of impulse conduction in myelinated fibers: Theoretical basis of the velocity-diameter relation. *Biophys. J.*, 8:596–607.

Grissmer, S. (1986). Properties of potassium and sodium channels in frog internode. *J. Physiol.*, 381:119–134.

Hahn, A. G., Chang, Y., and Webster, H. F. (1987). Development of myelinated nerve fibers in the sixth cranial nerve of the rat: A quantitative electron microscope study. *J. Comp. Neurol.*, 260:491–500.

Halter, J. A. and Clark, J. W., Jr. (1991). A distributed-parameter model of the myelinated nerve fiber. *J. Theor. Biol.*, 148:345–382.

Hardy, W. L. (1973a). Propagation speed in myelinated nerve: I. Experimental dependence on external Na^+ and on temperature. *Biophys. J.*, 13:1054–1070.

Hardy, W. L. (1973b). Propagation speed in myelinated nerve: II. Theoretical dependence on external Na^+ and on temperature. *Biophys. J.*, 13:1071–1089.

Hildebrand, C., Remahl, S., Persson, H., and Bjartmar, C. (1993). Myelinated nerve fibres in the CNS. *Prog. Neurobiol.*, 40:319–384.

Hodler, J., Stämpfli, R., and Tasaki, I. (1952). Role of potential wave spreading along myelinated nerve fiber in excitation and conduction. *Am. J. Physiol.*, 170:375–389.

Hursh, J. B. (1939). Conduction velocity and diameter of nerve fibers. *Am. J. Physiol.*, 127:131–139.

Hutchinson, N. A., Koles, Z. J., and Smith, R. S. (1970). Conduction velocity in myelinated nerve fibres of *Xenopus laevis. J. Physiol.*, 208:279–289.

Huxley, A. F. and Stämpfli, R. (1949a). Evidence for saltatory conduction in peripheral myelinated nerve fibres. *J. Physiol.*, 108:315–339.

Huxley, A. F. and Stämpfli, R. (1949b). Saltatory transmission of the nervous impulse. *Arch. Sci. Physiol.*, 3:435–448.

Huxley, A. F. and Stämpfli, R. (1951a). Direct determination of membrane resting potential and action potential in single myelinated nerve fibres. *J. Physiol.*, 112:476–495.

Huxley, A. F. and Stämpfli, R. (1951b). Effect of potassium and sodium on resting and action potentials of single myelinated nerve fibres. *J. Physiol.*, 112:496–508.

Jonas, P., Bräu, M. E., Hermsteiner, M., and Vogel, W. (1989). Single-channel recording in myelinated nerve fibers reveals one type of Na channel but different K channels. *Proc. Natl. Acad. Sci. U.S.A.*, 86:7238–7242.

Koles, Z. J. and Rasminsky, M. (1972). A computer simulation of conduction in demyelinated nerve fibres. *J. Physiol.*, 227:351–364.

Moore, J. W., Joyner, R. W., Brill, M. H., Waxman, S. G., and Najar-Joa, M. (1978). Simulations of conduction in uniform myelinated fibers: Relative sensitivity to changes in nodal and internodal parameters. *Biophys. J.*, 21:147–160.

Nonner, W. (1980). Relations between the inactivation of sodium channels and the immobilization of gating charge in frog myelinated nerve. *J. Physiol.*, 299:573–603.

Rasminsky, M. and Sears, T. A. (1972). Internodal conduction in undissected demyelinated nerve fibres. *J. Physiol.*, 227:323–350.

Ritchie, J. M. (1982). On the relation between fibre diameter and conduction velocity in myelinated fibres. *Proc. R. Soc. London, Ser. B*, 217:29–35.

Ritchie, J. M., Rang, H. P., and Pellegrino, R. (1981). Sodium and potassium channels in demyelinated and remyelinated mammalian nerve. *Nature*, 294:257–259.

Robertson, J. D. (1960). The molecular structure and contact relationships of cell membranes. *Prog. Biophys. Biophys. Chem.*, 10:344–418.

Rogart, R. B. and Ritchie, J. M. (1977). Physiological basis of conduction in myelinated nerve fibers. In Morell, P., ed., *Myelin,* 117–160. Plenum, New York.

Röper, J. and Schwarz, J. R. (1989). Heterogeneous distribution of fast and slow potassium channels in myelinated rate nerve fibres. *J. Physiol.,* 416:93–110.

Rushton, W. A. H. (1951). A theory of the effects of fiber size in medullated nerve. *J. Physiol.,* 115:101–122.

Shrager, P. (1987). The distribution of sodium and potassium channels in single demyelinated axons of the frog. *J. Physiol.,* 392:587–602.

Shrager, P., Chiu, S. Y., Ritchie, J. M., Zecevic, D., and Cohen, L. B. (1987). Optical recording of action potential propagation in demyelinated frog nerve. *Biophys. J.,* 51:351–355.

Shrager, P. and Rubinstein, C. T. (1990). Optical measurements of conduction in single myelinated axons. *J. Gen. Physiol.,* 115:867–890.

Tao-cheng, J.-H. and Rosenbluth, J. (1980). Nodal and paranodal membrane structure in complementary freeze-fracture replicas of amphibian peripheral nerves. *Brain Res.,* 199:249–265.

Tasaki, I. (1952). Conduction of impulses in the myelinated nerve fiber. In *Cold Spring Harbor Symposium on Quantitative Biology,* vol. 17, 37–41. Long Island Biological Association, Cold Spring Harbor, NY.

Tasaki, I. (1955). New measurements of the capacity and the resistance of the myelin sheath and the nodal membrane of the isolated frog nerve fiber. *Am. J. Physiol.,* 181:639–650.

Tasaki, I. (1956). Initiation and abolition of the action potential of a single node of Ranvier. *J. Gen. Physiol.,* 39:377–395.

Tasaki, I. and Bak, A. F. (1958). Current-voltage relations of single nodes of Ranvier as examined by voltage-clamp technique. *J. Neurophys.,* 21:124–137.

Tasaki, I. and Frank, K. (1955). Measurement of the action potential of myelinated nerve fiber. *Am. J. Physiol.,* 182:572–578.

Tasaki, I., Ishii, K., and Ito, H. (1943). On the relation between the conduction-rate, the fibre-diameter and the internodal distance of the medullated nerve fibre. *Jap. J. Med. Sci. Biophys.,* 9:189–199.

Tasaki, I. and Takeuchi, T. (1942). Weitere studien über den aktionsstrom der markhaltigen nervenfaser und über die elektrosaltatorische übertragung des nervenimpulses. *Pflügers Arch.,* 245:764–782.

Tasaki, I. and Tasaki, N. (1950). The electrical field which a transmitting nerve fiber produces in the fluid medium. *Biochim. Biophys. Acta,* 5:335–342.

Tuisku, F. and Hildebrand, C. (1992). Nodes of Ranvier and myelin sheath dimensions along exceptionally thin myelinated vertebrate PNS axons. *J. Neurocytol.,* 21:796–806.

Utzschneider, D. A., Archer, D. R., Kocsis, J. D., and Waxman, S. G. (1994). Transplantation of glial cells enhances action potential conduction of amyelinated spinal cord axons in the myelin-deficient rat. *Proc. Natl. Acad. Sci. U.S.A.,* 91:53–57.

Utzschneider, D. A., Thio, C., Sontheimer, H., Ritchie, J. M., Waxman, S. G., and Kocsis, J. D. (1993). Action potential conduction and sodium channel content in the optic nerve of the myelin-deficient rat. *Proc. R. Soc. London, Ser. B,* 254:245–250.

Waxman, S. G. and Bennett, M. V. L. (1972). Relative conduction velocities of small myelinated and non-myelinated fibres in the central nervous system. *Nature*, 238:217–219.

Waxman, S. G. and Brill, M. H. (1978). Conduction through demyelinated plaques in multiple sclerosis: Computer simulations of facilitation by short internodes. *J. Neurol., Neurosurg., Psychiatry*, 41:408–416.

Waxman, S. G. and Foster, R. E. (1980). Ionic channel distribution and heterogeneity of the axon membrane in myelinated fibers. *Brain Res. Rev.*, 2:205–234.

Waxman, S. G. and Ritchie, J. M. (1985). Organization of ion channels in the myelinated nerve fiber. *Science*, 228:1502–1507.

Waxman, S. G. and Wood, S. L. (1984). Impulse conduction in inhomogeneous axons: Effects of variation in voltage-sensitive ionic conductances on invasion of demyelinated axon segments and preterminal fibers. *Brain Res.*, 294:111–122.

Voltage-Gated Ion Channels

When Bert Sakmann and I started measurements by placing pipettes onto the surface of denervated muscle fibers, we soon realized that it was not easy to obtain a satisfactory seal. Although Bert Sakmann was very experienced in enzymatically treating cell surfaces through his work in B. Katz's laboratory, and although the work of Katz and Miledi (1972) and our own voltage-clamp measurements had shown that denervated muscle should have an appropriate density of diffusely dispersed ACh channels, our initial attempts failed. Our seal resistances were just about 10-20 MΩ, two orders of magnitude lower than desired. However, by reducing the pipette size and by optimizing its shape, we slowly arrived at a point where signals emerged from the background—first some characteristic noise, later on blips that resembled square pulses, as expected. In 1976 we published records (Neher and Sakmann, 1976) that, with good confidence, could be interpreted as single-channel currents. The fact that similar records could be obtained both in our Göttingen laboratory and in the laboratory of Charles F. Stevens at Yale (where I spent parts of 1975 and 1976) gave us confidence that they were not the result of some local demon, but rather were signals of biological significance. The square-wave nature of the signals was proof of the hypothesis that channels in biological membranes open and close stochastically in an all-or-none manner. For the first time one could watch conformational changes in biological macromolecules in situ and in real time. However the measurement was far from perfect. There still was excessive background noise, concealing small and more short-lived contributions of other channel types. Besides, the amplitudes of single-channel currents had a wide distribution, since the majority of channels were located under the rim of the pipette, such that their current contributions were recorded only partially.

We made many systematic attempts to overcome the seal problem (manipulating and cleaning cell surfaces, coating pipette surfaces, reversing charges on the glass surface, etc.) with little success.

By about 1980 we had almost given up on attempts to improve the seal, when we noticed by chance that the seal suddenly increased by more than two orders of magnitude when slight suction was applied to the pipette. The resulting seal was in the gigaohm range, the so-called "gigaseal." It turned out that a gigaseal could be obtained reproducibly when suction was combined with some simple measures to provide for clean surfaces, such as using a fresh pipette for each approach and using filtered solutions. The improved seal resulted in much improved background noise (Hamill et al., 1981).

—Neher, 1992

6.1   Historical Perspective

The Hodgkin-Huxley model, published in 1952, proved to be a turning point in physiology. In retrospect, it seems clear that this study culminated an era in which biophysicists strove to understand the basis of the electrical excitability of cells. The Hodgkin-Huxley model explained the electrical excitability of the squid giant axon in terms of the properties of the voltage- and time-dependent sodium and potassium conductances. The characteristic differences in kinetics of these two conductances explained the generation of the action potential which had been in question for more than a century. This study also initiated a new focus on the membrane mechanisms that give rise to the kinetics of the sodium and potassium conductances. Thus, the work ushered in a new era in which the molecular basis of electrical excitability was sought.

6.1.1   New Electrical Recording Techniques

During the period following 1952, many new techniques were developed to extend the types of electrical events that could be recorded from cells as well as the range of cell types from which electrical recordings could be obtained. Methods were developed for recording the membrane potential with extracellular electrodes—the *gap techniques* described in Chapter 5. Prior to the 1970s, intracellular electrical recording from single cells was accomplished either via capillary electrodes inserted longitudinally into very large cells such as the giant axon of the squid or via micropipette electrodes used to penetrate cells transversely by puncturing the membranes (Weiss, 1996, Figure 7.2). The penetrating micropipettes were widely used, but clearly had deleterious effects on small cells. During the 1970s and 1980s, new techniques were developed for recording from single cells and from isolated patches of cellular membrane that contained individual ion channels (Neher and Sakmann, 1976; Hamill et al., 1981; Sakmann and Neher, 1983). In the new technique, a micropipette, called a *patch pipette*, with a smooth (fire-polished) tip with a diameter of about 1 μm is placed in contact with a cell membrane as shown schematically in Figure 6.1. Initial success was obtained with cells grown in tissue culture under aseptic conditions, but a wide variety of cell types were subsequently used. If a slight suction is applied to the pipette, the glass seals to the cell with a very high-resistance seal (typically in the 10–100 GΩ range)

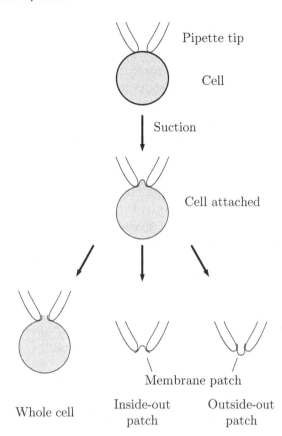

Figure 6.1 Illustration of method for recording from cells and small patches of membrane (adapted from Hamill et al., 1981, Figure 9).

so that current flow through the micropipette is confined to the patch of membrane in the lumen of the pipette. Electrical properties of this configuration, called the *cell-attached configuration*, have been recorded, but more useful results are obtained by further manipulations of the cell-attached configuration.

If sufficient suction is applied to suck out the patch of membrane in the tip while the pipette remains in contact with the cell, the result is the *whole-cell recording configuration*. In this configuration, current is recorded from the whole cell with an electrode of relatively low source impedance. These 1 μm patch pipettes have tip diameters that are a factor of ten larger than pipettes used to penetrate cellular membranes. The lower source resistance of the patch pipettes produces less noise; hence, smaller signals can be resolved. In addition, the lower source impedance results in a higher recording bandwidth, which enables measurement of rapid changes in membrane current in response to rapid changes in membrane potential. The solution within the patch pipette exchanges fairly quickly with that of the cell; hence, the composition of the cytoplasm of the cell can be changed. The external concentration can be controlled by placing the cell in a bath of known composition. In this way, it is possible to study the membranes of relatively small cells and to approach the type of control of membrane electrical variables and solution compositions that was previously possible only for very large cells. Furthermore, this method of recording from single cells seems to be less traumatic to a cell than impaling it with a micropipette, even one that has a tip diameter that is a factor of ten smaller than that of the patch pipette. The tight seal between the patch pipette and the membrane reduces the incidence of large shunt leakage paths between the inside and the outside of the cell that are an important limitation of measurements with penetrating micropipettes. Development of this method has enabled investigation of currents under voltage clamp for a wider range of smaller cells than is possible with either the large capillary electrodes or the penetrating micropipettes. In whole-cell recording, the cytoplasm is in diffusive contact with the solution in the patch pipette (Pusch and Neher, 1988). In order to reduce changes in concentration of large solutes in the cytoplasm, the antibiotic Nyastatin can be placed in the lumen of the patch pipette, now called a *perforated patch pipette* (Horn and Marty, 1988). The Nyastatin barrier is permeable to small solutes, but not to large solutes such as metabolites and proteins.

If the patch of membrane is separated from the cell, the resulting recording method is called *patch recording*. In this method, a small patch of membrane, in one of two configurations—outside out or inside out—is lodged in the tip of the pipette. In an outside-out patch, the external surface of the membrane faces the outside of the pipette; in an inside-out patch, the cytoplasmic surface of the membrane faces the outside of the pipette. With either configuration, it is possible to control the concentration of the solution in the pipette and in the bath as well as the potential across the membrane in the tip of the pipette (Figure 6.2). The current through a single ionic channel in the patch can be measured under voltage clamp. This technique is called the *patch-clamp technique.*

6.1.2 Macroscopic Ionic Currents

Following the publication of the studies of Hodgkin and Huxley on the squid giant axon, it became clear that the study of the membrane ionic current components under voltage-clamp conditions yielded important information about membrane mechanisms. Thus, these voltage-clamp studies were refined and generalized in several distinct directions. We call currents measured this way

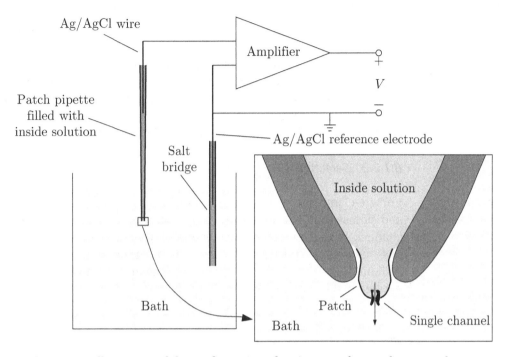

Figure 6.2 Illustration of the configuration of a pipette and a membrane patch.

macroscopic currents to distinguish them from the *single-channel currents* that flow through single ion channels.

6.1.2.1 Selective Blocking of Channels by Pharmacological Agents

Pharmacological agents, often natural toxins known to affect electrically ex-citable cells, were applied to cells to determine which components of the membrane current were affected. Often the motivation was simply to better define the toxic action of the agent. But there was also interest in determining the implications for membrane mechanisms of the action of these agents. A large number of different substances were investigated, of which the most im-portant were tetrodotoxin (TTX) and tetraethylammonium (TEA), which were shown to selectively block sodium and potassium currents, respectively. The discovery of these two agents had immediate technical implications for the investigation of ionic currents. These agents enabled investigators to block ei-ther the sodium current or the potassium current through the membrane in a reversible manner. Therefore, it was possible to study the individual ionic currents more conveniently and selectively than had been possible previously. Subsequently, these agents have been used routinely to study one or another component of the ionic current. More important, these studies indicated that not only are the mechanisms that give rise to the sodium and potassium conductances kinetically distinct, but they are also pharmacologically distinct. Thus, it was assumed that these conductances resulted from distinct structures in membranes, which came to be called *channels*.

6.1.2.2 Selectivity of Channels

Voltage-clamp currents were measured after ions were substituted for sodium and potassium. The purpose was to determine which ions permeated each channel. It became clear that channels, though selectively permeable, could pass many ionic species. Thus, the *sodium channel* is a channel through which sodium normally permeates the membrane and through which other ions can permeate the membrane. Similarly, the potassium channel also allows ions other than potassium to pass. These studies provided important information on the physicochemical properties of the channels and also identified ions to which the channels were impermeant. The latter finding has also proved to be important technically in the separation of ion currents carried by different channels. The sequence of ions to which a channel is permeant ordered according to the permeability, the *selectivity sequence*, is a characteristic property of a channel.

6.1.2.3 Profusion of Ion Channels

Voltage-clamp techniques were applied to different peripheral nerve fibers and later to cell bodies, as well as to nonneuronal cells. The initial motivation for many of these studies was to determine whether electrically excitable cells all behaved similarly to the giant axon of the squid. The studies on peripheral nerve fibers revealed the presence of sodium and potassium channels whose kinetic and pharmacological properties were very similar to those described for the squid giant axon. It appeared for a time as if the mechanism of generation of action potentials in electrically excitable cells could be explained universally by the two channels described by Hodgkin and Huxley. However, as the new techniques enabled the study of a wider variety of cell types, it became clear that the results could not all be accounted for by these same two channels. For example, certain types of nerve cell bodies and muscle cells were found to generate action potentials that were not blocked by TTX as are the action potentials of peripheral axons. Furthermore, after applying both

TTX and TEA so as to block both the sodium and potassium channels, an appreciable ionic current remained. This remaining ionic current could be resolved into distinct ionic currents whose properties differed from those of the sodium and potassium channels found in peripheral axons. From such studies of macroscopic ionic currents under voltage-clamp conditions, new channels were identified on the basis of their kinetic and pharmacological properties.

6.1.3 Gating Currents

Although most of the attention was initially focused on refining knowledge of the ionic currents, investigation of the capacitance current gave important new insights into membrane mechanisms. The capacitance current results from the redistribution of charge on the membrane surface, which reflects redistribution of charge in the membrane. It has been found that a portion of this capacitance current, called the *gating current,* behaves as if it were generated by the motion of charges in the membrane that accompany the opening of ionic channels. The gating currents have provided information about channels that is complementary to the information obtained from ionic currents.

6.1.4 Single-Channel Currents

In the 1970s, the method of recording currents through single ionic membrane channels was developed by Neher and Sakmann. This development revolutionized the study of ionic transport through membranes, and Neher and Sakmann were awarded the Nobel Prize for this development in 1991. This technique demonstrated that the previously hypothetical ionic channels in fact existed as distinct sites in the membrane. Furthermore, a number of new channels were identified. With the *single-channel recording method,* it is possible to identify even those channels that are present at low density or that pass only small ionic currents. The current transported by such channels is normally swamped by the current produced by more numerous channels that pass large currents. It is now clear that membranes contain a variety of types of channels with distinct kinetic and pharmacological properties. Channels are categorized according to the identity of the physical variable, called the *gating variable,* that opens the channel, as well as by the ionic species that normally flows through an open channel. So, for example, channels that are opened by a change in the membrane potential are called *voltage-gated channels,* whereas channels that are opened by the binding of a ligand on the surface of the membrane are called *ligand-gated channels,* and so on. Thus, the *voltage-gated sodium channel* refers to a channel that is gated by a change in membrane potential and normally allows transport of sodium.

A principal finding is that single channels have a discrete number of states of conduction—often only one—when the channel is open and a transmembrane ionic current flows. Channels have multiple closed or nonconducting states. A channel switches randomly between its open and closed states, with the probability of being in one state or another dependent upon the gating variable. Combined with results obtained from macroscopic currents, a view of channels began to emerge (Figure 6.3). The channel is formed by an integral membrane macromolecule that spans the membrane. The voltage-current

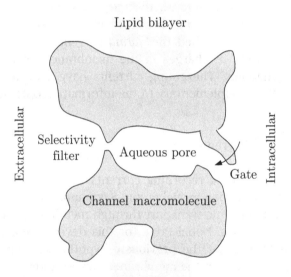

Figure 6.3 Schematic drawing of a hypothetical membrane ionic channel (adapted from Hille, 1988, Figure 1). The channel is formed by a proteinaceus macromolecule that spans the lipid bilayer and contains an aqueous pore. The gate determines whether the channel is open, and the selectivity filter determines which ions permeate an open channel. In a voltage-gated channel, the gate state is determined by the potential across the membrane, because the channel macromolecule contains charged groups that experience a force of electric origin when the membrane potential changes. This force is conveyed to the gate portion of the molecule. The resulting gate motion opens and closes the channel.

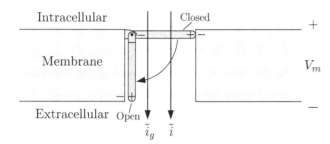

Figure 6.4 Schematic diagram of a channel with one two-state activation gate. The ionic current through the channel is a random variable denoted by $\tilde{\imath}$. The gating current resulting from motion of the charged gate is a random variable denoted by $\tilde{\imath}_g$.

characteristics of the channel are determined both by a gate that determines the extent to which the channel is opened or closed and by a selectivity filter that largely determines the current-voltage relation of an open channel.

6.1.5 Development of Kinetic Models of Channel Gating

The advent of refined data on macroscopic currents obtained under voltage clamp, the discovery of gating currents, and the measurement of currents through single ion channels led to the development of kinetic models to encompass the emerging measurements. The simplest such model is that of a hypothetical channel with one two-state gate as shown in Figure 6.4. The gate is charged so that a depolarization of the membrane potential tends, in the case shown, to open the channel by displacing the gate to its open conformation. Since the gate is of molecular dimensions, it is subject to thermal forces. Hence, the gate opens and closes at random, but its probability of being in the open conformation is increased as the membrane potential is increased. When the gate opens, the positive gate charge moves from the inner surface of the membrane to the outer surface. The motion of this gating charge constitutes a gating current. When the gate is open, ionic currents flow through the channel. This model is the simplest of a class of models of gating mechanisms and is qualitatively consistent with the behavior of single channels in membranes. Later in this chapter, we shall analyze this model in some detail and compare its predictions with measurements on membranes. Although no known channels behave precisely as predicted by this model, this model gives a great deal of insight into channel gating and leads directly to multistate models that represent channels more faithfully.

6.1.6 Channel Macromolecules

Progress in recording from single channels has been accompanied by progress in the molecular biology of channels. Channel proteins have been isolated, their amino acid sequences have been determined, and the genes specifying the proteins have been identified. Modification of channel proteins has allowed for studies that relate the biophysical properties of the channels to their molecular architecture. The availability of such information on a variety of channels from a variety of species has enabled studies of the phylogenetic development of channels.

6.1.7 Summary

In this chapter, we shall discuss voltage-gated channels. We shall start by examining properties of channels gleaned from measurements of macroscopic ionic currents obtained under voltage clamp. We shall then discuss measurements of gating currents and of ionic currents through single channels. We shall then analyze the simplest possible model of a voltage-gated ion channel: a channel with one two-state gate that gates an ohmic channel. This discussion is a springboard for a more general discussion of multiple-state models of voltage-gated ion channels. Finally, we shall briefly discuss the molecular biology of channel macromolecules. Throughout the chapter, we shall focus— where specificity is helpful—on the voltage-gated sodium channel.

6.2 Macroscopic Ionic Currents

In this section, we shall describe properties of ionic currents, focusing first on the sodium and potassium currents found in axons. We shall then briefly describe ionic currents through other types of ion channels.

6.2.1 Pharmacological Manipulations of Channels

Ionic channels are blocked by pharmacological agents called *channel blockers.* Since neurons are endowed with a rich selection of ion channels that are affected by channel blockers, and since many of these channel blockers are substances that are highly toxic to organisms, these blockers are commonly called *neurotoxins.* A large number of neurotoxins have been identified (Adams and Swanson, 1994). The role of these neurotoxins has been critical in the study of channels. We shall illustrate experiments on channel blockers with important blockers of sodium and potassium channels.

The closely related natural neurotoxins tetrodotoxin (TTX) and saxitoxin (STX) are highly toxic substances. TTX is found in several animals, including the *fugu* (puffer) fish, while STX is found in certain marine protozoa (dinoflagellates) that occasionally bloom to produce a reddish tint to the water, or a *red tide*. Since puffer fish have for generations been considered a delicacy in Japan, and since shellfish eaten by humans ingest the dinoflagellates, both TTX and STX are health hazards to humans. Thus, the toxic action of these compounds has been of interest to public health authorities. Both substances block action potentials when applied extracellularly to many electrically excitable cells. However, the specific site of action of these toxins was revealed in measurements of membrane current under voltage clamp (Narahashi et al., 1964; Nakamura et al., 1965), which showed that both TTX and STX completely, reversibly, repeatably, and selectively block the sodium current without affecting the potassium current (Figure 6.5). Only nanomolar concentrations of the toxins are required to block sodium currents.

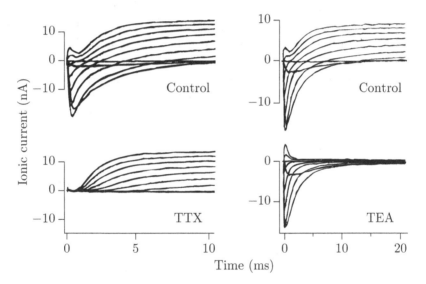

Figure 6.5 Effect of TTX and TEA on ionic currents obtained from a frog node of Ranvier (adapted from Hille, 1992, Figure 3-2). Each panel shows families of ionic currents obtained under voltage clamp for voltage sequences from −120 mV to a final potential that ranges from −60 to +60 mV for the different traces. The response to the depolarization is shown with leakage and capacitance currents removed, assuming linearity of both these components. The left panel shows responses in normal frog Ringer's solution (upper panel) and responses after 300 nmol/L TTX was added to the Ringer's. The temperature of the preparation was 13°C (Hille, 1966). The right panel shows responses from another node before (upper panel) and after (lower panel) the addition of 6 mmol/L TEA to the Ringer's solution. The temperature was 11°C (Hille, 1967b).

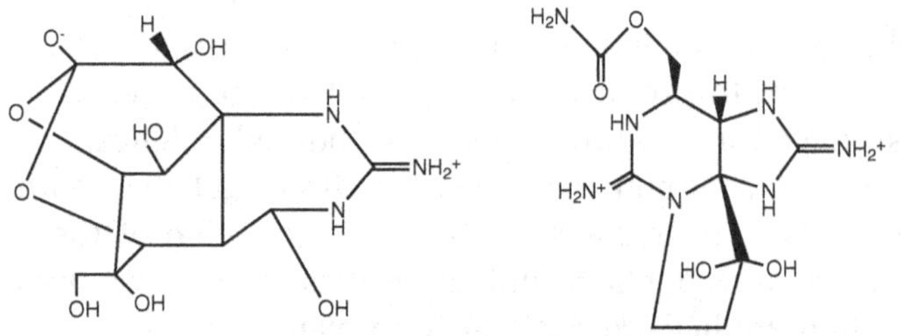

Figure 6.6 Chemical structures of tetrodotoxin and saxitoxin.

Tetrodotoxin   Saxitoxin

Since both tetrodotoxin and saxitoxin are highly specific blockers of the voltage-gated sodium channels, it is of interest to note that they have some similarities in molecular structure (Figure 6.6). For example, both contain the guanidinium group (a carbon with three amino groups seen on the right side of each molecule in Figure 6.6). One possible mechanism for blocking the sodium channel by TTX and STX is that the guanidinium group just fits into the mouth of the channel on the extracellular side of the membrane and plugs the channel.

The quaternary ammonium ions, which include tetraethylammonium (TEA), also have large effects on the action potentials of many types of electrically excitable cells. Studies of ionic currents under voltage clamp (Tasaki and Hagiwara, 1957; Hagiwara and Saito, 1959; Armstrong and Binstock, 1965) demonstrated that TEA selectively and reversibly blocks the potassium current without affecting the sodium current (Figure 6.5).

TTX and TEA are routinely used to block the sodium and/or the potassium currents in investigations of ionic and gating currents. These substances bind to cell membranes and to membrane fractions that contain the appropriate channels. These blockers may block channels either by lodging in the channel openings or by binding to receptor sites on the channels that result in conformational changes of the channels that block them.

Pharmacological manipulation can not only block conduction through channels, but also manipulate channel kinetics. For example, the proteolytic enzyme pronase, when applied intracellularly, removes inactivation of the sodium current (Figure 6.7). Thus, it seems that pronase enzymatically disables some part of the sodium channel macromolecule that is responsible for inactivation without affecting the part that is responsible for activation of the channel. This result suggests a segregation of function within the channel macromolecule.

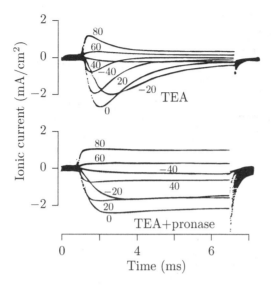

Figure 6.7 Effect of pronase on the sodium current of an axon (adapted from Armstrong, 1975, Figure 2). In the presence of TEA, the membrane ionic current contains no appreciable potassium current and consists predominantly of the sodium current. In the presence of both TEA and pronase in the intracellular solution, the sodium current exhibits little inactivation.

6.2.2 Selective Permeability of Channels

In the late nineteenth century, it was recognized that membranes were selectively permeable to different ions. By the 1950s, this selective permeability of *membranes* was explained by the independent-ion-channel idea. That is, membranes are selectively permeable because they contain ionic channels each of which is permeable to particular ions. If one channel is most prevalent in the membrane of a cell, that cell's membrane is predominantly permeable to that ion. At rest, cells are predominantly permeable to potassium because these channels are open when the membrane potential is at its resting value; other channels tend to be closed at this potential. With the advance of techniques for isolating macroscopic currents through a single population of channels (by pharmacological means), it became of interest to understand the mechanism of permeation through each type of channel.

One question that was raised was, How selectively permeable are the channels? Consider the experiment shown in Figure 6.8. A set of voltage-clamp ionic currents with the nerve fiber in an extracellular solution of normal composition are shown at the top of the figure. Each trace corresponds to the ionic current for a different value of membrane potential. On the time scale shown, these currents contain an early transient current that reverses its polarity near the sodium equilibrium potential. Therefore, this current is identified with the flow of sodium ions. If sodium ions in the extracellular Ringer's solution are replaced by lithium ions, there is still an early current component that re-

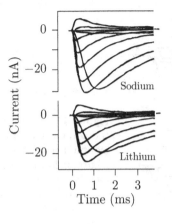

Figure 6.8 Traces that show voltage-clamp ionic currents obtained from a node of Ranvier of the frog for ten different values of membrane potential spaced 15 mV apart (adapted from Hille, 1972, Figure 1). Leakage and capacitance current have been subtracted. The time scale was chosen to show the early component of the ionic current. The upper traces show the voltage-clamp currents in normal Ringer's solution. The lower traces show these currents when lithium has replaced sodium in the Ringer's solution.

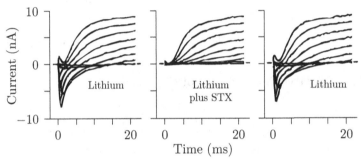

Figure 6.9 The traces show voltage-clamp ionic currents obtained from a node of Ranvier of the frog for membrane potentials spaced 15 mV apart in the range of −60 to +60 mV. (adapted from Hille, 1968, Figure 1). Leakage and capacitance current have been subtracted. The traces show the ionic currents obtained in Ringer's solution in which lithium has replaced sodium. The left and right panels show the responses in lithium-Ringer's solution both before and after the application of a lithium-Ringer's solution that contained 25 nmol/L STX.

verses at some reversal potential which acts as an effective equilibrium potential (lower trace in Figure 6.8). Since there are no sodium ions extracellularly and since the major cation is lithium, at least the inward component of the current in this experiment must be carried by lithium. Figure 6.9 shows that this early transient current carried by lithium ions is blocked by STX although the late maintained component of the ionic current is unaffected by STX. The interpretation of the results of these experiments is that (1) lithium can permeate the channel through which sodium normally flows; (2) STX blocks the channel and prevents permeation of the permeant ion whether it is sodium or lithium. Thus, the names assigned to the channels correspond to the names of

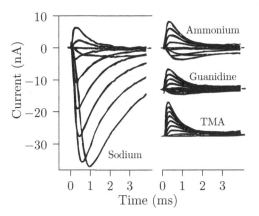

Figure 6.10 The traces show voltage-clamp ionic currents obtained from a node of Ranvier of the frog for membrane potentials spaced 15 mV apart in the range of −65 to +70 mV. Leakage and capacitance current have been subtracted. The traces show the ionic currents obtained in Ringer's solution in which the indicated cation has been substituted for sodium (adapted from Hille, 1971, Figure 4).

the ions that *normally* permeate these channels; they do not imply that these channels are exclusively permeable to that ion. For instance, under normal circumstances sodium is the predominant ion that flows through the sodium channel, but other ions can also flow through this channel.

Experiments such as the one illustrated in Figure 6.8 have been performed on numerous channels in several cell types. The question arises, Is there some quantitative method to assess the relative permeability of a channel to a given ion? In the absence of adequate microscopic theories of ion permeation through membranes, an ad hoc procedure has been widely used to assess the ion selectivity of channels from the macroscopic channel currents. The procedure is based on the observation that substitution of ions, such as lithium for sodium, can change the reversal potential of the ionic current (see Figure 6.10). This reversal potential is defined as that potential at which the ionic current is zero. Usually this reversal potential is obtained by interpolating between the two voltages that just cause inward and outward ionic currents, i.e., those that bracket the reversal potential. Clearly the reversal potential for the early component of the ionic current is not the sodium equilibrium potential when the external sodium has been replaced by some other permeant cation.

An empirical relation based on the Goldman theory (Weiss, 1996, Appendix 7.1) has been used to quantify the permeability of different ions through an ionic channel. Consider the following relation between the membrane potential and the concentrations and permeabilities of cations through a cationic channel:

$$V_j = \frac{RT}{F} \ln \left(\frac{\sum_k P_k c_k^o}{\sum_k P_k c_k^i} \right),$$

(6.1)

where V_j is the reversal potential for channel j and the P_ks are the permeabilities of channel j to ion species k. In the simplest instance of the use of Equation 6.1, if the channel is permeable to ion n only, Equation 6.1 reduces to $V_j = V_n$, where V_n is the Nernst equilibrium potential for ion n. If other permeant ions are present, their effect on the channel reversal potential depends on their permeabilities. For example, from the measurements shown in Figure 6.8, Equation 6.1 can be written

$$V_{Na} = \frac{RT}{F} \ln \frac{P_{Li} c_{Li}^o}{P_{Na} c_{Na}^i}$$

under the assumption that the only permeant ion in the external solution is lithium and the only permeant ion in the internal solution is sodium. From a measurement of the reversal potential of the early current component and the known concentrations of lithium and sodium, the ratio P_{Li}/P_{Na} can be estimated. If more than one permeant ion is present, the more complete Equation 6.1 can be used to estimate the permeability ratios from a sequence of measurements with different compositions of electrolytes.

Based on such measurements, the permeabilities of voltage-gated ion channels to a variety of cations have been determined as shown in Tables 6.1 and 6.2. Inspection of these tables reveals that a number of cations can pass through both the sodium channel and the potassium channel. In fact, potassium can traverse the sodium channel (as indicated in Figure 6.11), although with much lower permeability than sodium, and sodium can traverse the potassium channel, although with much lower permeability than potassium. Lithium traverses the sodium channel with a permeability that is close to that of sodium. Cesium permeates poorly through both the sodium and potassium channels. Hence, cesium is often used to substitute for both sodium and potassium when it is desirable to reduce the ionic currents normally carried by these ions.

6.2.3 Diversity of Ion Channels

Six years after the characterization of the sodium and potassium currents in squid giant axon by Hodgkin and Huxley, studies began to emerge on the characteristics of ionic currents in axons obtained from a variety of invertebrate and vertebrate species including frog myelinated fibers (Dodge and Frankenhaeuser, 1958), toad myelinated nerve fibers (Dodge and Frankenhaeuser, 1959), lobster giant axons (Julian et al., 1962), cockroach giant axons (Pichon and Boistel, 1967), marine worm giant axons (Goldman and Schauf, 1973), crayfish giant axons (Shrager, 1974), and rabbit myelinated nerve fibers

Table 6.1   Selectivity of sodium channels in four different cell types expressed as the ratio of permeabilities P_n/P_{Na} (adapted from Hille, 1992, Table 13-2).

Ion species	Permeability ratio			
	Frog node	Frog muscle	Squid axon	*Myxicola* axon
H^+	252	—	> 2	—
Na^+	1.0	1.0	1.0	1.0
$HONH_3^+$	0.94	0.94	—	—
Li^+	0.93	0.96	1.1	0.94
$H_2NNH_3^+$	0.59	0.31	—	0.85
Tl^+	0.33	—	—	—
NH_4^+	0.16	0.11	0.27	0.20
Foramidinium	0.14	—	—	0.13
Guanidinium	0.13	0.093	—	0.17
Hydroxyguanidinium	0.12	—	—	—
Ca^{++}	< 0.11	< 0.093	0.1	0.1
K^+	0.086	0.048	0.083	0.076
Aminoguanidinium	0.06	0.031	—	0.13
Rb^+	< 0.012	—	0.025	—
Cs^+	< 0.013	—	0.016	—
Methylammonium	< 0.007	< 0.009	—	—
TMA	< 0.005	< 0.008	—	—

(Chiu et al., 1979). While there are many variations in the sodium and potassium currents in these different preparations, they all share basic kinetic and pharmacological properties.

However, as methodology improved to allow for voltage-clamp studies of a larger variety of cell parts than simply axons, it became clear that cells contain a large variety of ion channels, although the sodium and potassium channels first identified in axons are among the most commonly found channels. A large number of different types of gated ion channels are known to exist. A summary of some of the more prominent ionic channels is shown in Table 6.3. A more complete listing is given elsewhere (Watson and Girdlestone, 1994) as is a comprehensive exposition on ion channels (Hille, 1992).

6.2.3.1   Functional Categorization of Ion Channels

Ion channels are categorized functionally on the basis of several different criteria. We shall discuss these briefly.

Table 6.2 Selectivity ratios for the delayed rectifier potassium channel for different cations expressed as the ratios of permeabilities P_n/P_K (adapted from Hille, 1992, Table 13-3).

| | Permeability ratio | | |
Ion species	Frog node	Frog muscle	Snail neuron
Tl^+	2.3	—	1.29
K^+	1.0	1.0	1.0
Rb^+	0.91	0.95	0.74
NH_4^+	0.13	—	0.15
Cs^+	<0.077	<0.11	0.18
Li^+	<0.018	<0.02	0.09
Na^+	<0.10	<0.03	0.07
$H_2NNH_3^+$	<0.29	—	—
Methylammonium	<0.021	—	—

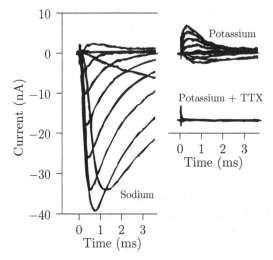

Figure 6.11 The traces show voltage-clamp ionic currents obtained from a node of Ranvier of the frog for ten different values of membrane potential spaced 15 mV apart (adapted from Hille, 1972, Figure 2). Leakage and capacitance current have been subtracted. The traces show the effect of 150 nmol/L TTX on the ionic currents obtained in Ringer's solution in which potassium has replaced sodium. TTX removes the current carried by potassium through the sodium channel.

Gating Variable

Ion channels are gated by one or more physicochemical variables. In this section, we are considering only voltage-gated channels, i.e., ion channels whose state of conduction is affected by the potential across the membrane. Other channels are gated by the binding of a ligand to either the extracellular or the intracellular surface of the membrane or by mechanical stress to the membrane.

Table 6.3 Selective list of ion channels, the ions they normally transport, and examples of specific pharmacological blockers. Receptors that act through second messengers to open channels or that modulate channel transport have not been included. The following abbreviations are used: AA, amino acid; GABA, γ-aminobutyric acid; NMDA, *N*-methyl-D-aspartate.

Name	Solute transported	Blockers
Voltage-gated channels		
Sodium	Sodium	Tetrodotoxin, saxitoxin
Delayed rectifier	Potassium	Tetraethylammonium
Fast transient	Potassium	4-aminopyridine
Inward rectifier	Potassium	Tetraethylammonium
Calcium	Calcium	
Chloride	Chloride	
Ligand-gated channels		
Calcium-activated potassium	Potassium	Tetraethylammonium
ATP-activated potassium	Potassium	
Nicotinic acetylcholine receptor	Sodium, potassium, calcium	d-tubocurarine, α-bungarotoxin
Excitatory AA (NMDA) receptor	Sodium, potassium, calcium	Dizocilipine, phencyclidine
Excitatory AA (non-NMDA) receptor	Sodium, potassium	
Serotonin receptor	Potassium	
Glycine receptor	Chloride	Strychnine
GABA$_A$ receptor	Chloride	Picrotoxin, bicuculine
Mechanically gated channels		
Hair-cell mechanical transducer	Cations	Aminoglycosides
Stress-gated	Cations	

Principal Permeant Ion

There are four principal ions transported by ion channels: sodium, potassium, calcium, and chloride. Some ion channels are predominantly permeant to one of these ions, whereas other ion channels have appreciable permeability to more than one of these ions. The principal permeant ion is used to classify ion channels. Hence, sodium channels are channels that are primarily permeant to sodium in vivo.

Selectivity Sequence

While many ion channels are principally permeant to one ion when surrounded by normal physiological solutions, they can be shown to be permeant to other ions in artificial solutions. Each ion channel has a characteristic selectivity sequence. For example, the voltage-gated sodium channel has the selectivity sequence (Table 6.1)

$$Na^+/Li^+/Tl^+/NH_4^+/Ca^{++}/K^+/Rb^+/Cs^+,$$

whereas the delayed rectifier potassium channel has the selectivity sequence (Table 6.2)

$$Tl^+/K^+/Rb^+/NH_4^+/Cs^+/Li^+/Na^+.$$

Rectification

Most voltage-gated ion channels exhibit some degree of electrical rectification, i.e., the current-voltage characteristic is nonlinear. The rectification is apparent in the voltage dependence of both the conductance of the channel and the current through the channel (Figure 6.12). Figure 6.12 illustrates behavior of the Hodgkin-Huxley model of the steady-state (delayed rectifier) potassium channel of the squid giant axon for different potassium ion concentrations. Even when the voltage dependence of the potassium conductance does not

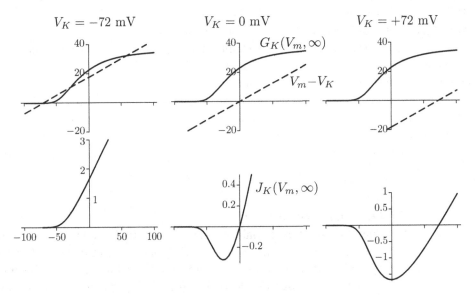

Figure 6.12 Steady-state rectification of a channel. The upper panels show the steady-state conductance as a function of the membrane potential (solid line; the ordinate scale is in mS/cm^2) for the (delayed rectifier) potassium channel of the squid giant axon, $G_K(V_m, \infty)$. Superimposed on the conductance is the quantity $V_m - V_K$ (dashed line; the ordinate scale multiplied by four is the potential in mV). The current $J_K(V_m, \infty) = G_K(V_m, \infty)(V_m - V_K)$ is plotted in the lower panels (ordinate scale in mA/cm^2). The columns are for different values of the potassium equilibrium potential. The labels on the curves in the center column apply to all three columns.

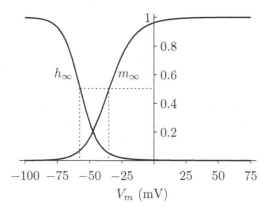

Figure 6.13 Voltage dependence of activation and inactivation variables, i.e., $m_\infty(V_m)$ and $h_\infty(V_m)$. The activation factor has a value of 0.5 at $V_m = -35$ mV, and the inactivation factor has a value of 0.5 at $V_m = -57$ mV.

change with ion concentration, the potassium ion current density does. The ionic current density is

$$J_K(V_m, \infty) = G_K(V_m, \infty)(V_m - V_K).$$

Since $G_K(V_m, \infty) \geq 0$, the sign of $J_K(V_m, \infty)$ depends only on the sign of $V_m - V_K$. Thus, for $V_K = -72$ mV, $J_K(V_m, \infty) > 0$ (outward) for the physiological range of V_m, since when $V_m < V_K$, $G_K(V_m, \infty) \approx 0$. For both $V_K = 0$ mV and $V_K = +72$ mV, $J_K(V_m, \infty) < 0$ (inward) for $V_m < V_K$, and $J_K(V_m, \infty) > 0$ for $V_m > V_K$. Also, $J_K(V_m, \infty) = 0$ for $V_m = V_K$. The delayed rectifier potassium channel exhibits *outward rectification,* since the conductance is large for depolarizations of the membrane for which the current is outward for normal potassium concentrations.

Voltage Dependence of Activation and Inactivation
Ion channels differ in the voltage range over which they are activated and inactivated. As shown in Figure 6.13 for the Hodgkin-Huxley model of sodium activation and inactivation, the membrane potential at which the (in)activation factor equals 0.5 is $V_{1/2} = -35$ mV for activation of sodium and $V_{1/2} = -57$ mV for inactivation. $V_{1/2}$ is a measure of the voltage at which the (in)activation variable is appreciably different from zero. The steepness of the voltage dependence can also be quantified by the slope of $m_\infty(V_m)$ versus V_m or by the slope of $\log m_\infty(V_m)$ versus V_m measured at $V_{1/2}$. The overlap of the activation and inactivation factors as a function of membrane potential gives an indication of the magnitude of the steady-state conductance of the channel. For example, for the Hodgkin-Huxley model of sodium in the squid giant axon, the activation and inactivation variables have approximately the same value near -47.5 mV. If the values of $V_{1/2}$ for m_∞ and h_∞ are moved further apart,

the intersection of these values gets smaller and the steady-state value of the conductance decreases. If these factors are sufficiently far apart in voltage, the channel makes no appreciable contribution to the resting potential of the cell.

Kinetics

The time course of activation and inactivation varies widely among different ion channels. Phenomenological models patterned after the Hodgkin-Huxley model have been developed for a number of different ion channels. The generic form of the conductance of such a model channel is

$$G_n(V_m, t) = \overline{G}_n m^\delta(V_m, t) h^\kappa(V_m, t),$$

where m and h are generic activation and inactivation factors, δ is an integer in the range of 1–4, and κ is either 0 or 1. The factors m and h satisfy first-order kinetics. Hence, the behavior of the onset of activation is dependent on the value of δ. If $\delta = 1$, the onset of activation rises linearly; if $\delta > 1$, the onset of activation is S shaped (parabolic). For channels that do not inactivate, $\kappa = 0$; for channels that do inactivate, $\kappa \neq 0$. The voltage-dependent time constants $\tau(V_m)$ and $\tau_h(V_m)$ determine the time course of activation and inactivation. These time constants vary from hundreds of microseconds for some ion channels to many seconds in others.

Pharmacology

As we have seen, TTX and TEA block voltage-gated sodium and (delayed rectifier) potassium channels. Similarly, for each type of voltage-gated ion channel there are an assortment of pharmacological agents that act on the channel (Adams and Swanson, 1994; Watson and Girdlestone, 1994). The pharmacological properties of a channel are important for unambiguous identification of the channel.

6.2.3.2   Common Ion Channels

The functional taxonomy of voltage-gated ion channels is somewhat murky. In part, the taxonomy depends upon the philosophical bent of the taxonomist; i.e., is the taxonomist a "lumper" or a "splitter"? A lumper will create a small number of categories, each of which will include ion channels with somewhat heterogeneous properties. A splitter will define the categories more narrowly so that each includes a more homogeneous population of channels at the cost of a large number of categories. In this classification, we shall take a decidedly lumper perspective, i.e., we shall ignore many distinctions to generate a small number of categories of ion channels. More detailed discussions can be found elsewhere (Jan and Jan, 1989; Hille, 1992; Johnston and Wu, 1995).

Sodium Channels

Voltage-gated sodium channels similar to those first discovered in squid giant axons are found widely among electrically excitable cells in the animal kingdom. Sodium channels in most cells show a rapid S-shaped activation and a slower exponential inactivation that follows $m^\delta h$ kinetics, where the integer δ that best fits the measurements varies somewhat (either two or three) across channels. Sodium channels are activated for relatively small depolarizations. There are some clear subcategories that have been defined on kinetic and pharmacological grounds. For example, one can distinguish sodium channels with relatively fast inactivation from those with slow inactivation. In addition, while most sodium channels are blocked by nanomolar extracellular concentrations of TTX, there are some sodium channels that are relatively insensitive to TTX. However, when compared to other ion channel types, sodium channels form a relatively homogeneous population.

Potassium Channels

In contrast to the comparative homogeneity of sodium channels, potassium channels show great diversity. Categorization and nomenclature for the different potassium channels are somewhat arcane. There are at least four major classes of potassium channels (Rudy, 1988; Pallotta and Wagoner, 1992) into which many potassium channels can be placed unambiguously, although there clearly are potassium channels that defy this categorization; they have properties that cut across the classes. Within each broad class, further subdivisions of properties into subclasses have been made. The four classes of potassium channels are indicated in Table 6.3, and they include three voltage-gated channels (delayed rectifier, fast transient, and inward rectifier) and one channel that is activated by intracellular calcium (the calcium-activated potassium channel). Examples of three potassium currents under voltage clamp based on a model of a single cell in the stomatogastric ganglion of the crab are shown in Figure 6.14. Since these currents are based on measurements in a single cell, they can be compared directly.

Delayed Rectifier Channels The potassium channel originally defined by Hodgkin and Huxley for the squid giant axon is referred to as the *delayed rectifier channel,* even though technically it does not exhibit a delay, but rather a lag in temporal response. Furthermore, all voltage-gated channels exhibit a lag and most exhibit rectification in their voltage-current characteristics. Never-

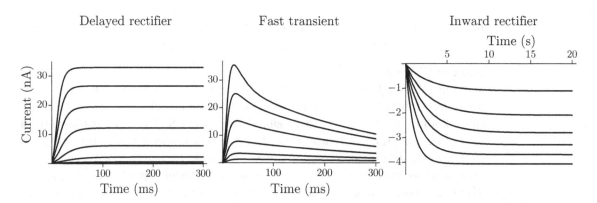

Figure 6.14 Examples of three potassium currents in a stomatogastric ganglion neuron in the rock crab *Cancer borealis* (Buchholtz et al., 1992). The traces were computed from models of the delayed rectifier, the fast transient, and the inward rectifier currents. The holding potential was −40 mV. The voltage ranges were (in 10 mV steps) −50 to +30 mV for the delayed rectifier; −40 to + 10 mV for the fast transient current; and −120 to −70 mV for the inward rectifier.

theless, the nomenclature has stuck. The delayed rectifier provides the major potassium current for many cells, both electrically excitable and electrically inexcitable cells. The conductance of the delayed rectifier has an S-shaped activation and an absent or very slow exponential inactivation. The inactivation in the squid axon is so slow that it was missed in the original work by Hodgkin and Huxley, because only short-duration depolarizations were used. The inactivation in squid giant axon has a time constant as long as seconds and is not observed with depolarizations that last tens of milliseconds. The family of delayed rectifier potassium channels exhibits somewhat diverse kinetics, voltage dependence, and pharmacology. Most delayed rectifier potassium currents are blocked by the external application of TEA at millimolar concentrations.

Fast Transient Channels The fast transient potassium channel, also called the *A channel,* is found widely in cells (Rogawski, 1985; Rudy, 1988). It differs from the delayed rectifier channel in its kinetics, voltage sensitivity, and pharmacology. The fast transient channel inactivates rapidly (time constant ≤ 100 ms) compared to the delayed rectifier, activates at a membrane potential near the resting potential of the cell, and is blocked by millimolar concentrations of 4-aminopyridine.

Inward Rectifier Channels The inward rectifier channel, also known as the *anomalous rectifier channel,* is also widely found in animal cells. The conductance of this channel increases with hyperpolarization of the membrane. Therefore, this channel produces an inward potassium current for hyperpolarizations below the potassium equilibrium potential. The channel shows no inactivation in response to maintained hyperpolarization of the membrane.

Calcium-Activated Channels The calcium-activated potassium channel is also widely found (Latorre et al., 1989; Marty, 1989). This channel is gated open by an increase in the internal concentration of ionized calcium in the micromolar range of concentration. Therefore, this type of channel is effective when in proximity to voltage-gated calcium channels. Several different populations of calcium-activated potassium channels have been described. Some populations are also voltage gated, while others show gating that is relatively independent of membrane potential.

Voltage-Gated Calcium Channels

Calcium channels show both activation and inactivation, are activated by depolarization, and have inward currents at normal physiological membrane potentials. Because the normal intracellular concentration of free (unbound) calcium is less than 1 μmol/L while the extracellular concentration of calcium is on the order of 1 mmol/L (in terrestrial vertebrates), the Nernst equilibrium potential for calcium is greater than +90 mV. This Nernst potential is not known accurately, in part because of the difficulty of measuring such low intracellular concentrations. Furthermore, intracellular calcium is subject to numerous cellular processes that either bind or transport calcium, making experiments on intracellular calcium concentration difficult to perform.

A number of different types of calcium channels have been found, and multiple types can occur in the same cell (Tsien et al., 1987; Hille, 1992). These channels differ in their kinetics, voltage dependence, inactivation, ion selectivity, and pharmacology. Many, but not all, calcium channels can be subclassified into four types: T-type, L-type, N-type, and P-type. The T-type channel is activated for small depolarizations ($V_m > -70$ mV), has rapid voltage-dependent inactivation, is blocked by low concentrations of Ni^{++} and octanol, and is found in cardiac and skeletal muscle. The L-type channel is activated for large depolarizations ($V_m > -10$ mV), inactivates rapidly, is sensitive to dihydropyridine, and is found in endocrine cells and neurons. The N-type channel is activated for large depolarizations ($V_m > -20$ mV), inactivates more slowly, is blocked by ω-conotoxin, and is found in neurons. The P-type channel is activated for large depolarizations, is insensitive to dihydropyridine and ω-conotoxin, and is found in neurons.

Voltage-Gated Chloride Channels

Chloride is the most abundant anion in plants and animals. Chloride channels have been found in virtually every type of cell investigated, both in the cellular membrane and in the membranes of organelles (Guggino, 1994; Peracchia, 1994). Progress in characterization of chloride channels lagged behind that of the cation channels, but has accelerated rapidly since the mid-1980s. A number of different types of chloride channels are now distinguished. Some are referred to as *background chloride channels* because they contribute appreciably to the resting potential of the cell; they tend to be open at rest. There are several types of background chloride channels, which are referred to as ClC-0, ClC-1, ClC-2, and CFTR. ClC-0 channels have been isolated from the electric organs of *Torpedo californica*, show outward rectification, and are activated by depolarization. ClC-1 channels are found in skeletal muscle and show inward rectification. ClC-2 channels are found widely in tissues, and they open slowly in response to large hyperpolarizations. CFTR (cystic fibrosis transmembrane regulator) channels are found in secretory epithelia, where they play a role in the transport of salts across the epithelium. Defects in the gene encoding this protein have been shown to be the genetic basis of cystic fibrosis.

6.2.3.3 Summary

Ion channels are found pervasively in the membranes of all types of cells and in the organelles of eukaryotic cells. Categorization based on biophysical and pharmacological properties reveals that there is a large number of categories of channels. Different cell types express different constellations of ion channels to serve different physiological needs. So, for example, a pacemaker neuron in the brain will express a different constellation of ion channels than an alga. The expression is dynamic so that the constellation of channels in a cell can change in response to signals received by the cell; ion channels can be mobilized for insertion into the membrane in response to a chemical signal. Channels may also show different spatial segregations on a cell surface. For example, the channel composition differs at the nodes and internodes of a myelinated fiber; the density of sodium channels is much higher at the nodes than at the internodes.

The physiological roles of the different ion channels are still being defined. Some roles are quite clear. For example, the primary role of the voltage-gated sodium channel is to produce action potentials in electrically excitable cells. However, that cannot be its sole function, since some electrically in-excitable cells also express sodium channels. In some cells, voltage-gated calcium channels are also involved in the production of (calcium) action potentials. Voltage-gated calcium channels also allow for the entry of calcium into cells, which triggers a number of cellular responses, e.g., responses to control calcium-based secretion, to open calcium-activated potassium channels, to control motility, and to control enzyme catalytic activity. The potassium channels are important in determining the resting potential of cells. The myriad types of potassium channels are involved in shaping the intrinsic electrical activity of neurons. These potassium channels shape the repolarization of individual action potentials, produce action potential trains with different temporal characteristics, and produce sustained action potential trains in pacemaker cells.

6.3 Gating Currents

To understand the physical basis of gating currents and the distinction between gating currents and ionic currents, it is helpful to first understand the distinction between conduction and displacement currents.

6.3.1 Conduction and Displacement Currents

In order to make the laws of electromagnetism consistent, Maxwell generalized the notion of current by introducing the *displacement current* into Ampère's law. Ampère's law, as amended by Maxwell, is

$$\nabla \times \mathcal{H} = \underbrace{\mathbf{J}_c}_{\text{Conduction current}} + \underbrace{\frac{\partial \mathcal{D}}{\partial t}}_{\text{Displacement current}} , \tag{6.2}$$

where $\mathcal{H}$ is the magnetic field intensity vector, $\mathbf{J}_c$ is the conduction current density vector, $\mathcal{D}$ is the displacement flux density vector, and $\nabla \times$ is the curl operator. The displacement current density is the rate of change of the displacement flux density with respect to time. This term is required to make Ampère's law consistent with both Gauss's law and conservation of charge as expressed in the continuity relation for conduction current. Gauss's law states that

$$\nabla \cdot \mathcal{D} = \rho_c, \tag{6.3}$$

where ρ_c is the free charge density, i.e., the density of charges that are free to move and lead to conduction current, and $\nabla\cdot$ is the divergence operator. We note that taking the divergence of Equation 6.2 yields

$$\nabla \cdot (\nabla \times \mathcal{H}) = \nabla \cdot \mathbf{J}_c + \frac{\partial(\nabla \cdot \mathcal{D})}{\partial t}.$$

We also note that the divergence of the curl of any vector is zero and use Gauss's law to eliminate $\mathcal{D}$, which yields the continuity relation for the conservation of charge,

$$\nabla \cdot \mathbf{J}_c = -\frac{\partial \rho_c}{\partial t}. \tag{6.4}$$

Omission of the displacement current term from Ampère's law would have led to the conclusion that $\nabla \cdot \mathbf{J}_c = 0$, which contradicts the continuity relation, and hence conflicts with conservation of charge.

To understand the displacement current density further, we need to investigate the displacement flux density, $\mathcal{D}$. This quantity can be expressed as $\mathcal{D} = \epsilon_o \mathcal{E} + \mathcal{P}$, where ϵ_o is the permittivity of free space, $\mathcal{E}$ is the electric field intensity, and $\mathcal{P}$ is the polarization density. $\mathcal{P}$ reflects the effect on the displacement flux density of the presence of electrically polarizable matter. Thus, the displacement flux density has contributions that are independent of the presence of matter and those due to the presence of polarizable matter whose effects are contained in the polarization density, $\mathcal{P}$. What is $\mathcal{P}$?

The relation between the polarization density and the electric field is a material property. Some materials have a permanent polarization density independent of the electric field; others have a polarization density that depends on the electric field. The application of an electric field typically results in the motion of charges in an electrical insulator. However, the motion of charges in an insulator differs appreciably from the motion of charges in a conductor. In a conductor, charges (e.g., electrons in a metal or ions in an electrolytic solution) are relatively free to move and can be transported over large distances. However, in an insulator charges move only microscopic distances. The point is illustrated schematically in Figure 6.15, which shows an insulator with microscopic charged domains. Each microscopic domain represents the material constituents at a microscopic level. For example, each domain might represent an individual atom that consists of a positively charged nucleus surrounded by a negatively charged electronic cloud. Alternatively, the representation might be at a molecular level, where each molecule shows a charge separation: one end is charged positively, the other end nega-

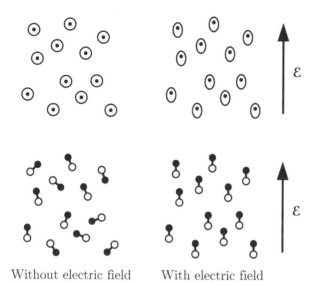

Figure 6.15 Illustration of the effect of an electric field on distribution of charge in an insulator. The black regions represent a positive charge, and the clear regions a negative charge. In the upper panel, application of the electric field shifts the center of gravity of the negative charge distribution with respect to the positive charge distribution. In the lower panel, application of the electric field orients the charges so that they are aligned with the electric field.

Figure 6.16 Definition of an electric dipole. The dipole moment $\mathbf{p} = \lim_{l \to 0, q \to \infty} q\mathbf{l}$ where $\mathbf{l}$ is a vector whose length (l) equals the distance between the charges and which points from the negative to the positive charge. The limit is taken as the product of q and $\mathbf{l}$ is held constant.

tively. In both representations, there are strong forces that prevent complete separation of the charged entities. However, application of an electric field can result in a microscopic separation of charge or a reorientation of the charge.

This separation of charge constitutes an electric dipole that has no net charge, but consists of two equal and opposite charges separated by a distance. The dipole moment $\mathbf{p}$ is defined as shown in Figure 6.16. While each dipole contains no net charge, the orientation of a population of dipoles in an electric field leads to a separation of charge. As can be seen in Figure 6.15, when the dipoles in a material are lined up, the center of gravity of all the negative charges is not identical with the center of gravity of all the positive charges. Hence, there exists a charge separation in the material that is reflected in the macroscopic polarization density. The polarization density is simply the density of dipole moments in a material. For example, if N is the density of dipoles in a material, each of which has a dipole moment $\mathbf{p}$, then $\mathcal{P} = N\mathbf{p}$.

Gauss's law (Equation 6.3) can be expressed in terms of the polarization density as

$$\nabla \cdot (\epsilon_o \mathcal{E} + \mathcal{P}) = \rho_c, \tag{6.5}$$

which can be rewritten as

$$\nabla \cdot (\epsilon_o \mathcal{E}) = \rho_c + \rho_p,$$

where ρ_p is the polarization charge density, defined as

$$\rho_p = -\nabla \cdot \mathcal{P}. \tag{6.6}$$

The sign of the relation between the polarization charge density and the polarization density is a consequence of the definition of **p** that points from $-q$ to $+q$, whereas the electric field of a pair of point charges points from the positive charge to the negative charge.

From Ampère's law (Equation 6.2), the component of the displacement current due to the polarization density is simply

$$\mathbf{J}_p = \frac{\partial \mathcal{P}}{\partial t}. \tag{6.7}$$

If we take the divergence of Equation 6.7 and combine this equation with Equation 6.6, we find that

$$\nabla \cdot \mathbf{J}_p = -\frac{\partial \rho_p}{\partial t},$$

which demonstrates that the polarization charge also obeys its own continuity relation.

What happens at a boundary between two materials? The boundary condition on the polarization density can be obtained directly by evaluating Equation 6.6 in a pillbox that encloses the boundary between the two materials, as shown in Figure 6.17. This volume integral on the pillbox is

$$-\int_V \nabla \cdot \mathcal{P} \, dv = \int_V \rho_P \, dv, \tag{6.8}$$

where dv is an incremental volume element and $\mathcal{V}$ is the volume of the pillbox. Using Gauss's integral theorem, the left-hand volume integral in Equation 6.8 can be replaced by a surface integral as follows:

$$-\oint_S \mathcal{P} \cdot d\mathbf{s} = \int_V \rho_P \, dv, \tag{6.9}$$

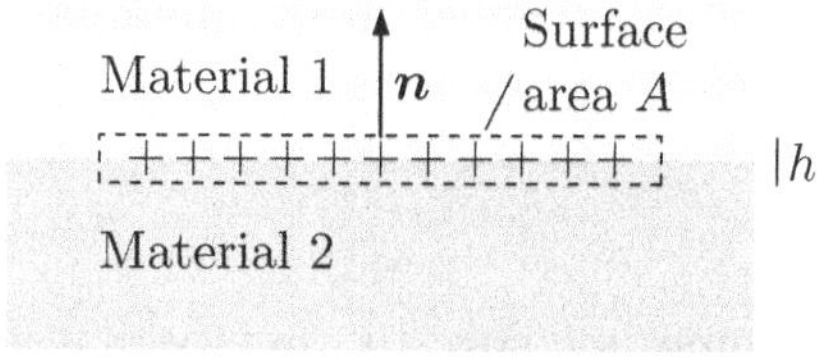

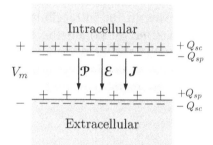

Figure 6.17   Pillbox through the boundary between two materials used to determine boundary conditions on polarization density. The boundary between the two materials extends in a plane perpendicular to the paper. A cross section of the pillbox is shown with a dashed line. The pillbox has two large surfaces of area A and a thickness of h. Here **n** is a unit normal pointing out of the surface of the pillbox.

Figure 6.18   Polarization density and both polarization and conduction surface charges on a membrane. The directions of the polarization density $\mathcal{P}$, electric field $\mathcal{E}$, and current density **J** are normal to the membrane and point outward. The magnitudes of these vectors are: $\mathcal{P}$, $\mathcal{E}$, and J.

where $d\mathbf{s}$ is an incremental surface area of the volume element and S is the surface of the pillbox. As the thickness of the pillbox $h \rightarrow 0$, the contribution of the edges of the pillbox becomes negligible and the left-hand term in Equation 6.9 approaches $-(\mathcal{P}_1 \cdot \mathbf{n}A - \mathcal{P}_2 \cdot \mathbf{n}A)$. The right-hand term is simply the charge enclosed in the infinitesimal pillbox, which we can express as $Q_{sp}A$, where Q_{sp} is the surface polarization charge density. Substitution of these terms into Equation 6.9 yields

$$-\mathbf{n} \cdot (\mathcal{P}_1 - \mathcal{P}_2) = Q_{sp}. \tag{6.10}$$

Further insight can be obtained into the polarization current and its measurement from an analysis of the simple capacitance structure shown in Figure 6.18. A membrane separates two good conductors, here indicated by the intracellular and extracellular media. The membrane is a pure dielectric and allows no conduction current. The membrane is assumed to have a polarization density normal to the membrane. The polarization density gives rise to a polarization surface charge on each membrane : solution interface as shown.

The polarization current density component in the membrane J_p can be expressed as

$$J_p = \frac{d\mathcal{P}}{dt} = \frac{dQ_{sp}}{dt}.$$ (6.11)

Thus, the component of the membrane current density that is due to polarizable matter in the membrane can be expressed in terms of either the polarization surface charge or the polarization density in the membrane.

The only current that can be detected with electrodes not located in the membrane proper is the conduction current detected in the intracellular or the extracellular solution. Can the membrane polarization current be detected from a measurement of the conduction current in the solutions? To determine this, we shall compute current flowing in the intracellular and extracellular conductors, which we shall obtain after we obtain the electric field in the membrane. From the geometry, the electric field in the membrane has the magnitude (V_m/d). Integration of Gauss's law in the form of Equation 6.5 on a pillbox over the intracellular solution: membrane interface yields

$$\epsilon_o \frac{V_m}{d} + \mathcal{P} = Q_{sc}.$$

An identical argument can be applied to the other interface, which shows that there is a surface conduction charge that equals $-Q_{sc}$. The conduction current density in the intracellular and extracellular media is evaluated from integrating the continuity equation for the conduction current in a suitable pillbox, e.g., on the intracellular solution: membrane interface, to yield

$$J_c = \frac{dQ_{sc}}{dt} = \frac{\epsilon_o}{d}\frac{dV_m}{dt} + \frac{d\mathcal{P}}{dt},$$

which is simply the total displacement current in the membrane.

What happens when the polarization density in the membrane changes? A change in polarization density leads to a change in polarization charge density, which contributes to the displacement current in the membrane and changes the surface conduction charge on the membrane, which leads to a change in the conduction current density in the intracellular and extracellular solutions. Thus, there is a component of the conduction current in the solutions that is due to the polarization current in the membrane. The surface conduction charge on the membrane is the intermediary between the polarization current in the membrane and the conduction current in the solutions.

6.3.2 Production of Gating Currents by Voltage-Gated Channels

By definition, voltage-gated channels, such as the voltage-gated sodium and potassium channels, open in response to a change in membrane potential. What is the mechanism that links the change in potential to the opening of such a channel? One plausible possibility is that the channel macromolecule contains charged groups. The macromolecule need contain not a net charge, only an asymmetric charge distribution that can interact with the electric field in the membrane. For example, the molecule might have no net charge, but might have a nonzero dipole moment. In the closed conformation, the charges are distributed in one (minimum-energy) configuration; in the open conformation this distribution is different. When the channel opens in response to a change in membrane potential, the membrane macromolecule undergoes a conformational change in which the charges redistribute themselves in a minimum-energy conformation at the new potential. The time rate of change of the charge redistribution in response to a membrane potential change constitutes a displacement current through the membrane. Membranes may contain a number of sources of displacement current that are unrelated to the gating of ionic channels. Hence, the term *gating current* is reserved for those components of the displacement current through a membrane that are caused by the gating of ionic channels. This gating current is an electrical sign of the change in conformation of the channel macromolecule. Thus, it appears that gating currents are inevitable, but a priori it is not clear that they are detectable.

6.3.3 Components of Membrane Current

The membrane current detectable by external electrodes on the two sides of the membrane contains two types of currents: conduction or ionic currents and displacement or capacitance currents. Ionic currents represent ions that flow through the membrane. Capacitance currents represent redistribution of charge on the surface of the membrane, which reflects redistribution of charge in the interior of the membrane. The ionic and capacitance current components are shown schematically in Figure 6.19 in response to a membrane depolarization. Ionic currents are shown for a voltage-gated sodium channel (f) and a voltage-gated potassium channel (i). The capacitance current results from the net movement of surface charge on the two membrane solution interfaces (a). Sources of this change in surface charge reflect redistribution of charge within the lipid bilayer (b); either net movement of charge in the membrane (c) or reorientation of dipoles (d) for macromolecules in the membrane that

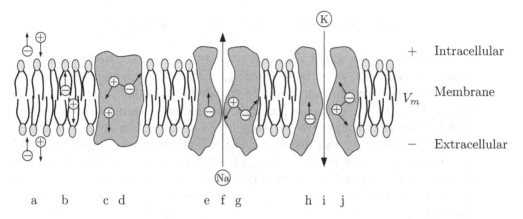

Figure 6.19 Components of membrane current (adapted from Bezanilla, 1982, Figure 1).

do not gate ionic channels; and charge redistribution associated with voltage-gated ionic channels (e, g, h, and j). This list does not exhaust all the possible mechanisms of generation of capacitance current.

We can express the membrane current density, $J_m(t)$, as

$$J_m(t) = J_g(t) + J_{ng}(t) + J_{ion}(t), \tag{6.12}$$

where $J_g(t)$ is the capacitance current associated with the gating macromolecules or the gating current, $J_{ng}(t)$ is the capacitance current not associated with the gating macromolecules, and $J_{ion}(t)$ is the total ionic current density through the membrane. The two capacitance currents can be expressed in terms of the rate of change of the two charge densities

$$J_g(t) = \frac{dQ_g(t)}{dt} \quad \text{and} \quad J_{ng}(t) = \frac{dQ_{ng}(t)}{dt}, \tag{6.13}$$

where $Q_g(t)$ is the integral of $J_g(t)$ and is called the *gating charge density* and $Q_{ng}(t)$ is the integral of $J_{ng}(t)$ and is a charge density associated with nongating capacitance current.

6.3.4 Estimation of Gating Current as an Asymmetrical Capacitance Current

To estimate the gating current, other membrane current components, such as other capacitance current components and ionic currents, must be reduced in magnitude or estimated separately. The relatively large ionic currents can be reduced by substituting impermeant ion species (such as cesium) for highly permeant ions (such as sodium and potassium) and by using channel-blocking molecules such as TTX and TEA. These two pharmacological channel blockers prevent permeation of ions through the channel, but do not eliminate the voltage-induced charge redistribution in the membrane. This difference in effect has implications for the site of action of these blockers.

Thus, the ionic currents can be reduced, but how can the two capacitance currents be separated? There is at present no a priori way to achieve such a separation. However, the capacitance current can be separated into two components: the linear capacitance current and the nonlinear capacitance current. Because there are a discrete number of channels in the membrane, we expect the charge carried by the gating macromolecule to be a nonlinear function of membrane potential. For example, suppose that depolarization opens a channel. Then if a depolarization is large enough to open all the channels a further depolarization will open no more. Thus, we expect the voltage dependence of the gating charge to saturate at large depolarizations. Therefore, the gating current is expected to contribute an appreciable component to the nonlinear capacitance current.

Let the capacitance current, J_c, be parsed into linear and nonlinear components as follows:

$$J_c(t) = J_{cl}(t) + J_{cn}(t), \tag{6.14}$$

where $J_{cl}(t)$ is the linear component of the capacitance current and $J_{cn}(t)$ is the nonlinear component of the capacitance current. Equation 6.14 can be written as

$$J_c(t) = \frac{dQ_{cl}(t)}{dt} + \frac{dQ_{cn}(t)}{dt} = \frac{dQ_{cl}(t)}{dV_m}\frac{dV_m(t)}{dt} + \frac{dQ_{cn}(t)}{dV_m}\frac{dV_m(t)}{dt}, \tag{6.15}$$

where Q_{cl} represents the charge density associated with the linear component of the capacitance current and Q_{cn} represents the charge density associated with the nonlinear (voltage-dependent) component of the capacitance current. Equation 6.15 can also be expressed as

$$J_c(t) = C_l\frac{dV_m(t)}{dt} + C_n(V_m)\frac{dV_m(t)}{dt}, \tag{6.16}$$

where C_l represents the capacitance associated with the linear component of the capacitance current and C_n represents the capacitance associated with the nonlinear (voltage-dependent) component of the capacitance current.

The linear and nonlinear components of capacitance current are shown schematically in response to changes in membrane potential in Figure 6.20.

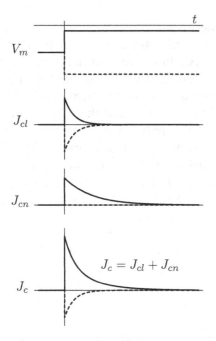

Figure 6.20 Schematic diagram of the linear and nonlinear components of capacitance currents. All the capacitance currents are shown as decaying transients, in part because the resistance in series with the membrane is assumed not to be zero.

During the holding potential at -100 mV, the membrane is hyperpolarized, and let us presume that all the channels are closed. A step of hyperpolarization to -200 mV will open no further channels; hence, we might expect the current due to the gating charge to be zero and all the capacitance current to be due to the nongating charge. However, for a step of depolarization of 100 mV to a membrane potential of 0 mV, there should be a large gating current as well as a capacitance current due to other charges. The assumption that the other sources produce a capacitance current that is related linearly to the membrane potential and that for sufficiently large hyperpolarizations the gating current is zero allows for a unique separation of the gating current from the total capacitance current. If these assumptions are valid, addition of the current due to the hyperpolarization to that due to depolarization yields the gating current (Figure 6.20). Because large hyperpolarizations are not well tolerated by cells, the method of separation that is used takes advantage of the assumption of linearity. For example, the current in response to a hyperpolarization of 20 mV is measured and multiplied by five, then added to the current in response to a depolarization of 100 mV. Measurements show that the capacitance current is related approximately linearly to the membrane potential for hyperpolarizations of different amplitudes.

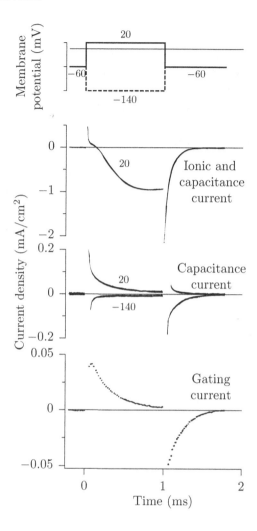

Figure 6.21   Measurement of gating current in a voltage-clamped squid giant axon (adapted from Keynes and Rojas, 1976, Figure 9). The axon was internally perfused with a solution containing cesium fluoride and no sodium or potassium. The upper trace shows the membrane potential profiles that produced the current densities shown below. The trace labeled *Ionic and capacitance current* shows the current density measured in response to an 80 mV depolarization when the axon was bathed in an artificial seawater solution that contained 1/4 the normal concentration of sodium. The trace labeled *Capacitance current* shows the response to both a depolarization and a hyperpolarization of 80 mV when the axon was bathed in a solution that contained no sodium, but 300 nmol/L of TTX. The sum of the responses to the depolarization and the hyperpolarization is defined as the *Gating current*. The lowest panel shows the average of 300 gating currents.

Figure 6.21 shows measurements of membrane currents under voltage clamp for a perfused giant axon of the squid. The membrane potential (top trace) is a step of ±80 mV starting at the holding potential of −60 mV. The second trace shows the membrane current that occurs in response to a depolarization of 80 mV when the external solution is seawater containing 1/4 of the normal sodium concentration (to reduce the sodium current) and the internal solution contains a solution containing cesium ions (to reduce the potassium current). The membrane current contains a rapid capacitance current at the onset of the voltage step followed by an inward current that is carried by sodium ions. Only the onset of the sodium current occurs on this brief time scale. If an impermeant ion is substituted for the remaining sodium ions in the external solution and if TTX is added to the external solution, the remaining ionic current is greatly reduced. Under these conditions, responses to 80 mV depolarizations and 80 mV hyperpolarizations (Figure 6.21) are presumed to reflect the capacitance current only. The current in response to depolarizations is not simply the mirror image of the response to hyperpolarizations; the capacitance current is asymmetrical. The sum of these two capacitance currents is an outward capacitance current (lowest trace in Figure 6.21) which is an estimate of the gating current. Because the gating current is relatively small compared to both the ionic current and the total capacitance current, signal averaging is used to reduce the measurement noise. For example, to obtain the results shown in Figure 6.21, responses to 300 depolarizing and 300 hyperpolarizing steps were averaged and then added to obtain the gating current shown in the bottom trace. Figure 6.22 indicates that the magnitude of the gating current is small compared with the ionic current carried by sodium.

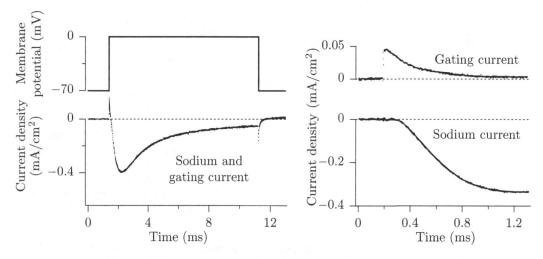

Figure 6.22 Comparison of gating and sodium currents in squid giant axon in response to a 70 mV pulse of depolarization (adapted from Bezanilla, 1985, Figure 3). The axon was perfused internally with solutions containing impermeant cations. The *Sodium and gating current* was obtained from an axon in a bathing solution that contained 1/5 the normal sodium concentration. Linear leakage and capacitance current were subtracted by the addition of an appropriately scaled response to a hyperpolarization. The *Gating current* was measured in a bathing solution that contained no sodium, but contained 300 nmol/L of TTX. The *Sodium current* was obtained by subtracting the *Gating current* trace from the *Sodium and gating current trace.*

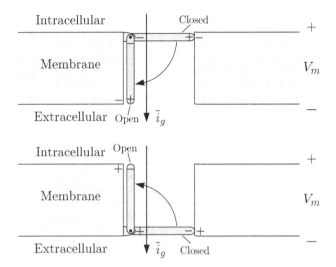

Figure 6.23 Schematic diagram illustrating gating currents for positively or negatively charged gates. For both gates, a depolarization of the membrane will tend to open the gate. For the gate whose free end is positively charged, a depolarization results in an outward motion of a positive charge. For the gate whose free end is negatively charged, a depolarization results in an inward displacement of a negative charge. Thus, for both gates a depolarization results in an outward capacitance current.

In these measurements, the potassium current has been reduced by the use of TEA and the nongating capacitance current has been subtracted as in Figure 6.21.

6.3.5 Properties of the Gating Current

6.3.5.1 Polarity of the Gating Current

We have seen (Figure 6.21) that the gating current in response to a step depolarization is an outward current. How can we interpret this result? Since the sodium and potassium channels described in Chapter 4 open when the membrane is depolarized, the membrane potential-induced charge redistribution should correspond to an outward movement of positive charge or an inward movement of negative charge (as indicated schematically in Figure 6.23). Both of these possibilities correspond to an outward gating current.

6.3.5.2 Kinetics of Gating Current

Figure 6.22 shows the gating current in response to a membrane potential pulse. Initial results (Keynes and Rojas, 1974) suggested that the gating current decline was characterized by an exponential function of time, suggesting first-order kinetics of the gating current. However, more recent work suggests that the gating current has several components that behave differently as a

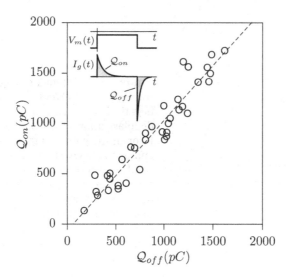

Figure 6.24 Relation of gating charge at onset and offset of a depolarization (adapted from Keynes and Rojas, 1974, Figure 8). $I_g(t)$ is the total gating current through the membrane. The dashed regression line has the equation $Q_{on} = 1.07 Q_{off} - 47$ with a correlation coefficient of 0.95.

function of membrane potential and time. The dissection of gating current kinetics into different components is ongoing, and when completed should provide information on the motion of charges in the membrane that occur when the ionic channel moves among its allowable discrete states.

6.3.5.3 Response of On-Transient and Off-Transient Gating Charges to Brief Pulses

In response to brief pulses of membrane potential, the total gating charge, Q_{on}, estimated by integrating the gating current at the onset of the membrane potential pulse is equal to the total gating charge, Q_{off}, estimated from the offset of the pulse (Figure 6.24). This result suggests that for such brief pulses, the charge redistribution in the membrane is reversible.

6.3.5.4 Saturation of Gating Charge

If the total gating charge is measured at the offset of a depolarization of the membrane potential, Q_{off}, this charge saturates at large depolarizations (Figure 6.25). This result is consistent with the notion that the gating charge represents a finite density of membrane-bound charges that are displaced in response to a depolarization. For a sufficiently large depolarization, all the gating charges are displaced so that all the gates are in their open conformations. Further depolarizations do not result in any further displacement of charge in the membrane.

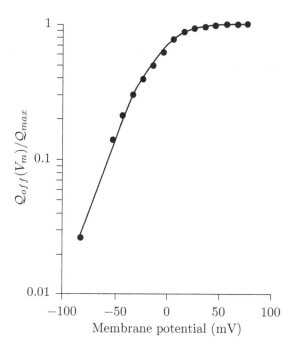

Figure 6.25 Dependence of gating charge on membrane potential for a squid giant axon (adapted from Keynes and Rojas, 1974, Figure 9). The gating charge normalized to the maximum gating charge is plotted versus the membrane potential in semilogarithmic coordinates.

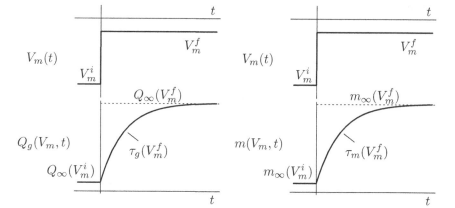

Figure 6.26 Comparison of time course of gating charge with that of the sodium activation factor.

6.3.5.5 *Relation of Gating Charge to Sodium Activation*

As indicated in Figure 6.26, the properties of the gating charge shown in Figures 6.21 through 6.25 have some qualitative similarities to the Hodgkin-Huxley activation parameter m. Both show approximately exponential changes in time in response to a depolarization. Both have final values that saturate with increasing membrane potential, and their time constants have similar dependence on membrane potential. However, quantitative comparisons between sodium activation and gating current are deferred until after discussions of two-state (Section 6.5) and multiple-state (Section 6.6) models of ion channels.

6.3.6 Summary and Conclusions

The gating currents, which are estimated from the asymmetry of the capacitance current, appear to have the properties one would expect of membrane potential-induced charge redistributions of the sodium channel. We have summarized some of the simpler properties of these gating currents. Gating currents related to charge distribution of the potassium channel have also been obtained (Bezanilla et al., 1991).

As we shall see, gating currents can help investigators to determine the kinetic behavior of channels. However, with only a few exceptions (Mika and Palti, 1994), most measurements of gating currents have been restricted by their poor spatial resolution. Measurements from a whole cell represent the gating currents from a large population of channels. Furthermore, there is no guarantee that measurements of the gating currents result from a homogeneous population of channels. Difficulties with the spatial resolution of the measurements, which limit the usefulness of measurements of gating currents, have been obviated for measurements of ionic currents that can be obtained from single channels.

6.4   Ionic Currents in Single Channels

A major advance in the study of transport through biological membranes occurred in the 1970s, when techniques were developed (Neher and Sakmann, 1976) to record ionic currents through individual ionic channels. This technical development allowed for the investigation of the kinetic properties of channels, such as the voltage-gated sodium channel, many of whose properties were already well known, with a new spatial resolution. These measurements allowed for much more direct studies of channel mechanisms.

6.4.1   Properties of Single-Channel Currents

A priori one might have expected that the currents through single channels were simply scaled replicas of the macroscopic currents, as shown schemat-

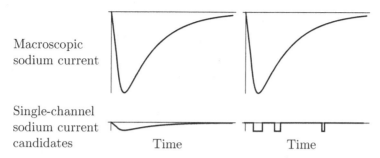

Figure 6.27 Two possible relations between macroscopic and single-channel currents. In one relation (left panel), the single-channel current is a scaled version of the macroscopic current. In another (right panel), the waveform of the single-channel current differs appreciably from the waveform of the macroscopic current.

ically in the left panel of Figure 6.27. In this conception, we can imagine that the gate position is a continuous variable that lets more or less current through depending on the gate position. This conception is analogous to a faucet that lets more or less water through depending on the position of its valve. However, this expectation has proved wrong in several respects. The waveform of the current through a single ion channel does not resemble the waveform of the macroscopic current.

6.4.1.1   Rapid Transitions of Single Channels between Discrete Conducting States

As shown in Figure 6.28, single-channel currents exhibit rapid transitions between discrete conducting states. Most channels described to date have one conducting or open state and multiple closed states. In the open state, current flows through the channel, and in the closed states it does not. However, there are some channels with more than one conducting state.

6.4.1.2   Probabilistic Opening and Closing of Single Channels

The transitions between open and closed channel states are not only rapid, but they occur at random. The rates of transitions among states depend upon the gating variable. So, for example, the voltage-gated channel shows a larger probability of opening and a larger probability of remaining open for $V = -190$ mV than for $V = -110$ mV (Figure 6.28).

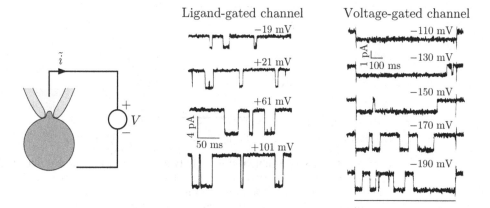

Figure 6.28 Measurements of currents through voltage-clamped patches of membrane in two different cell types illustrating two different types of single channels. The left panel shows a schematic diagram of the recording arrangement. The center panel shows results obtained from a ligand-gated channel of a cutaneous pectoris muscle fiber in response to 50 nmol/L suberyldicholine present in the patch pipette (adapted from Hamill et al., 1981, Figure 8). The right panel shows currents recorded from a voltage-gated potassium channel in tunicate egg (adapted from Fukushima, 1982, Figure 1). Both measurements were obtained with the *cell-attached configuration* (Figure 6.1). The parameter for each trace is the potential applied to the patch pipette with respect to the bath, V. The potential is applied continuously to the ligand-gated channel and for 900 ms (indicated by the horizontal bar at the bottom of the traces) for the voltage-gated channel.

6.4.1.3 Two Distinct Effects of Membrane Potential on Single-Channel Currents

A change in the membrane potential across a single channel affects the single open-channel current magnitude for all single channels, whether they are voltage gated or not. For example, note that the magnitude of the current when the channel is open increases as the magnitude of the potential increases for the ligand-gated channel in Figure 6.28. However, the rate of opening or closing of the channel does not change appreciably. Careful statistical analysis would need to be done to check the validity of this assertion in a quantitative manner. This channel is gated by a ligand, suberyldicholine, which is an acetylcholine agonist whose concentration is not changed during the course of these measurements. These results are distinctly different from those shown for the voltage-gated channel. For this channel, the magnitude of the open-channel current also increases as the magnitude of the potential increases, which is consistent with the ligand-gated channel. However, for the voltage-gated chan-

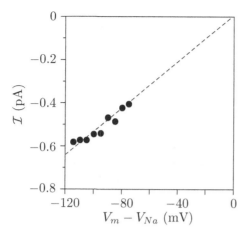

Figure 6.29   Voltage-current characteristic of a single open channel from cardiac muscle of a rat (adapted from Saint et al., 1994, Figure 5). The amplitude of the single open channel current is plotted versus the difference in potential from the sodium equilibrium potential across the patch.

nel, the probability that the channel is open also increases markedly as the potential V is made more negative.

6.4.1.4   *Voltage-Current Characteristic of an Open Channel*

A measurement of the amplitude of the single-channel current as a function of membrane potential is shown in Figure 6.29. The current amplitude is plotted versus $V_m - V_{Na}$, and the results suggest that the current is proportional to the difference in potential from the sodium equilibrium potential, i.e., $I = \gamma(V_m - V_{Na})$. The slope of the line is the single open-channel conductance, which is $\gamma = 5.3$ pS for the data shown.

6.4.1.5   *Macroscopic Membrane Currents Resulting from the Summation of Single-Channel Currents*

A summation of the current through one single channel over a sequence of identical depolarizations resembles the macroscopic current through a population of channels of the same type (Figure 6.30). This result is consistent with the notion that the macroscopic current represents the sum of currents through a population of channels.

6.4.1.6   *Channel Kinetic Properties Revealed by Statistics of Open and Closed Times*

Measurements of single-channel currents under voltage-clamp conditions reveal properties of channel gating in a quite direct manner. For exam-

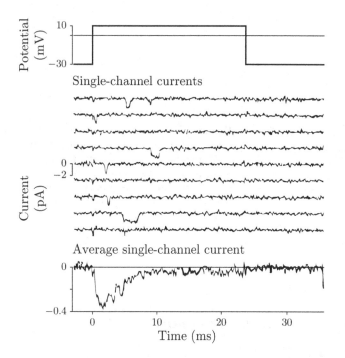

Figure 6.30 and associated plots.

Figure 6.30 Time course of single sodium channels under voltage clamp (adapted from Sigworth and Neher, 1980, Figure 2). The upper panel shows the potential across the patch. The next panel shows the current through the patch for nine successive pulses of membrane potential identical to the one shown in the top panel. Leakage and capacitance currents were subtracted from each of these traces. The lowest trace shows the average current at each instant of time following the voltage pulse obtained from 300 records, of which the nine traces shown are a subset.

ple, inactivation of sodium is manifested in macroscopic current as a reduction in sodium current for a constant membrane potential. Prior to single-channel recording, this might have been interpreted as due to single channels, each of whose currents decreased with time during a constant membrane potential. This point was investigated quite directly by recording from single channels before and after the application of substances that block sodium inactivation. Application of pronase (Figure 6.7) and N-bromoacetamide to the inside of the membrane can virtually eliminate sodium inactivation. Measurements of single-channel currents reveal that both of these blockers of inactivation affect the duration of the time that a channel remains open and not the current amplitude (Figure 6.31). So the possibility that the single-channel currents simply mirror the macroscopic currents is shown to be incorrect by such experiments. Furthermore, the effect of inactivation blockers can be quantified from measurements of statistical properties of the dwell times of open channels. Measurements of the durations for which a channel is open can be summarized by a histogram that shows the fraction of intervals of a given length as a function of the interval length (Figure 6.32). Histograms in the presence and absence of NBA reveal that the durations of

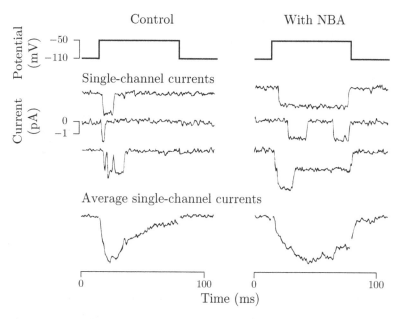

Figure 6.31   Effect of NBA on single-channel currents (adapted from Patlak and Horn, 1982, Figure 1). The left and right panels show the response of a patch obtained from tissue-cultured embryonic muscle cells of the rat both before (left panels) and after (right panels) addition of 300 μmol/L of NBA. Both individual responses to the change in membrane potential and averages of the responses to 144 (for the control) and 96 (for the NBA) voltage pulses are shown. The average currents have been normalized to have the same peak amplitude.

channel-open dwell times are an order of magnitude longer in the presence of NBA than in a control solution.

6.4.1.7   Conduction through Open Channels

Most voltage-gated ionic channels investigated contain one open state whose conductance is relatively independent of voltage. The conductances of these single channels tend to be in the range of 1–10^2 pS (Hille, 1992, Tables 12-3 and 12-4).

6.4.2   Density of Ion Channels

A number of different methods have been used to estimate the density of ionic channels in membranes. Some of these methods involve the use of radio-

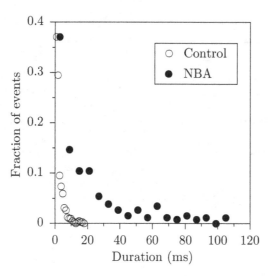

Figure 6.32 Effect of NBA on single-channel open times (adapted from Patlak and Horn, 1982, Figure 3). The durations of open intervals were measured from several patches, and the fraction of intervals having a given duration is plotted versus the duration. The mean durations were 3.2 ms for the control measurements and 29.1 ms after treatment with NBA.

actively labeled channel blockers that bind to channels. Based on the assumption that one molecule of blocker blocks one channel, one can obtain an upper bound on the density of channels. Some of these and other estimates are summarized in Table 6.4 for the voltage-gated sodium channel. These estimates range from a few channels to hundreds of channels per square micrometer of membrane. Thus, the channels are separated by an average distance of between 10 nm and 1 μm.

We saw in Chapter 5 that the density of voltage-gated sodium channels varies in myelinated fibers; the density is much higher at the nodes of Ranvier than in the internodes. The density is also much higher at the trigger zone of a neuron than over the rest of the surface of the neuron. Furthermore, the distribution of channels in myelinated fibers changes when the fibers are demyelinated. Thus, channel densities need not be uniform over the surface of a cell or stable in time.

6.4.3 Summary

Currents measured from single ion channels are random rectangular waves that switch rapidly and randomly between a discrete set of conduction states. The current through an open channel depends upon the membrane potential. The probability of switching depends upon a gating variable. In voltage-gated channels, this probability depends upon the membrane potential. Macroscopic currents are sums of single-channel currents. The probabilistic properties of single-channel currents reveals kinetic properties of the channel.

Table 6.4 Estimates of sodium channel density based on several different techniques. (adapted from Hille, 1992, Tables 12-1, 12-2, and 12-3). The toxin density method is based on estimates of the density of TTX and STX binding sites; the gating charge method is based on measurements of the maximum gating charge, an assumption that this charge is due to gating the sodium channel and an assumed valence of the sodium gate of 6; the conductance method is based on comparison of measurements of the macroscopic conductance of membranes with the single-channel conductance estimated either from single-channel recordings or from noise analysis.

| Preparation | Channel density (channels·μm^{-2}) | | |
	Toxin density	Gating charge	Conductance
Squid giant axon	166–533	300	330
Frog node of Ranvier	—	3000	400–1900
Rat node of Ranvier	—	2100	700
Lobster walking leg nerve	90	—	—
Crayfish giant axon	—	367	—
Myxicola giant axon	—	105	—
Garfish olfactory nerve	35	—	—
Frog sartorius muscle	195–380	—	—
Frog twitch muscle	—	650	—
Rabbit vagus nerve (unmyelinated)	110	—	—
Rabbit myelinated nerve	23,000	—	—
Neuroblastoma cells	78	—	—
Rat diaphragm, soleus, and EDL muscles	209–557	—	—
Rat ventricle	—	43	—
Dog purkinje fiber	—	200	—
Mouse skeletal muscle	—	—	65
Bovine chromaffin cells	—	—	1.5–10

6.5 Model of a Voltage-Gated Channel with One Two-State Molecular Gate

In this section, we shall analyze the predictions of a model of a voltage-gated ionic channel that has one two-state molecular gate that tends to open when the membrane is depolarized, i.e., the channel has an activation gate. This model is the simplest of a class of models of voltage-gated ionic channels, yet this model captures the flavor of such models without the analytic complexity of more elaborate models. This model is explored in detail in this section; the next section deals with generalizations to more realistic models of the gating of single channels.

Consider a membrane containing a population of independent and identical macromolecules that form channels through the membrane. The density of channels is $\mathcal{N}$ channels/cm^2. Each channel is controlled by a charged molecular gate. At each instant in time, each gate is found in one of two conformations, open or closed, as shown schematically in Figure 6.4. When the gate is in its open position, the channel is open and an ionic current flows through the channel; when the gate is closed, no ionic current flows through the channel. The channel is perfectly selectively permeable, so that only one ionic species, n, flows through an open channel; all other ions are rejected. Since the gate is charged, its state depends upon the membrane potential. Because of thermal energy, the gate opens and closes at random, but the fraction of time it spends in one state or the other depends on the membrane potential. As the charged gate opens and closes, its charge moves in the membrane. The motion of the gate charge constitutes a gating current. In Figure 6.4, the moving end of the gate has been arbitrarily assumed to have a positive charge, so that a depolarization will tend to open the channel. That is, the model describes an *activation gate*. In this section we shall derive the properties of such a gated channel as well as the properties of a population of identical statistically independent channels, each having one two-state gate.

6.5.1 General Considerations

Derivation of the macroscopic and single-channel variables associated with a channel with one two-state gate involves a number of different variables, which are summarized in Table 6.5. The single-channel variables associated with a gated ion channel are shown in Figure 6.33. The gate has two states, open (O) and closed (C). Under the influence of its thermal energy, the gate makes transitions between its two states at random times. Thus, the time sequence of gate states, or state occupancies, is a random rectangular wave, denoted by $\tilde{s}$, that jumps back and forth between the closed and open states.

For a given membrane potential, the channel has three attributes that depend only upon its state—the state conductance, the state gating charge, and the state ionic current. Specification of the voltage-current characteristic of the open channel relates the state conductance and the state ionic current. The gating current depends upon state transitions as well as upon the state gating charges. Hence, it is not a state attribute.

When the channel is open, the conductance of the open channel, called the *single open-channel conductance*, is γ. When the channel is closed, it has a

Table 6.5 Definitions of symbols for macroscopic variables, single-channel variables, and state attributes.

Symbol	Units	Definition
Macroscopic variables		
J	A/m^2	Average macroscopic ionic current density
$\tilde{J}$	A/m^2	Macroscopic ionic current density random variable
J_g	A/m^2	Average macroscopic gating current density
$\tilde{J}_g$	A/m^2	Macroscopic gating current density random variable
$\tilde{n}_J$	A/m^2	Macroscopic ionic current density noise
$\tilde{n}_{Jg}$	A/m^2	Macroscopic gating current density noise
$\mathcal{Q}_g$	C	Average macroscopic gating charge
Q_g	C/m^2	Average macroscopic gating charge density
G	S/m^2	Average macroscopic specific conductance
Single-channel variables		
$\tilde{s}$		State occupancy random variable
i	A	Average single-channel ionic current
$\tilde{i}$	A	Single-channel ionic current random variable
q_g	C	Average single-channel gating charge
$\tilde{q}_g$	C	Single-channel gating charge random variable
i_g	A	Average single-channel gating current
$\tilde{i}_g$	A	Single-channel gating current random variable
g	S	Average single-channel conductance
$\tilde{g}$	S	Single-channel conductance random variable
State attributes		
$\mathcal{I}$	A	Single open-channel ionic current
$\mathcal{Q}$	C	Single open-channel gating charge
γ	S	Single open-channel conductance

conductance of zero. Thus, as the state occupancy jumps from C to O, the conductance jumps from 0 to γ. Therefore, the conductance is also a random rectangular wave, which we call the single-channel conductance random variable $\tilde{g}$. When the channel is open, an ionic current called the *single open-channel ionic current* $\mathcal{I}$ flows through the channel. When the channel is closed, no ionic current flows through the channel. Thus, the single-channel current random variable $\tilde{i}$ is a rectangular wave that jumps between the values 0 and $\mathcal{I}$. When the gate is closed, the component of the polarization vector perpendicular to the plane of the membrane is zero, whereas when the channel is open this component is nonzero. Therefore, a measure of the polarization is

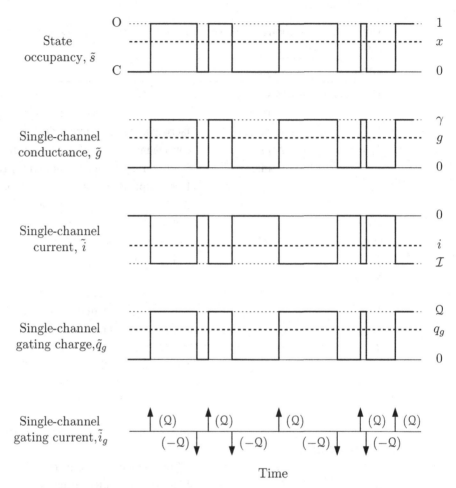

Figure 6.33 Variables associated with a channel with one two-state gate with a time-independent occupancy probability. The state occupancy, single-channel conductance, single-channel current, and single-channel gating charge random variables are indicated with solid rectangular waves. The single-channel gating current random variable $\tilde{i}_g$ is indicated with a sequence of impulses whose areas are indicated. The state occupancy probability x, average single-channel conductance g, average single-channel current i, and average single-channel gating charge q_g are indicated with dashed lines. The current through an open channel has been drawn arbitrarily for a negative value of $\mathcal{I}$.

simply the charge of the gate in the open position. Thus, the single-channel gating charge random variable $\tilde{q}_g$ has a waveform that is identical to that of the other variables and shows transitions between 0 and Q, which is called the *single open-channel gating charge*. Since the single-channel gating current is the derivative of the single-channel gating charge, the single-channel gating current $\tilde{i}_g$ is also a random variable that consists of a sequence of impulses (Dirac delta functions) of area $\pm Q$ at the state transition times.

The probability that the gate is open is x, where $0 \leq x \leq 1$. If $x = 0$, the gate is almost always closed; if $x = 1$, the gate is almost always open. If $x = 0.5$, on average the gate is open half the time and closed half the time. If the probability that the gate is open is known, and if other attributes of the open gate (such as the ionic conductance, ionic current, and gating charge when the gate is open) are known, then the expected value or average value of the conductance, the ionic current through the channel, and the gating charge can be determined.

To define the expected value of a discrete random variable, consider a generic single-channel random variable $\tilde{v}(t)$ that at each instant in time t takes on one of M values (generic state attributes) $\{a_1, a_2, \ldots a_M\}$. Let the probability that the random variable has value a_m at time t be x_m for m from 1 to M. Then the expected value of $\tilde{v}(t)$ is defined as

$$v = E[\tilde{v}(t)] = \sum_1^M a_m x_m.$$

Since in the two-state gate model the conductance takes on only one of two values, 0 and y, with probability $(1 - x)$ and x, respectively, we have

$$g = E[\tilde{g}(t)] = 0 \cdot (1 - x) + y \cdot x = yx.$$

We can visualize the computation of the expected value by imagining an ensemble of independent channels, all with the same statistical properties as shown schematically in Figure 6.34. Each channel has a conductance random variable that is a rectangular wave, with transitions at random times. Since the channels are independent, the transitions in different channels occur independently. We can compute an estimate of the expected value at each instant of time t from N such conductance random variables. Let the number of realizations whose conductance is y be n_y. Then an estimate of the expected value is simply the fraction of channels open at each instant in time times the conductance of an open channel, or yn_y/N. Therefore, we can imagine an infinite ensemble of statistically independent channels with identical statistical properties; therefore,

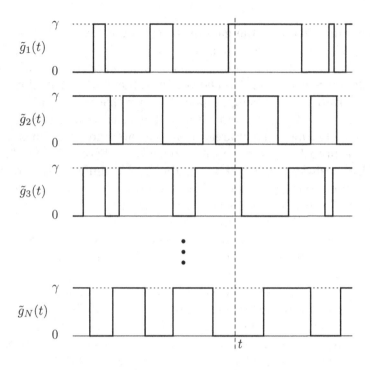

Figure 6.34 Illustration of method for computing the ensemble average of a random variable.

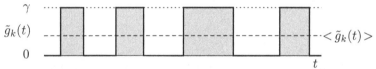

Figure 6.35 Illustration of method for computing the time average of a random variable.

$$g = E[\tilde{g}(t)] = \lim_{N \to \infty} \frac{n_\gamma}{N} \gamma.$$

If the conductance is independent of time, the average value of the conductance can also be computed as a time average from a single conductance record as shown schematically in Figure 6.35. The time average is defined as the area under the conductance curve divided by the time interval of the area calculation, or

$$g = <\tilde{g}_k(t)> = \lim_{T \to \infty} \frac{1}{T} \int_{T/2}^{T/2} \tilde{g}_k(t)\, dt.$$

To summarize, the average single-channel conductance, g, the average single-channel current, i, and the average single-channel gating charge, q_g, are

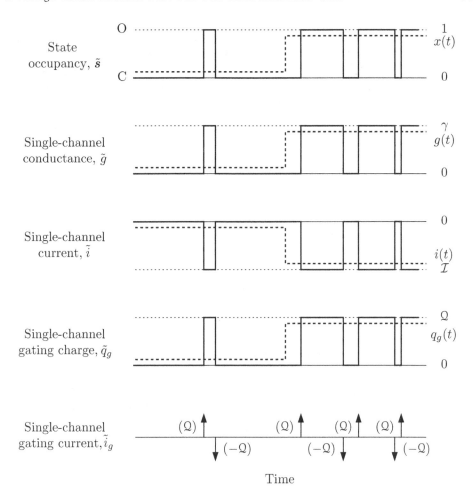

Figure 6.36 Variables associated with a channel with one two-state gate with a time-dependent occupancy probability.

$$g = E[\tilde{g}] = \gamma x,$$
$$i = E[\tilde{i}] = \mathcal{I}x, \tag{6.17}$$
$$q_g = E[\tilde{q}_g] = \mathcal{Q}x.$$

Thus far in the description of the random variables, we have implicitly assumed that the occupancy probability x is constant. However, we shall need to be concerned with time-varying occupancy probabilities. As shown in Figure 6.36, a change in the occupancy probability changes the fraction of the time the gate is found in its two states, and hence changes the fraction of the time that each of the electrical variables spends at one value. The magnitudes of the random variables remain the same.

Therefore, we can compute the expected value at each instant of time to define the time-varying expected values of the conductance and current variables, as follows:

$$g(t) = E[\tilde{g}(t)] = \gamma x(t),$$

$$i(t) = E[\tilde{i}(t)] = \mathcal{I}x(t), \tag{6.18}$$

$$q_g(t) = E[\tilde{q}_g(t)] = \mathcal{Q}x(t).$$

Therefore, for this simple model all single-channel random variables have average values that can be computed simply from $x(t)$. Therefore, we shall proceed to develop a model that allows for computation of $x(t)$ for the two-state gate model.

6.5.2 Single-Channel Variables

6.5.2.1 Expected Values and Probabilities

Permeation through an Open Channel

The simplest possible permeation mechanism is one in which the current through an open channel, the single open-channel ionic current, $\mathcal{I}$, is related linearly to the electrochemical potential difference across the membrane as follows:

$$\mathcal{I} = \gamma(V_m - V_n), \tag{6.19}$$

where V_n is the Nernst equilibrium potential for ion n. The reference directions for current and potential are indicated in Figure 6.4.

Kinetics of State Occupancy Probability

As shown schematically in Figure 6.37, each gate is free to move between its open and closed state by first-order kinetics, where C and O are the closed and open states, respectively, and α and β are the opening and closing rate constants,

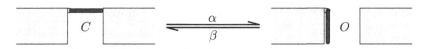

Figure 6.37 Schematic kinetic diagram of a channel with one two-state gate.

$$C \underset{\beta}{\overset{\alpha}{\rightleftharpoons}} O.$$

Let $\mathfrak{N}(t)$ be the average number of open channels per unit area of membrane at time t. Therefore, the net rate of openings/cm^2 is

$$\frac{d\mathfrak{N}(t)}{dt} = \alpha \left(\mathcal{N} - \mathfrak{N}(t) \right) - \beta \mathfrak{N}(t), \tag{6.20}$$

where $\alpha(\mathcal{N} - \mathfrak{N})$ is the rate of opening of closed states and $\beta\mathfrak{N}$ is the rate of closing of open states. Then $x(t) = \mathfrak{N}(t)/\mathcal{N}$, where $x(t)$ is the mean fraction of open channels or the probability that an individual channel is open. Since $x(t)$ is the probability that the open state is occupied at time t, $x(t)$ is called the *state occupancy probability*. Then

$$\frac{dx(t)}{dt} + (\alpha + \beta) \, x(t) = \alpha. \tag{6.21}$$

If α and β are constant, Equation 6.21 has the solution

$$x(t) = x_\infty - (x_\infty - x_0)e^{-t/\tau_x}, \tag{6.22}$$

where

$$x_\infty = \frac{\alpha}{\alpha + \beta} \tag{6.23}$$

and

$$\tau_x = \frac{1}{\alpha + \beta}. \tag{6.24}$$

Note that if the opening rate constant is much larger than the closing rate constant ($\alpha \gg \beta$), then $x_\infty \approx 1$, i.e., all the gates are open at equilibrium. Alternatively, if $\alpha \ll \beta$, then $x_\infty \approx 0$ and all the gates are closed at equilibrium.

Voltage Gating

We can now relate the rate constants to the conformational energy of the gate in its open and closed conformations as shown in Figure 6.38. Let E_O be the energy of the gate in its open conformation, and let E_C be the energy in the closed conformation. E_B is the barrier energy. E_C and E_O are local minima of energy that represent stable conformations of the gate—namely, the closed and the open conformations. The theory of absolute reaction rates (Weiss, 1996, Chapter 6, Section 6.3) gives a relation between the rate constants and the conformational energy as follows:

$$\alpha = Ae^{(E_C - E_B)/kT}, \tag{6.25}$$

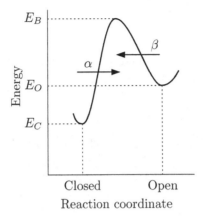

Figure 6.38 Energy diagram for a two-state gate.

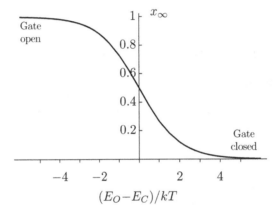

Figure 6.39 Occupancy probability as a function of energy.

$$\beta = Ae^{(E_O - E_B)/kT}, \tag{6.26}$$

where k is Boltzmann's constant. This relation indicates that if the barrier energy is extremely large, the rate constants for opening or closing are small. The equilibrium fraction of open channels or the probability that a channel is open can be expressed in terms of these energies by substituting Equations 6.25 and 6.26 into 6.23 to yield

$$x_\infty = \frac{e^{(E_C - E_B)/kT}}{e^{(E_C - E_B)/kT} + e^{(E_O - E_B)/kT}} = \frac{1}{1 + e^{(E_O - E_C)/kT}}. \tag{6.27}$$

As indicated in Equation 6.27 and shown in Figure 6.39, the equilibrium probability that a gate is open depends only on the ratio of the difference in energies between the open and closed states to the thermal energy (kT). If

$E_C/kT \gg E_O/kT$, then $x_\infty \to 1$, i.e., the gate is open. If $E_O/kT \gg E_C/kT$, then $x_\infty \to 0$ and the gate is closed. If $(E_O - E_C)/kT = 0$, then $x_\infty = 0.5$, i.e., the gate is open or closed with equal probability. Physically, the thermal energy is so great compared to the difference in potential energies of the closed and open states that the gate is equally likely to be found in either state.

The ratio of the probability that the gate is open to that of the gate's being closed is

$$\frac{x_\infty}{1 - x_\infty} = \frac{1}{\frac{1}{x_\infty} - 1} = \frac{1}{1 + e^{(E_O - E_C)/kT} - 1} = e^{(E_C - E_O)/kT}, \tag{6.28}$$

which is the ratio of Boltzmann factors![1] Equation 6.28 shows again that if $|E_C - E_O| \ll kT$, then the probability that the gate is open is equal to the probability that it is closed. Thus, on average, if we have a population of channels, half the channels will be open and the other half will be closed. The thermal energy will continually bounce the channels at random between the two conformations. Equation 6.28 also shows that the smaller is E_O, the larger will be the number of gates in the open conformation. Thus, if the open conformation has a lower energy state than the closed conformation, at equilibrium more channels will be in the open state than in the closed state.

Because the gates are assumed to be charged, a change in the membrane potential will change the conformational potential energy of the gate. We can imagine the potential energy diagram changing as shown in Figure 6.40, where we have arbitrarily made all three energy profiles coincide at E_C since only energy differences affect the rate constants. Note that the diagram is constructed so that for a depolarization, $E_C - E_B$ increases and $E_O - E_B$ decreases. Hence, a depolarization of the membrane will increase the opening rate constant, α, and decrease the closing rate constant, β. Hyperpolarization produces the opposite effect. Thus, a depolarization favors opening of a gate, and hyperpolarization favors gate closing, i.e., the gate acts as an activation gate.

Figure 6.40 illustrates in a qualitative manner how the the energy differences must change for an activation gate. Next we shall develop a quantitative relation between energy differences and the membrane potential. As a guide to determining the form of this relation, we shall review the electrostatic stored energy for simple charge configurations. By definition of the potential, the

---

1. This result is a special case of a general result in statistical mechanics. Suppose we have a small system of particles in contact with a heat source. Then the probability of finding the system in a particular energy state ϵ is proportional to $e^{-\epsilon/kT}$.

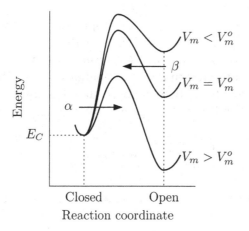

Figure 6.40 Voltage dependence of energy diagram for a two-state gate.

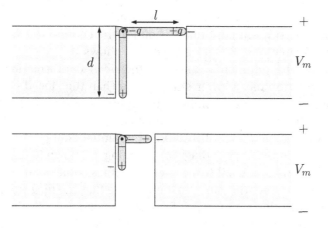

Figure 6.41 Two different schemes for a two-state gate. In the upper scheme, the gate spans the thickness of the membrane; in the lower scheme, it does not.

electrostatic stored energy of a charge q in an electric field with electrostatic potential ψ is $q\psi$. For a dipole that consists of two charges, q and $-q$, separated by distance l as shown in Figure 6.16, the electrostatic energy is found by summing the stored electrostatic energy of the two charges which, after finding that the limit of the energy stored as the distance between the charges vanishes if the dipole moment is kept constant, yields an electrostatic stored energy of $\mathbf{p} \cdot \nabla\psi$.

Now let us apply these results to the two-state gates shown in Figure 6.41. Suppose we make the simple assumption that the electric field in the channel is uniform and has the same dependence on membrane potential both when the channel is open and closed. Hence, the gradient of the electric potential is V_m/d. Let us assume that the gate has a dipole moment $\mathbf{p} = q\mathbf{l}$, where $\mathbf{l}$ is a vector that points from the negative to the positive charge, each of which has a charge of q. Then, in the closed conformation, the electrostatic stored energy in the gate is zero, because the dipole moment is perpendicular to the gradient of the potential. When the gate is open, the stored electrostatic energy is $-qlV_m/d$. If the gate spans the membrane, $l/d = 1$; if it does not, ql/d acts as an effective gating charge, $\mathcal{Q} = ql/d$. Therefore, in general we can write the electrostatic stored energy in the open configuration as $-\mathcal{Q}V_m$.

With this interpretation of the stored energy in mind, we can proceed to determine a more general relation between stored energy and membrane potential. We expand each energy difference in a Taylor's series in the potential as follows:

$$E_C - E_B = \sum_n a_n (V_m)^n, \quad \text{and} \quad E_O - E_B = \sum_n b_n (V_m)^n.$$

Next we assume that the higher-order terms are negligible and retain only the constant and linear term as follows:

$$E_C - E_B = a_0 + a_1 V_m, \quad \text{and} \quad E_O - E_B = b_0 + b_1 V_m.$$

The constant terms represent the energy differences that do not depend upon the membrane potential, i.e., they represent mechanical and chemical potential energy differences between conformations. Note that since the linear term is a potential energy, the coefficient of V_m must be a quantity that has the units of charge. Therefore, we choose to rewrite the energy in the following manner:

$$E_C - E_B = \mathcal{Q}(0.5 + \zeta)(V_m - V_C),$$

and

$$E_O - E_B = -\mathcal{Q}(0.5 - \zeta)(V_m - V_O),$$

where $\mathcal{Q}$ is the equivalent gating charge and ζ is an asymmetry factor that lies in the range of -0.5 to 0.5. The factor ζ allows the change in stored energy that results from a change in voltage to be different for the open and closed states. V_C and V_O are constants that are independent of V_m.

The most important factor introduced into the relation is the choice of the sign of the linear term. For an activation gate, $E_C - E_B$ increases with increasing V_m, and $E_O - E_B$ decreases with increasing V_m. We can also express the gating charge as $\mathcal{Q} = ze$, where z is the equivalent valence of the gating charge and e is the charge of an electron. Since $ze/kT = zF/RT$, the rate constants can be expressed as follows

$$\alpha = Ae^{z(0.5+\zeta)(F/RT)(V_m - V_C)},$$ (6.29)

$$\beta = Ae^{-z(0.5-\zeta)(F/RT)(V_m - V_O)}.$$ (6.30)

Hence, upon depolarizing the membrane, α increases and β decreases. The different parameters affect the rate constants differently. Consider α first. Both of the quantities A and V_C can be thought of either as acting to multiply the rate constants or as factors that shift the voltage dependence of the rate constant along the voltage axis. The constants A and V_O act similarly for β. The parameter z determines the magnitude of the rate of voltage dependence of the rate constants, and the factor ζ determines the asymmetry in the rate of voltage dependence of α and β. If $\zeta = 0$, then the magnitude of the rate of voltage dependence of the two rates constants is the same.

If Equations 6.29 and 6.30 are substituted into 6.23, then

$$x_\infty = \frac{1}{1 + e^{-z(F/RT)(V_m - V_{1/2})}},$$ (6.31)

where $V_{1/2} = (0.5 - \zeta)V_O + (0.5 + \zeta)V_C$. Thus, when $V_m = V_{1/2}$, then $x_\infty = 1/2$. Equation 6.31 implies that if x_∞ is plotted versus $z(F/RT)(V_m - V_{1/2})$, the resultant curve, shown in Figure 6.42, is independent of all parameters. For large V_m, $x_\infty \to 1$ and the gate is open. For small V_m, $x_\infty \to 0$ and the gate is closed.

It is also useful to plot x_∞ in semilogarithmic coordinates as shown in Figure 6.42. The asymptotic values are easy to determine. When the exponential term in Equation 6.31 is large, which occurs when V_m is sufficiently positive, then $\ln x_\infty \to 0$. When this exponential is small, which occurs when V_m is sufficiently negative, then $\ln x_\infty \to (zF/RT)(V_m - V_{1/2})$. That is, for sufficiently negative values of V_m, $\ln x_\infty$ plotted versus the normalized potential is a straight line with a slope of one. Therefore, the asymptotic slope of $\ln x_\infty$ plotted versus V_m is $(zF)/(RT)$. Therefore, the quantity z can be estimated directly from the slope of $\ln x_\infty$ plotted versus V_m at sufficiently negative values of the membrane potential.

The dependence of the time constant on membrane potential is

$$\tau_x = \frac{1}{\alpha + \beta} = \frac{1}{Ae^{z(0.5+\zeta)(F/RT)(V_m - V_C)} + Ae^{-z(0.5-\zeta)(F/RT)(V_m - V_O)}},$$

which, after some algebraic manipulations can be written as

$$\tau_x = \frac{1}{2Ae^{(zF/RT)(\zeta(V_m - V_{1/2}) + (\zeta^2 - 0.25)(V_C - V_O))}\cosh\left((zF/2RT)(V_m - V_{1/2})\right)}.$$ (6.32)

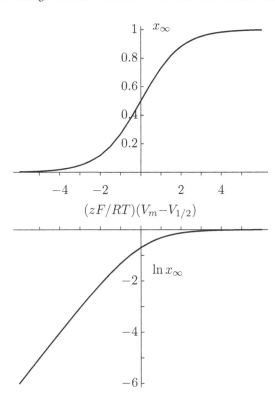

Figure 6.42 Steady-state probability that a channel with a two-state gate is open plotted in normalized coordinates on both a linear (upper panel) and a logarithmic (lower panel) ordinate scale.

The voltage dependence of the time constant can be seen most clearly in normalized coordinates. The normalized time constant is defined as

$$\tau_{xn} = \tau_x 2A e^{(zF/RT)(\zeta^2 - 0.25)(V_C - V_O)} \tag{6.33}$$

$$= \frac{1}{e^{(zF/RT)(\zeta(V_m - V_{1/2})} \cosh\left((zF/2RT)(V_m - V_{1/2})\right)}.$$

In Figure 6.43, τ_{xn} is plotted versus V_m. For $\zeta = 0$, the dependence of τ_x on V_m is an even function of $V_m - V_{1/2}$, i.e., the function is symmetrical. For nonzero values of ζ, the dependence of τ_x on $V_m - V_{1/2}$ is asymmetrical. For $\zeta < 0$, the rate of decline of τ_x with voltage is steeper for $V_m < V_{1/2}$ than for $V_m > V_{1/2}$. In fact, for $\zeta = -0.5$, τ_x is a saturating exponential function of V_m and does not decline at all for $V_m > V_{1/2}$. For $\zeta > 0$, the voltage dependence of τ_x is the reverse of that for $\zeta < 0$. The asymptotic form of the voltage dependence is seen more clearly in semilogarithmic coordinates. As indicated in Equation 6.32, the voltage dependence is exponential for large positive and negative values of V_m. Therefore, when τ_x is plotted in semilogarithmic

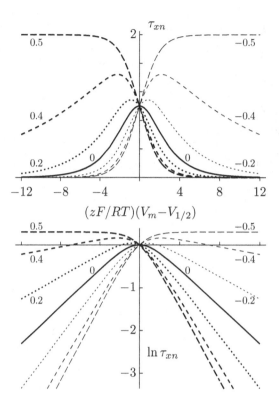

Figure 6.43 Time constant for the opening of a channel with a two-state gate plotted in normalized coordinates on both a linear (upper panel) and a logarithmic (lower panel) ordinate scale. The parameter is ζ.

coordinates, the asymptotes are linear in V_m. The asymptotes can be found directly from Equation 6.32 and are

$$\lim_{V_m \to -\infty} \ln \tau_x = -\ln A - \frac{zF}{RT}(\zeta - 0.5)(V_m - V_O),$$

$$\lim_{V_m \to \infty} \ln \tau_x = -\ln A - \frac{zF}{RT}(\zeta + 0.5)(V_m - V_C).$$

Thus, ζ determines the ratio of the asymptotic slopes of $\ln \tau_x$ plotted versus V_m.

Ionic and Gating Variables

For a voltage-gated channel, the state occupancy probability is a function of both membrane potential and time, so we write it as $x(V_m, t)$. Specification of $x(V_m, t)$ determines the expected value of each of the single-channel variables. The expected value of the conductance is

$$g(V_m, t) = E[\tilde{g}(V_m, t)] = \gamma x(V_m, t), \tag{6.34}$$

i.e., the conductance of the open channel γ times the probability that the channel is open. The current through an open channel is given by Equation 6.19. The expected value of the current through a channel, $i(V_m, t)$, is the product of the current through an open channel and the probability that the channel is open,

$$i(V_m, t) = E[\tilde{i}(V_m, t)] = \mathcal{I}x(V_m, t) = \gamma x(V_m, t)(V_m - V_n), \tag{6.35}$$

where V_n is the Nernst equilibrium potential for ion n.

When the gate opens, the gating charge goes from the closed to the open conformation, and then $\mathcal{Q}$ units of charge are displaced in the membrane. Hence, the expected value of the gating charge is

$$q_g(V_m, t) = E[\tilde{q}_g(V_m, t)] = \mathcal{Q}x(V_m, t), \tag{6.36}$$

and the gating current is

$$i_g(V_m, t) = \frac{dq_g(V_m, t)}{dt} = \mathcal{Q}\frac{dx(V_m, t)}{dt}. \tag{6.37}$$

Thus, the ionic conductance, ionic current, gating charge, and gating current are determined by the state occupancy probability $x(V_m, t)$. In response to a step of voltage, $x(V_m, t)$ is an exponential function of time with time constant, τ_x, which is a function of membrane potential (Figure 6.43), and with final value, x_∞, which is also a function of membrane potential (Figure 6.42). Hence, $g(V_m, t)$, and $q_g(V_m, t)$ are also exponential functions of time whose time constant and final value are proportional to those of $x(V_m, t)$ (Figure 6.44). These variables are continuous functions of time in response to a discontinuous pulse of membrane potential. However, the expected value of the gating current, $i_g(V_m, t)$, is proportional to the derivative of $x(V_m, t)$ with respect to t (Figure 6.44). Hence, the $i_g(V_m, t)$ is a discontinuous function of time. The expected value of the ionic current $i(V_m, t) = g(V_m, t)(V_m - V_n)$. Hence, the ionic current is discontinuous when the membrane potential is discontinuous. The discontinuity in the current is proportional to the value of $g(V_m, t)$. Thus, when this conductance is small, the discontinuity in current is small. When this conductance is large, as it is before the offset of the membrane potential, the discontinuity in current is large and gives rise to the prominent *tail current* at the offset of the membrane potential. The magnitude of this discontinuity of the tail current is proportional to the value of $g(V_m, t)$.

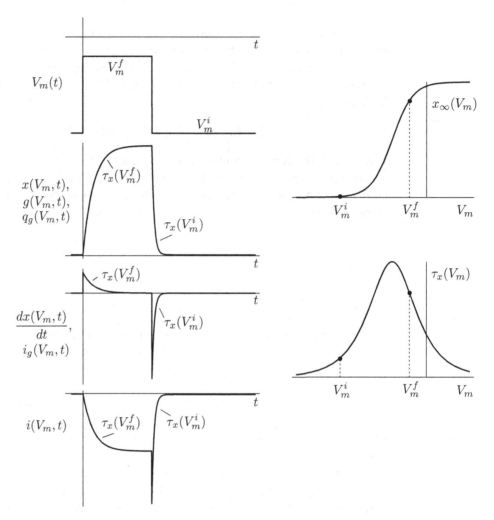

Figure 6.44 Variables that determine the expected values of the ionic conductance, ionic current, gating charge, and gating current for a channel with one two-state gate in response to a pulse of membrane potential. We can see that $g(V_m, t)$, $i(V_m, t)$, and $q_g(V_m, t)$ are proportional to $x(V_m, t)$, and $i_g(V_m, t)$ is proportional to $dx(V_m, t)/dt$. The right panels show the dependence of the parameters x_∞ and τ_x on membrane potential.

6.5.2.2   Fluctuations of Single-Channel Variables

Relation between State Occupancy and Rate Constants

At any time t, the channel is either in its open (O) or closed (C) state. We shall assume that the state of the channel at some later time depends only on the present state of the channel and not on the occurrence of past states. Such a process is called a *discrete-state, continuous-time Markov process* (see Appendix 6.1). We can define the state transition probabilities of such a Markov process as

$$y_{CO}(t, t + \tau) = \text{Prob}\{\tilde{s}(t + \tau) = O \,|\, \tilde{s}(t) = C\}, \tag{6.38}$$

$$y_{OC}(t, t + \tau) = \text{Prob}\{\tilde{s}(t + \tau) = C \,|\, \tilde{s}(t) = O\},$$

where y_{CO} is the probability that the channel is open at time $t + \tau$ given that the channel was closed at time t, and where y_{OC} is the probability that the channel is closed at time $t + \tau$ given that the channel was open at time t.

The probability that a transition occurs between the two states of the channel during a brief time interval $(t, t + \Delta t)$ can be computed from the rate constants. We assume that for a sufficiently brief interval the probability of a transition in the interval will be proportional to the length of the interval, so that

$$y_{CO}(t, t + \Delta t) = \alpha(t)\Delta t + o(\Delta t),$$

$$y_{OC}(t, t + \Delta t) = \beta(t)\Delta t + o(\Delta t), \tag{6.39}$$

where $\lim_{\Delta t \to 0} o(\Delta t)/\Delta t = 0$ and where $\alpha(t)$ and $\beta(t)$ are the rates of transitions at time t from closed to open and from open to closed states, respectively. Since the channel is always in one of its two states, the probability that the channel is closed and does not open in the interval Δt is simply $1 - \alpha(t)\Delta t + o(\Delta t)$, and the probability that the channel is open and does not close in the interval is $1 - \beta(t)\Delta t + o(\Delta t)$.

Thus, the state occupancies can be computed numerically if the rate constants are given. Since the state occupancies are random variables, the computation necessarily involves a simulation that has access to the computation of random numbers. Results of such simulations are shown in Figure 6.45 to illustrate the effect of the rate constants on the state occupancies. The traces all show the state occupancies for a fixed time interval for different values of rate constants, all of which are constants over the time interval of each computation. As the ratio α/β is increased (from the top to the bottom row), the open state is occupied a larger fraction of the time. For the three values shown,

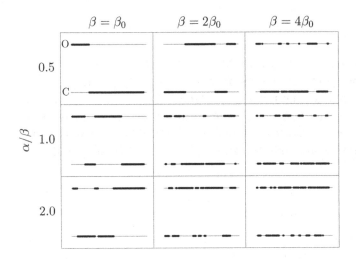

Figure 6.45 Dependence of state occupancy on rate constants. Each panel shows the the state occupancy as a function of time for a different combination of rate constants. The state occupancy was computed at 400 points in time with a base rate such that the probability of a transition from closed to open in an interval of duration Δt was chosen to be $\beta_0 \Delta t = 0.01$. The columns show the effect of changing this base rate, and the rows show the effect of changing the ratio of rate constants α/β.

the probability that the state will be open is 1/3 ($\alpha/\beta = 0.5$), 1/2 ($\alpha/\beta = 1$), and 2/3 ($\alpha/\beta = 2$). As both rate constants are increased (from the left to the right column), the rate of transitions from closed to open and from open to closed is increased, although the fraction of time that the channel is open is not changed.

The Variance

A simple measure of the variation of the single-channel random variables is the standard deviation of the variable, which can also be interpreted as the root-mean-squared difference between the variable and its mean value. To compute this quantity, we first compute the second moment of each of the single-channel random variables. Since each variable takes on only two values, the value of the second moment is easily computed. For example, the conductance random variable has the value y with probability $x(V_m, t)$ and 0 with probability $1 - x(V_m, t)$ at each instant in time. Hence, the second moment is simply $y^2 x(V_m, t) + 0(1 - x(V_m, t)) = y^2 x(V_m, t)$. The other single-channel variables are computed similarly to yield

$$E[\tilde{g}^2(V_m, t)] = \gamma^2 x(V_m, t),$$

$$E[\tilde{\imath}^2(V_m, t)] = \mathcal{I}^2 x(V_m, t), \tag{6.40}$$

$$E[\tilde{q}_g^2(V_m, t)] = \mathcal{Q}^2 x(V_m, t).$$

The variance of a variable is the difference between the second moment and the mean value squared, so the variances are

$$\sigma_{\tilde{g}}^2(V_m, t) = E[\tilde{g}^2(V_m, t)] - E^2[\tilde{g}(V_m, t)] = \gamma^2 x(V_m, t) - \gamma^2 x^2(V_m, t)$$

$$= \gamma^2 x(V_m, t)\,(1 - x(V_m, t)),$$

$$\sigma_{\tilde{i}}^2(V_m, t) = E[\tilde{i}^2(V_m, t)] - E^2[\tilde{i}(V_m, t)] = \mathcal{I}^2 x(V_m, t) - \mathcal{I}^2 x^2(V_m, t)$$

$$= \mathcal{I}^2 x(V_m, t)\,(1 - x(V_m, t)),\qquad(6.41)$$

$$\sigma_{\tilde{q}_g}^2(V_m, t) = E[\tilde{q}_g^2(V_m, t)] - E^2[\tilde{q}_g(V_m, t)] = \mathcal{Q}^2 x(V_m, t) - \mathcal{Q}^2 x^2(V_m, t)$$

$$= \mathcal{Q}^2 x(V_m, t)\,(1 - x(V_m, t)).$$

The standard deviation is the square root of the variance, so

$$\sigma_g(V_m, t) = \gamma\sqrt{x(V_m, t)\,(1 - x(V_m, t))},$$

$$\sigma_i(V_m, t) = \mathcal{I}\sqrt{x(V_m, t)\,(1 - x(V_m, t))}\qquad(6.42)$$

$$\sigma_{q_g}(V_m, t) = \mathcal{Q}\sqrt{x(V_m, t)\,(1 - x(V_m, t))}.$$

All of the variances have the form

$$\sigma^2 = x(1 - x)$$

except for scale factors. This expression can be rewritten as

$$\sigma^2 + x^2 - x = 0,$$

or, after completing the square, as

$$\sigma^2 + \left(x - \frac{1}{2}\right)^2 = \frac{1}{4}.$$

Hence, the relation between σ and x is that of a circle centered at $\sigma = 0$ and $x = 1/2$ with a radius of $1/2$, as shown in Figure 6.46. Thus, the standard deviation is zero when $x = 0$ and $x = 1$ and is a maximum when $x = 1/2$. This makes sense, since when $x = 0$ or $x = 1$, the channel is either closed or

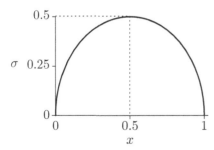

Figure 6.46 Relation between standard deviation and state occupancy probability.

open essentially all of the time; hence, there are no fluctuations in any of the channel variables. When $x = 1/2$, the channel spends half its time open and half its time closed, so that fluctuations are maximized at the value $\sigma = 1/2$.

The standard deviations of the single-channel random variables have a dependence on x of the form shown in Figure 6.46 except for scale factors. The maximum values of the standard deviations are $\sigma_g(V_m, t)_{max} = \gamma/2$, $\sigma_i(V_m, t)_{max} = \mathcal{I}/2$, and $\sigma_{q_g}(V_m, t)_{max} = \mathcal{Q}/2$.

The Autocovariance Function and the Power Density Spectrum

As indicated in Figure 6.45, as the rate constants are increased, the channel flickers more rapidly between the open and closed states. The expected time variation of any of the single-channel random variables can be quantified by means of the autocovariance function or power density spectrum (see Appendix 6.1). We shall summarize these results for time-independent rate constants. The autocovariance function for the conductance fluctuations of a single channel with single open-channel conductance γ and constant rate constants is

$$c_g(\tau) = \gamma^2 \frac{\alpha^2}{(\alpha + \beta)^2} + \gamma^2 \frac{\alpha\beta}{(\alpha + \beta)^2} e^{-(\alpha+\beta)|\tau|}. \tag{6.43}$$

Since $c_g(0) = E[\tilde{g}^2]$ and $\lim_{\tau \to \infty} c_g(\tau) = E^2[\tilde{g}]$, we can check the consistency of the covariance function with previous computations of the mean and the variance of the conductance random variable. From Equation 6.43, we can see that

$$\lim_{|\tau| \to \infty} c_g(\tau) = \gamma^2 \frac{\alpha^2}{(\alpha + \beta)^2},$$

which shows that

$$E[\tilde{g}] = \gamma \frac{\alpha}{\alpha + \beta}.$$

When the rate constants are independent of time, x takes on its equilibrium value $x_\infty = \alpha/(\alpha + \beta)$, as can be seen from Equation 6.23. Therefore, from the covariance function we see that $E[\tilde{g}] = \gamma x_\infty$, which is consistent with direct calculation of the expected value of the conductance. From Equation 6.43, we can see that

$$c_g(0) = \gamma^2 \frac{\alpha^2}{(\alpha + \beta)^2} + \gamma^2 \frac{\alpha\beta}{(\alpha + \beta)^2}.$$

Combining these results yields

$$\sigma_g^2 = \gamma^2 \frac{\alpha\beta}{(\alpha+\beta)^2}.$$

Since, $x_\infty = \alpha/(\alpha+\beta)$, then $1 - x_\infty = \beta/(\alpha+\beta)$, and

$$\sigma_g^2 = \gamma^2 x_\infty(1 - x_\infty),$$

which is the same expression that was derived directly in Equation 6.41.

The power density spectrum is defined as the Fourier transform of the autocovariance function and is

$$C_g(f) = \gamma^2 \frac{\alpha^2}{(\alpha+\beta)^2}\delta(f) + 2\gamma^2\frac{\alpha\beta}{(\alpha+\beta)^2}\left(\frac{1}{(\alpha+\beta)^2 + (2\pi f)^2}\right), \qquad (6.44)$$

where $\delta(f)$ is the unit impulse function. The impulse results from the constant value of the covariance function and reflects the square of the expected value of the conductance.

The autocovariance function and the power density function are plotted for both slow and fast rate constants in Figure 6.47. For the fast rate constants, the conductance random variable switches more rapidly between its two values, 0 and γ, than for the slow rate constants. This behavior is reflected in the more rapid decay of the exponential covariance function and in the greater power at high frequencies in the power density spectrum for the results from the channel with fast rate constants. The covariance functions and power density spectra of both the ionic current and the gating charge are identical to those of the conductance with γ replaced by $\mathcal{I}$ and $\mathcal{Q}$, respectively.

6.5.2.3 *State Occupancy Dwell Times*

Single-channel currents for a channel with one two-state gate are rectangular waves that switch between zero current when the channel is closed, and $\mathcal{I}$ when the current is open (Figure 6.48). The durations of the open (τ_O) and closed (τ_C) states clearly depend upon the rate constants. For example, if the rate constants increase, the channel will flicker more rapidly between its open and closed states; hence, both the open and closed durations will decrease. In this section, we shall derive the relation between the distribution of the open and closed durations and the rate constants.

Let $P_O(T > \tau)$ be the probability that the interval T, for which the channel is open, exceeds τ. Then we can write

$$P_O(T > \tau + \Delta\tau) = P_O(T > \tau)(1 - \beta\Delta\tau), \qquad (6.45)$$

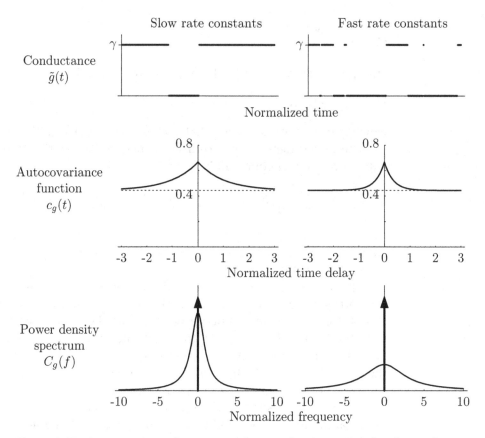

Figure 6.47 Autocovariance functions and power density spectra for the conductance fluctuations of a channel with one two-state gate. The upper panels show the conductance random variables for two different sets of rate constants. The rate constants were chosen so that $\beta/\alpha = 2$, so closed durations are on average twice as long as open durations. The right panels (fast rate constants) show the results for a channel whose rate constants are three times larger than those for the channel whose results are shown in the left panels (slow rate constants).

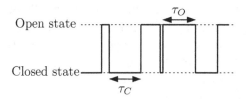

Figure 6.48 Schematic diagram illustrating the definitions of the closed and open durations.

because the probability that the interval is greater than $\tau + \Delta\tau$ is simply the probability that the interval is greater than τ times the probability that no transition from open to closed occurs in the interval $(\tau, \tau + \Delta\tau)$. The latter probability is gotten from Equation 6.39. Equation 6.45 can be rearranged as follows:

$$\frac{P_O(T > \tau + \Delta\tau) - P_O(T > \tau)}{\Delta\tau} = -\beta P_O(T > \tau),$$

and then in the limit as $\Delta\tau \to 0$ this becomes

$$\frac{dP_O(T > \tau)}{d\tau} = -\beta P_O(T > \tau).$$

This equation can be integrated to yield

$$\ln P_O(T > \tau)|_{\tau=0}^{\tau=\tau_O} = -\beta\tau\,|_{\tau=0}^{\tau=\tau_O},$$

and since $P_O(T > 0) = 1$, we have

$$P_O(T > \tau_O) = e^{-\beta\tau_O}.$$

We also have that the probability that the interval is less than or equal to duration τ_O is simply

$$P_O(T \leq \tau_O) = 1 - e^{-\beta\tau_O}.$$

From these two relations, it is easy to determine the probability density function that the open intervals are of duration τ_O, as follows:

$$p_O(\tau_O) = \lim_{\Delta\tau \to 0} \frac{P_O(\tau_O \leq T < \tau_O + \Delta\tau)}{\Delta\tau}$$

and

$$p_O(\tau_O) = \beta e^{-\beta\tau_O}. \tag{6.46}$$

In an analogous manner, the probability density function of closed times is

$$p_C(\tau_C) = \alpha e^{-\alpha\tau_C}. \tag{6.47}$$

Thus, the density functions of both the open and closed times, called *dwell times*, are exponential (Figure 6.49).

From the distribution of these times, we can compute the expected value of the open and closed times as follows:

$$E[\tau_O] = \int_0^\infty \tau_O p_O(\tau_O) d\tau_O \quad \text{and} \quad E[\tau_C] = \int_0^\infty \tau_C p_C(\tau_C) d\tau_C. \tag{6.48}$$

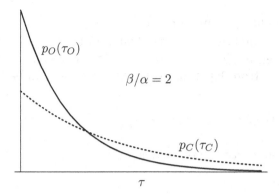

Figure 6.49 Distributions of dwell times for a channel with one two-state gate. The probability density functions of the open and closed durations are plotted on a common normalized time scale (τ). The rate constants were chosen so that $\beta/\alpha = 2$ so that closed durations are on average twice as long as open durations.

Substitution of Equations 6.46 and 6.47 into Equation 6.48 yields

$$E[\tau_O] = \frac{1}{\beta} \quad \text{and} \quad E[\tau_C] = \frac{1}{\alpha}. \tag{6.49}$$

Thus, we have the result that the average duration of time the channel is open, the average open dwell time, is inversely proportional to the rate of closing; i.e., increasing the rate at which the channel closes decreases the average time the channel is open. Similarly, the average duration of time the channel is closed is inversely proportional to the rate at which the channel opens.

These results show that the closing rate constant can be determined from the mean value of the distribution of open times and that the opening rate constant can be determined from the mean value of the distribution of closed times. By repeating measurements of these distributions at constant values of the membrane potential, the dependence of rate constants on membrane potential can be determined.

6.5.3 Relation of Macroscopic and Single-Channel Variables

6.5.3.1 *Expected Values of Macroscopic Variables*

Suppose we have a cell membrane that contains a population of independent but otherwise identical channels, each having a single two-state gate. The number of such channels is N, the surface area of the membrane is A, and the density of these channels is $\mathcal{N} = N/A$ channels per unit area of membrane. Then the conductance of a unit area of membrane is

$$\tilde{G}(V_m, t) = \frac{1}{A} \sum_{1}^{N} \tilde{g}(V_m, t),$$

whose expected value is the average macroscopic conductance

$$G(V_m, t) = E[\tilde{G}(V_m, t)] = E\left[\frac{1}{A}\sum_{1}^{N}\tilde{g}(V_m, t)\right]$$

$$= \frac{1}{A}\sum_{1}^{N}E[\tilde{g}(V_m, t)] = \gamma \mathcal{N}x(V_m, t). \tag{6.50}$$

In an analogous manner, the average macroscopic current density (Figure 6.50) is

$$J(V_m, t) = E[\tilde{J}(V_m, t)] = E\left[\frac{1}{A}\sum_{1}^{N}\tilde{i}(V_m, t)\right]$$

$$= \frac{1}{A}\sum_{1}^{N}E[\tilde{i}(V_m, t)] = \gamma \mathcal{N}x(V_m, t)(V_m - V_n). \tag{6.51}$$

The gating charge density is

$$Q_g(V_m, t) = E[\tilde{Q}_g(V_m, t)] = E\left[\frac{1}{A}\sum_{1}^{N}\tilde{q}_g(V_m, t)\right]$$

$$= \frac{1}{A}\sum_{1}^{N}E[\tilde{q}_g(V_m, t)] = \mathcal{Q}\mathcal{N}x(V_m, t), \tag{6.52}$$

and the gating current density is

$$J_g(V_m, t) = \frac{dQ_g(V_m, t)}{dt} = \mathcal{Q}\mathcal{N}\frac{dx(V_m, t)}{dt}. \tag{6.53}$$

As we have already seen, in response to a step of voltage, $x(V_m, t)$ is an exponential function of time with time constant, τ_x, which is a function of membrane potential (Figure 6.43) and with final value, x_∞, which is also a function of membrane potential (Figure 6.42). Hence, the macroscopic conductance, $G(V_m, t)$, the macroscopic ionic current, $J(V_m, t)$ and the macroscopic gating charge are exponential functions of time whose time constant and final value are proportional to those of $x(V_m, t)$ (Figure 6.44). The macroscopic gating current is proportional to the derivative of $x(V_m, t)$ with respect to t (Figure 6.44).

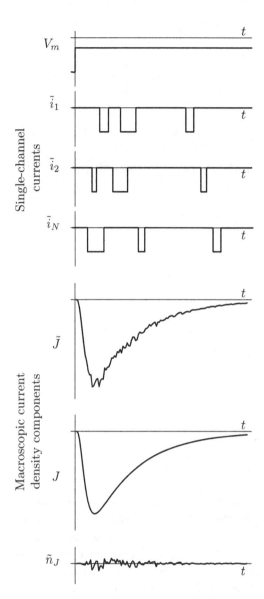

Figure 6.50 A schematic diagram illustrating the relation between single-channel and macroscopic ionic currents. The single-channel currents for three channels—channels 1, 2, and N—are shown. Each is a random rectangular wave of fixed amplitude. The sum of a collection of N such channels in a unit area of membrane is the macroscopic current density random variable $\tilde{J}$, whose expected value is J. The current noise is $\tilde{n}_J = \tilde{J} - J$.

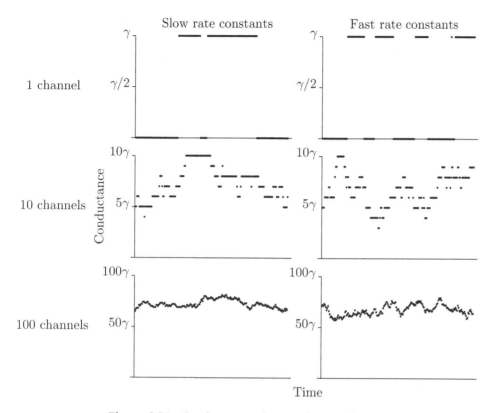

Figure 6.51   Conductance of a population of independent channels. The parameters are the same as in Figure 6.47.

6.5.3.2   *Fluctuations*

A population of statistically independent, but otherwise identical channels will produce a macroscopic conductance that is the sum of the single-channel conductances as shown in Figure 6.51. The total conductance is just the sum of the individual conductances. Whereas the average values add linearly, the fluctuations do not. Therefore, the relative amount of fluctuation decreases as the density of channels increases.

To find the variance of the macroscopic variables, we make use of the result that the variance of the sum of statistically independent random variables is the sum of the variances. Therefore, scaling Equation 6.41 yields

$$\sigma_G^2(V_m, t) = \mathcal{N}\gamma^2 x(V_m, t)\left(1 - x(V_m, t)\right),$$

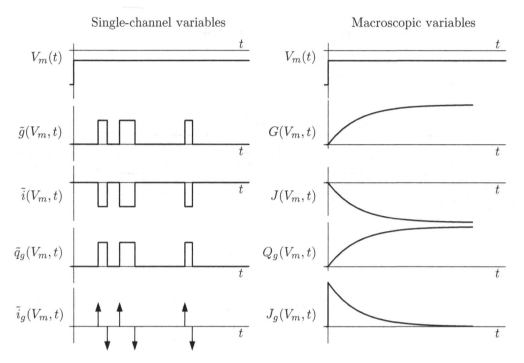

Figure 6.52 Schematic diagram showing the relation of single-channel to macroscopic variables for a channel model with one two-state gate.

and the standard deviation is

$$\sigma_G(V_m, t) = \sqrt{\mathcal{N} \gamma^2 x(V_m, t) \left(1 - x(V_m, t)\right)}.$$

Similar results can be found for the other macroscopic variables. In addition, the autocovariance function and the power density function of the macroscopic variables also scale with the channel density.

6.5.4 Summary and Conclusion

Figure 6.52 summarizes the relations among single-channel and macroscopic variables predicted by the simple model of a channel with one two-state gate that gates a simple ohmic channel. The single-channel ionic conductance, ionic current, and gating charge random variables are all random rectangular waves with the same transition times. The single-channel gating current random variables are impulses that occur at the transition times. The rates of transitions between states are exponential functions of membrane potential. As these rates increase, the rates of transition increase. Since the model has first-order kinetics, the expected values of the single-channel ionic conductance, ionic current, gating charge, and gating current have first-order kinetics in response to step changes in membrane potential. The expected values of the single-channel random variables are proportional to the macroscopic currents. Therefore, the macroscopic variables are exponential functions of time when the membrane potential is constant. The kinetics of all the variables are equivalent. For example, the kinetics of the ionic current are equivalent to the kinetics of the gating current and vice versa; gating and ionic currents give equivalent information about channel kinetics. Both the equilibrium values of the currents and the time constants are functions of membrane potential. The equilibrium values are sigmoidal functions of membrane potential whose rate of voltage dependence is proportional to the gating charge. The dependence of the time constant on the membrane potential depends on two parameters: the gating charge, which determines the rate of voltage dependence, and the asymmetry factor, which determines the asymmetry of the dependence of the time constant on the membrane potential. Examination of the dwell times of single channels gives similar information to examination of the kinetics of ionic currents or gating currents. Furthermore, examination of the fluctuations of single-channel currents also gives the same information. The two-state gate model provides strong constraints on the interrelations of many types of measurements, including both single-channel and macroscopic variables associated with ionic and gating variables.

6.6   Models of Multiple-State Channels

Properties of the channel model in which each channel contains one two-state gate show many qualitative similarities to those of the sodium and potassium channels of cellular membranes. The channel model has a voltage-dependent ionic conductance kinetics, a voltage dependent gating charge, etc. However, the model shows many quantitative differences from the measured characteristics of the sodium and potassium channels. For example, the kinetics of the conductance change following a step change in potential do not agree, as indicated in Figure 6.53. The two-state gate model has exponential onset kinetics, whereas the sodium and potassium conductances show S-shaped activations and the sodium conductance shows an exponential inactivation.

In this section, we shall examine more general channel models that can more faithfully represent the measured properties of ionic channels in mem-

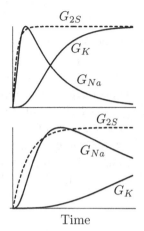

Time

Figure 6.53 Comparison of the kinetics of conductance in response to a step of membrane potential for the two-state gate model with those of the sodium and potassium channels. The conductance of the two-state gate model (G_{2S}) has an exponential activation and does not inactivate. The sodium conductance (G_{Na}) has an S-shaped activation and an exponential inactivation. The potassium conductance (G_K) has an S-shaped activation and no inactivation. The conductances have been normalized to have the same peak value and are shown on two time scales.

branes. We shall begin by synthesizing a channel model with four independent two-state gates. This model gives channel kinetics that are identical to those predicted by the Hodgkin-Huxley model. Comparison of the predictions of these models with measurements of ionic and gating currents indicates the inadequacy of the Hodgkin-Huxley model to fit these measurements. Then we shall investigate more general channel models.

6.6.1   The Hodgkin-Huxley Model: A Molecular Interpretation

6.6.1.1   A Kinetic Scheme Consistent with the Hodgkin-Huxley Model

In the Hodgkin-Huxley model, the sodium conductance is proportional to the product of m^3h. Now consider a channel with four independent gates, each of which has to be open for the channel to be open (conducting). Three of these gates are activation gates; they open upon depolarization of the membrane. One of these gates is an inactivation gate and opens upon hyperpolarization of the membrane. Let m be the probability that an activation gate is open, and let h be the probability that the inactivation gate is open. The probability that the channel is open equals the probability that all four gates are open. If the gates are independent, the probability that all four gates are open is m^3h. Thus, the average macroscopic conductance for the sodium channel is simply $G_{Na} = \gamma_{Na}\mathcal{N}m^3h = \overline{G}_{Na}m^3h$, where $\overline{G}_{Na} = \gamma_{Na}\mathcal{N}$.

Consider the activation gates first. Each activation gate has first-order kinetics, i.e.,

$$C_g \underset{\beta_m}{\overset{\alpha_m}{\rightleftharpoons}} O_g,$$

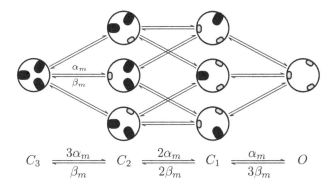

$$C_3 \underset{\beta_m}{\overset{3\alpha_m}{\rightleftharpoons}} C_2 \underset{2\beta_m}{\overset{2\alpha_m}{\rightleftharpoons}} C_1 \underset{3\beta_m}{\overset{\alpha_m}{\rightleftharpoons}} O$$

Figure 6.54 Schematic diagram showing the states of a channel with three activation gates. Closed gates are shown black and large; open gates are shown small and gray. The rate constants for all transitions are the same; only one pair of rate constants is shown.

and, as shown in Figure 6.54, each channel contains three activation gates. Thus, we now need to distinguish between the state of the gate and the states of the channel. Assuming that the gates are indistinguishable, the collection of three gates leads to four possible states of the activation gates: three closed states, C_3, C_2, and C_1, and one open state, O. If each activation gate opens with a rate constant α_m and closes with a rate constant β_m, then the transition rate from state C_3 to state C_2 is $3\alpha_m$, because any of the three independent gates can open to produce this transition. Thus, the state transition occurs three times as fast as the rate of opening of one gate. The transition rate from state C_2 to state C_3 is β_m, because there is only one gate that can close for this transition to occur. It is clear from this analysis that the kinetic diagram that corresponds to the notion of three independent activation gates is the one given in Figure 6.54. This state diagram indicates why the m^3 kinetics are slower than exponential, as indicated in Figure 6.53. Clearly, if the channel starts in state C_3, it must traverse states C_2 and C_1 before it can get to the open state. Each transition acts as a first-order system, which accounts for the S-shaped onset kinetics.

To achieve inactivation and activation with four independent two-state gates, it is necessary to add only one additional gate, which is an inactivation gate, to the kinetic scheme given in Figure 6.54. This model has the kinetic diagram shown in Figure 6.55. This kinetic diagram corresponds to the m^3h kinetics of the Hodgkin-Huxley model. The essential features of this model are shown more abstractly in Figure 6.56. This kinetic diagram shows that the model has eight states, of which only one, state S_4, is conducting, and all the other states are nonconducting. The essential feature that leads to the kinetic scheme of the Hodgkin-Huxley model for the sodium conductance is the specific ratios for forward and backward rate constants that link the states.

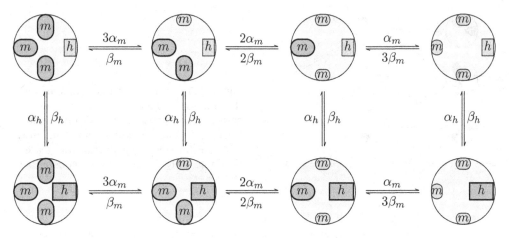

Figure 6.55 Kinetic diagram showing the states of a channel with three activation gates and one inactivation gate (adapted from French and Horn, 1983, Figure 1). Each state is illustrated by one of its configurations. For example, there are three distinct configurations of the gates, for which there are two closed m gates and a closed h gate.

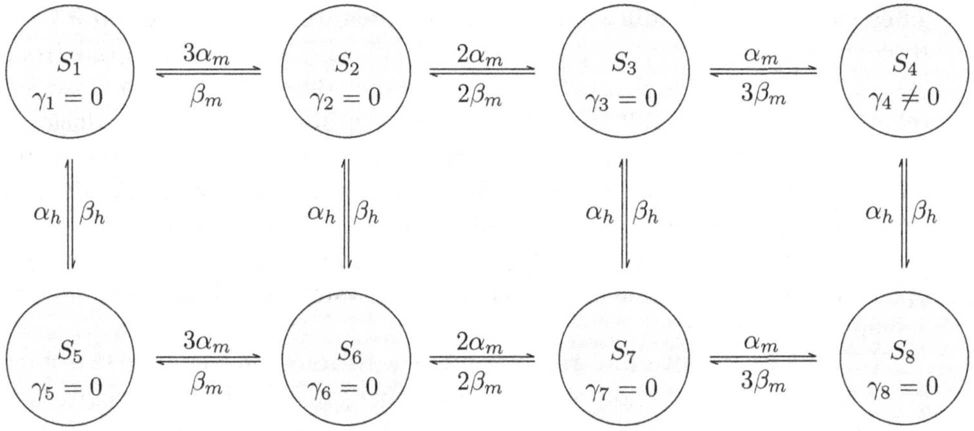

Figure 6.56 Kinetic diagram showing the states of a channel with three activation gates and one inactivation gate. This eight-state model has kinetics that are equivalent to the Hodgkin-Huxley model for the sodium channel.

6.6.1.2 Comparison of Measurements with Predictions of Channel Models that Have Hodgkin-Huxley Kinetics

With the advent of new techniques, the Hodgkin-Huxley model for the sodium conductance, with three independent activation gates as shown in Figure 6.54, has been subject to more extensive testing, including measurements of macroscopic ionic currents, gating currents, and single-channel currents from single channels. One important issue is to determine whether this model can predict all the measurements. Careful comparisons between these kinetics and measurements are given elsewhere (Patlak, 1991; Keynes, 1994b). In this section, we shall give two examples of such comparisons.

Voltage Dependence of Equilibrium Values
The independent gate model for activation makes strong predictions regarding the relation between ionic conductances and gating charges. We shall explore one of these. First, let us compute the steady-state value of the gating charge density. Suppose we assume that $m_\infty(V_m)$ represents the probability that a gate is open when the membrane potential has been at V_m for a long time. Suppose that the gating charge associated with an open gate is Q. Then the average gating charge for a channel that contains three independent gates is $3Qm_\infty(V_m)$. If there are $\mathcal{N}$ channels per unit area of membrane, the macroscopic surface charge density of the gating charge is $Q_g(V_m, \infty) = 3\mathcal{N}Qm_\infty(V_m)$.

Next we compute the steady-state value of the activation of the ionic conductance. If all three gates must be open for the channel to be open, the probability that the channel is open is $m_\infty^3(V_m)$ if the inactivation gate is open. The conductance of an open channel is γ. If there are $\mathcal{N}$ channels per unit area of membrane, the average steady-state conductance density of a population of such channels is $G_{Na}(V_m, \infty) = \mathcal{N}\gamma m_\infty^3(V_m)$.

Therefore, the kinetic scheme that gives rise to the Hodgkin-Huxley model for sodium conductance also predicts that the steady-state value of the gating charge is proportional to $m_\infty(V_m)$ and that the steady-state value of the conductance activation is proportional to $m_\infty^3(V_m)$. To test such ideas, it is important to measure both the steady-state conductance and the gating charge of the same preparation. Measurements of the steady-state value of the conductance activation can be obtained from voltage-clamp currents, but these are more accurate if inactivation is removed, i.e., with the use of pronase. The gating charge can be measured by integrating the gating current. The results of such an experiment are shown in Figure 6.57. Clearly, the activation of the sodium conductance has a much steeper dependence on membrane po-

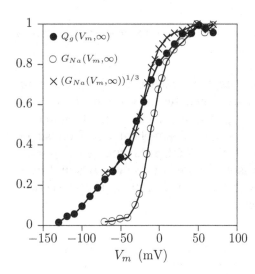

Figure 6.57 Comparison of the voltage dependence of sodium conductance with that of gating charge (adapted from Stimers et al., 1985, Figure 4B). Measurements were obtained in voltage-clamped, perfused giant axons of squid (*Loligo pealii*). Cesium ions replaced the intracellular potassium ions to reduce the effect of potassium currents on the measurements. Intracellular solutions also contained pronase, which removed sodium inactivation. To measure gating currents, sodium-free extracellular solutions containing TTX were used to minimize the effect of sodium currents on gating currents. Gating charge was obtained by integrating gating currents. The temperature was 7.5°. All measurements have been normalized to their maximum values.

tential than does the gating charge. The voltage dependence of the cube root of the conductance is almost in line with the voltage dependence of the gating charge, although some relatively small differences are seen near 0 mV. On the whole, the relation between the voltage dependence of the conductance and that of the gating charge is in accord with the Hodgkin-Huxley model with three independent activation gates.

Offset Time Constants

The Hodgkin-Huxley model makes specific predictions about the time constants of $m(V_m, t)$, $G_{Na}(V_m, t)$, and $J_{Na}(V_m, t)$ as shown in Figure 6.58. Consider the response to a pulse of membrane potential that starts at V_m^i, increases to V_m^f for a sufficient time interval, T, to allow m to reach its steady-state value of $m_\infty(V_m^f)$, and then returns to V_m^i. The value of V_m^i is chosen to be sufficiently hyperpolarized so that $m_\infty(V_m^i) \approx 0$. Therefore, the solution for m during the offset of the membrane potential is

$$m(V_m, t) = m_\infty(V_m^f)e^{-(t-T)/\tau_m(V_m^i)} \text{ for } t > T.$$

If inactivation were removed so that $h = 1$, then the sodium conductance would be

$$G_{Na}(V_m, t) = \overline{G}_{Na}m^3(V_m, t) \text{ for } t > T,$$

$$G_{Na}(V_m, t) = \overline{G}_{Na}m_\infty^3(V_m^f)e^{-3(t-T)/\tau_m(V_m^i)} \text{ for } t > T,$$

and the current would be

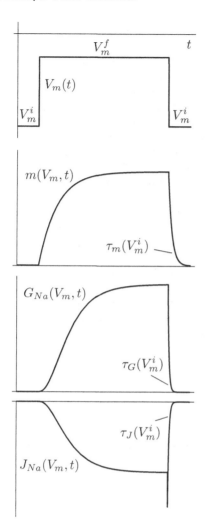

Figure 6.58 Predictions of the Hodgkin-Huxley model for m, G_{Na}, and J_{Na} in response to a pulse of membrane potential. Provided that $m_\infty \approx 0$, all three variables are exponential functions of time at the offset of the membrane potential pulse, but the time constants are $\tau_m(V_m^i)$ for m and $\tau_G(V_m^i) = \tau_J(V_m^i) = \tau_m(V_m^i)/3$ for G_{Na} and J_{Na}.

$$J_{Na}(V_m, t) = \overline{G}_{Na} m_\infty^3(V_m^f)(V_m^i - V_{Na})e^{-3(t-T)/\tau_m(V_m^i)} \text{ for } t > T.$$

Because G_{Na} is proportional to the cube of m, and because an exponential cubed is another exponential that decays three times faster, the Hodgkin-Huxley model predicts that the offset time constant of both the conductance and the current is $\tau_G(V_m^i) = \tau_J(V_m^i) = \tau_m(V_m^i)/3$.

Measurements of the so-called *tail currents* have been known for some time to be inconsistent with this prediction. Convincing evidence of the inconsistency of the model with measurements of tail currents in squid giant

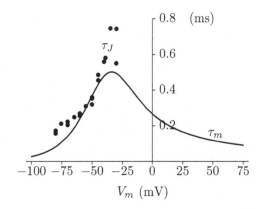

Figure 6.59 Comparison of the voltage dependence of the time constant of the tail current with τ_m (adapted from Oxford, 1981, Figure 2). The solid line shows τ_m according to the Hodgkin-Huxley model for a temperature of 6.3°. The points are from measurements of the time constant of tail currents obtained from voltage-clamped, perfused giant axons of squids (*Loligo pealii*). Cesium ions replaced the intracellular potassium ions to reduce the effect of potassium currents on the measurements. The intracellular solutions also contained pronase, which removed sodium inactivation.

axons is shown in Figure 6.59. In these experiments, cesium was substituted for potassium intracellularly to reduce the potassium current, and inactivation was blocked by the use of pronase or NBA. Capacitance and leakage currents were subtracted to minimize their effects on measurements of the sodium current. The time constant of the offset current was measured and compared to the offset current expected of m. Note that the time constant of the offset current, τ_J, is much closer in value to τ_m than to $\tau_m/3$, which is the value required by Hodgkin-Huxley kinetics.

Summary

The predictions of the Hodgkin-Huxley model do not accurately fit all the measurements of voltage-clamp currents, gating charge, and single-channel currents; that is, the model of the sodium channel consisting of four independent two-state gates does not fit exactly. Therefore, we shall consider more general channel models.

6.6.2 The Theory of Multistate Channels

We can now generalize the notion of a channel with a discrete set of states. Suppose we have a channel that has M discrete states that we enumerate as $\{S_1, S_2, S_3, \ldots S_M\}$. Each state has state attributes such as the state conductance, state ionic current, and state gating charge. We can enumerate these as follows: the state conductances are $\{\gamma_1, \gamma_2, \gamma_3, \ldots \gamma_M\}$, the state ionic currents are $\{\mathcal{I}_1, \mathcal{I}_2, \mathcal{I}_3, \ldots \mathcal{I}_M\}$, and the state gating charges are $\{\mathcal{Q}_1, \mathcal{Q}_2, \mathcal{Q}_3, \ldots \mathcal{Q}_M\}$. When the channel is in state S_j, the state conductance is γ_j, the state ionic current is $\mathcal{I}_j$, and the state gating charge is $\mathcal{Q}_j$.

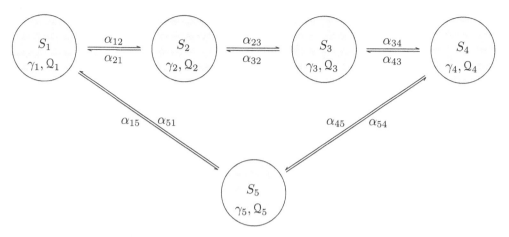

Figure 6.60   Kinetic diagram of a channel with five states.

6.6.2.1   *Description of Multistate Channels*

Random Variables

The kinetics of a multistate channel are governed by a state diagram such as that shown in Figure 6.60, which shows a state diagram for a five-state kinetic model. The state diagram indicates the allowed state transitions as well as the rate constant of the transition. For example, for the kinetic diagram shown in Figure 6.60 transitions can occur directly between states S_1 and S_2, S_2 and S_3, S_3 and S_4, S_4 and S_5, and S_1 and S_5, but not directly between S_1 and S_3.

The channel switches randomly among its states according to the transition rate constants so that the state occupancy is a random variable $\tilde{s}$ as shown schematically in Figure 6.61 for a channel that has five states. As the channel switches among its states, the single-channel random variables (the channel conductance, channel ionic current, channel gating charge, and channel gating current) change. Thus, these are all random variables as well. In particular, the channel conductance $\tilde{g}(V_m, t)$ is a random, piecewise constant random variable that switches among the state conductances. In the example shown in Figure 6.61, it has been arbitrarily assumed that states S_1, S_2, and S_3 are closed (nonconducting) states so that $\gamma_1 = \gamma_2 = \gamma_3 = 0$. State S_4 and S_5 are assumed to be conducting states, but with different state conductances. Therefore, the channel conductance switches between the values 0, γ_4, and γ_5. It has also been assumed that the current is related simply to the conductance by the relation $\mathcal{I}_j = \gamma_j(V_m - V_n)$ and that $V_m < V_n$. Therefore, current flows only in time intervals when the channel is in state four or state five and the

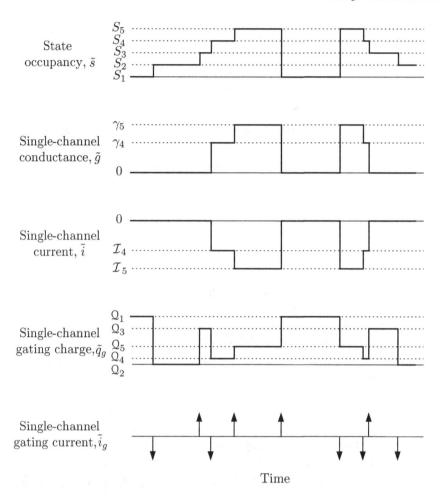

Figure 6.61 Variables associated with a channel that has five states with time-independent occupancy probabilities. States S_1–S_3 are nonconducting or closed states, i.e., $\gamma_1 = \gamma_2 = \gamma_3 = 0$. States S_4 and S_5 are conducting states with nonzero state conductances of γ_4 and γ_5. All five states have distinct gating charges as shown. The gating current is a sequence of impulses that have areas that equal the amplitude of the discontinuity in the gating charge. The areas of the impulses are (from left to right) $Q_2 - Q_1$, $Q_3 - Q_2$, $Q_4 - Q_3$, $Q_5 - Q_4$, $Q_1 - Q_5$, $Q_5 - Q_1$, $Q_4 - Q_5$, $Q_3 - Q_4$, and $Q_2 - Q_3$.

current is negative in these intervals. For the sake of illustration, it has been assumed that all the states have distinct gating charges so that a transition between any two states produces a gating current.

In a channel with one two-state gate (Figure 6.33), the channel conductance, channel ionic current, and channel gating charge all have the same waveforms in time. In contrast, for a channel with multiple states, it is apparent from Figure 6.61 that these variables can have quite different waveforms. Therefore, measurements of ionic currents and gating currents can give distinct information about channel kinetics.

State Occupancy Probabilities

The state occupancy probabilities determine the expected values of all the single-channel random variables—the single channel conductance, current, gating charge, and gating current. Therefore, we shall explore the state occupancy probabilities first. The probability that at time t the channel is in state j is $x_j(V_m, t)$. Since the channel must be in some state at each instant in time, the state occupancies must satisfy the relation $\sum_j x_j(V_m, t) = 1$ at each instant in time t. To indicate the method for determining the state occupancy probabilities, we shall consider a simple three-state channel model defined by the kinetic scheme

$$S_1 \underset{\alpha_{21}}{\overset{\alpha_{12}}{\rightleftharpoons}} S_2 \underset{\alpha_{32}}{\overset{\alpha_{23}}{\rightleftharpoons}} S_3,$$

where the αs are functions of V_m. The kinetic equations are

$$\frac{dx_1(V_m, t)}{dt} = -\alpha_{12}x_1(V_m, t) + \alpha_{21}x_2(V_m, t),$$

$$\frac{dx_2(V_m, t)}{dt} = \alpha_{12}x_1(V_m, t) - (\alpha_{21} + \alpha_{23})x_2(V_m, t) + \alpha_{32}x_3(V_m, t),$$

$$\frac{dx_3(V_m, t)}{dt} = \alpha_{23}x_2(V_m, t) - \alpha_{32}x_3(V_m, t).$$

These equations are not independent, i.e., the sum of equations 1 and 3 equals -1 times equation 2. Thus, one of these equations can be removed and the reduced equations augmented by the relation $x_1(V_m, t) + x_2(V_m, t) + x_3(V_m, t) = 1$ to yield a set of independent equations whose number equals the number of unknowns.

There are many methods for finding the solutions to such equations. One method is to use a trial solution of the form $x_j = A_j e^{-\lambda t}$, substitute this into the coupled, first-order linear differential equations, and solve for the values of λ—which are called the eigenvalues, characteristic frequencies, or natural frequencies of the system—that satisfy the differential equations. Another method is to use the Laplace transform to reduce the coupled differential equations to coupled algebraic equations. Alternatively, the set of equations can be written as a matrix differential equation

$$\frac{d\mathbf{x}(V_m, t)}{dt} = \mathbf{x}(V_m, t)\,\boldsymbol{\alpha},$$

where

$$\boldsymbol{\alpha} = \begin{bmatrix} -\alpha_{12} & \alpha_{12} & 0 \\ \alpha_{21} & -\alpha_{21} - \alpha_{23} & \alpha_{23} \\ 0 & \alpha_{32} & -\alpha_{32} \end{bmatrix}.$$

The matrix equation can be solved using the methods outlined in Appendix 6.1. Whichever method is chosen, ultimately a polynomial in λ, the characteristic polynomial, is obtained and must be solved to find the eigenvalues. The eigenvalues are algebraic functions of the rate constants of the kinetic scheme (see Appendix 6.1). For an M-state channel, there will be M eigenvalues, of which one will be $\lambda = 0$. If the eigenvalues are distinct, then in general

$$x_j(V_m, t) = \sum_{k=1}^{M} W_{jk} e^{\lambda_k t},$$

where λ_k is an algebraic function of the α_{mn}s and the W_{jk}s are weights of the natural response components.

6.6.2.2 *Properties of Channel Behavior*

In this section, we shall explore properties of multiple-state voltage-gated channels. Figures 6.62-6.66 show the responses of such channels to a step of membrane potential applied at $t = 0$. For $t < 0$, the membrane potential produces rate constants such that the channel occupies state 1 with probability 1, i.e., $x_1(t) = 1$ for $t < 0$. For $t > 0$, the membrane potential produces the rate constants indicated in the state diagrams given in the upper panel of each waveform. In this way, this section focuses on the dependence of the response on the number of states, on the distribution of state attributes, and on the rate constants, and not on the relation between the rate constants and the membrane potential.

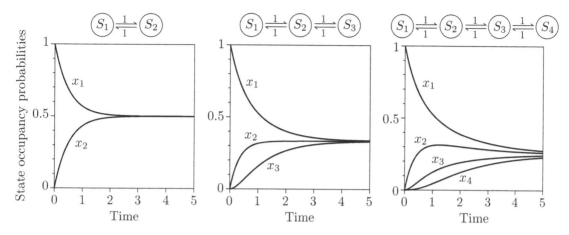

Figure 6.62 State occupancy probabilities for a two-state, a three-state, and a four-state channel, each channel having a linear array of states. All the rate constants are arbitrarily set to one. In each case, the channel was assumed to be in state 1 at $t = 0$.

Dependence on the Number of States

How do the state occupancy probabilities depend upon the number of states? Figure 6.62 shows the state occupancies for a two-state, a three-state, and a four-state channel where all the channels are in a linear array. All the rate constants were arbitrarily set to one, and for each channel the initial state was assumed to be state 1, so that the state occupancy probability $x_1(V_m, 0) = 1$ and the state occupancy probability of all other states was zero. Since all the rate constants are the same, all states are equally likely at equilibrium. For each channel model, $x_1(V_m, t)$ decays to its equilibrium state where all the other state occupancy probabilities increase to their equilibrium values. For the two-state models, the time courses of both state occupancies are exponential (as was shown in Section 6.5. However, for channel models with a number of states greater than two, the state occupancies of later states show decidedly S-shaped onsets. Thus, the S-shaped onset represents a lag incurred by the state occupancy having to traverse earlier stages. For example, in the four-state model, $x_3(V_m, t)$ shows a lag at the onset because state 2 must be occupied before state 3 can be occupied. Similarly, there is greater lag in the state occupancy of state 4 because both state 2 and state 3 must be occupied before state 4 can be occupied. The lag in the onset of the state occupancies of later states results from the time taken to traverse earlier states.

Channel state diagram

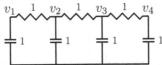

Electrical network

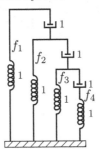

Mechanical system

Figure 6.63 Channel state diagram, electrical network, and mechanical system that have the same kinetics. The channel is characterized by its state occupancy probabilities, $\{x_1, x_2, x_3, x_4\}$; the electrical network by its voltages, $\{v_1, v_2, v_3, v_4\}$; and the mechanical system by its forces, $\{f_1, f_2, f_3, f_4\}$. The voltages are all referred to the common node to which all the capacitances are connected; the forces are on each spring, and all springs are attached to a common rigid plane. The differential equations that describe the state occupancy probabilities, voltages, and forces are the same for these systems.

Analogy to Other Multistate Systems

The equations of state occupancy for the kinetic model shown in the top of Figure 6.63 are

$$\frac{dx_1(t)}{dt} = -x_1(t) + x_2(t),$$

$$\frac{dx_2(t)}{dt} = +x_1(t) - 2x_2(t) + x_3(t),$$

$$\frac{dx_3(t)}{dt} = + x_2(t) - 2x_3(t) + x_4(t),$$

$$\frac{dx_4(t)}{dt} = + x_3(t) - x_4(t).$$

The differential equations that describe the state occupancy probabilities in the channel state diagram in Figure 6.63 are the same as those for the node voltages in the electrical network and the spring forces in the mechanical system except for relabeling the variables. Therefore, each system exhibits identical kinetic behavior. If each system is initiated by making variable number 1 (x_1, v_1, and f_1) nonzero while all other variables are zero, each system will change transiently in a manner identical to that shown for the four-state system in Figure 6.62. This analogy allows us to summarize the kinetics of a multistate channel system in another manner. Each stage of the system acts as a low-pass filter (either an electrical RC network or a mechanical dashpot/spring system) so that state occupancy probability 2 is a low-pass filtered version of state occupancy probability 1 and state occupancy probability 3 is a low-pass filtered version of state occupancy probability 2, etc. These successive stages of low-pass filtering result in the increasing lag at the onset of subsequent state occupancy probabilities in a cascade of states.

6.6.2.3 *Expected Values of Conductance, Ionic Current, Gating Charge, and Gating Current*

The expected values of the single-channel random variables are directly obtained from the state occupancy probabilities,

$$g(V_m, t) = E[\tilde{g}(V_m, t)] = \sum_{j=1}^{M} \gamma_j x_j(V_m, t),$$

$$i(V_m, t) = E[\tilde{i}(V_m, t)] = \sum_{j=1}^{M} \mathcal{I}_j x_j(V_m, t), \qquad (6.54)$$

$$q_g(V_m, t) = E[\tilde{q}_g(V_m, t)] = \sum_{j=1}^{M} \mathcal{Q}_j x_j(V_m, t),$$

$$i_g(V_m, t) = E[\tilde{i}_g(V_m, t)] = \sum_{j=1}^{M} \mathcal{Q}_j \frac{dx_j(V_m, t)}{dt}.$$

Thus, we can now examine how conduction and gating of a channel depend on both the number of states and on the state attributes.

Effect of Transition Rates

Sufficiently large changes in the transition rates will produce large changes in the kinetics of ion conduction and gating in channel models. However, as indicated in Figure 6.64, large changes in the constellation of transition rates need not be accompanied by large changes in channel kinetics. In this figure, the state attributes are fixed, but the transition rates have been varied. The transition rates have been chosen so that the lowest nonzero rate eigenvalue equals one. Even though the transition rates differ appreciably, the kinetics of both the gating current and the ionic current do not differ a great deal. This example illustrates that it is likely to be difficult to determine the individual

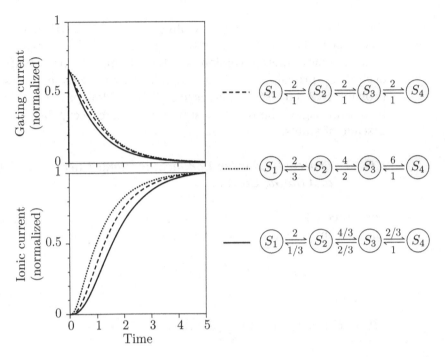

Figure 6.64 Comparison of ionic and gating currents for three four-state channel models with different rate constants. Each model has the same state conductances, $\{0, 0, 0, 1\}$, and state gating charges, $\{0, 1/3, 2/3, 1\}$. The rate constants were chosen so that for each channel model, the lowest rate nonzero eigenvalue was one. In each case, the channel was assumed to be in state 1 at $t = 0$.

transition rates between states from measurements of the kinetics of gating and ionic currents alone.

Effect of State Attributes
By fixing the transition rates and varying the state attributes, the effect of state attributes on the conduction and gating variables can be examined. Figure 6.65 shows results for a four-state channel model with a fixed set of transition rates. The resulting state occupancy probabilities are shown in the left panel. The two right panels show the effect of changes in the state conductances. In the center panel, all states are nonconducting except state 4, which has a state conductance of $\gamma_4 = 15$. Therefore, the expected value of the conductance is $g(V_m, t) = 15x_4(V_m, t)$. For this choice of state attributes, the expected value of the conductance rises according to an S-shaped curve to a final value at

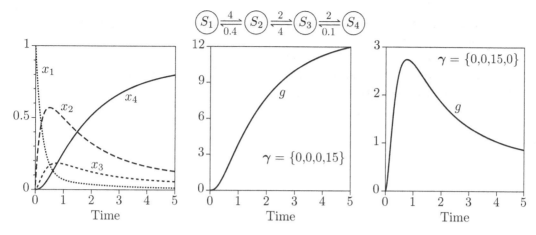

Figure 6.65   Comparison of expected values of ionic conductances for two four-state channel models with the same rate constants, but different state conductances. Each model had the rate constants shown in the kinetic diagram at the top and yielded the state occupancy probabilities shown in the left panel. The channel was assumed to be in state 1 at $t = 0$. The other panels show the expected conductance for the indicated state conductances.

which it is maintained. This type of response is similar to that found for the potassium conductance in the Hodgkin-Huxley model. In contrast, the right panel is identical to the center panel, except that now all states are nonconducting except state 3, which has a conductance $y_3 = 15$. Therefore, $g(V_m, t) = 15x_3(V_m, t)$. For this choice of state attributes, the expected value of the conductance rises according to an S-shaped curve to a peak value and then decreases. This type of response is similar to that of an inactivating channel such as that found for the sodium conductance in the Hodgkin-Huxley model. These two examples are simple illustrations of the differences between channels that do not inactivate and those that do.

Relation of Gating Charge to Ionic Conductance

Equations 6.54 demonstrate that the expected values of the gating charge ($q_g(V_m, t)$) and the ionic conductance ($g(V_m, t)$) are weighted sums of the state occupancy probabilities, but the weights differ. For the expected value of the gating charge, the weights are the state gating charges. For the expected value of the ionic conductance, the weights are the state conductances. Since these weights are in general different, the expected values of the gating charge and the ionic conductance will differ.

Relation of State Gating Charge to Expected Value of Gating Charge and Gating Current

Mathematically, we might assign state gating charges arbitrarily to different states. However, physically there must be constraints on this assignment. For example, if an increase in the membrane potential opens a channel, it must do so by displacing a positive charge outward through the membrane or a negative charge inward. The effect of state gating charge on the expected value of both the gating charge and the gating current is shown in Figure 6.66. The

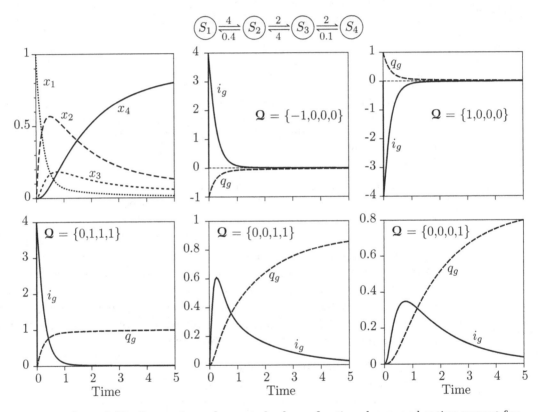

Figure 6.66 Comparison of expected values of gating charge and gating current for different four-state channel models with the same rate constants, but different state gating charges. Each model had the rate constants shown in the kinetic diagram at the top and yielded the state occupancy probabilities shown in the upper left panel. The channel was assumed to be in state 1 at $t = 0$. The other panels show the expected gating charge, $q_g(V_m, t)$, and gating current, $i_g(V_m, t) = dq_g(V_m, t)/dt$, for the state gating charges indicated.

same four-state channel model used in Figure 6.65 is used again with the same transition rates. The upper left panel shows the resulting state occupancy probabilities. The remaining panels show the effects of different assignments of gating charge to the different states. If only the first state has a gating charge and if it is negative, the channel gating charge is negative, with a time course equal to that of the occupancy probability of state 1. Under these conditions, the channel gating current is positive or outward. If the sign of the state 1 gating charge is reversed, both the channel gating charge and the channel gating current are reversed.

6.6.3 Summary

There has been a great deal of progress in the development of models of the gating of ion channels. The Markov process type of model assumes that each channel has a discrete number of states that are connected by a kinetic diagram that defines the allowable transitions between these states and the rates at which these transitions occur. The channel has attributes assigned to the states, such as the conductance and the gating charge. The probability of a transition from an initial state to a final state in an incremental interval of time is assumed to be proportional to the product of that time interval and a transition rate that depends only on the initial state. This basic set of assumptions defines a rich class of models from which one can compute the average values and fluctuations of all macroscopic as well as single-channel electrical variables associated with single ion channels, including both ionic current and gating current variables. This constitutes a set of very strong constraints. Thus, it should in principle be possible to determine whether these types of kinetic models are consistent with all measured variables on a single type of channel. This has not yet been achieved for any ion channel. In part, as new information about channel properties appears, the kinetic diagrams have become exceedingly complex, making estimation of kinetic parameters more and more difficult. Furthermore, without the ability to study directly the individual kinetic steps, it is in principle not possible to determine a kinetic model uniquely.

There is a different type of difficulty inherent in the class of channel models we have explored. In these models, gating and permeation are independent processes. It is unlikely that motions of charges that result from changes in the electric field in the membrane will not themselves change that field to affect permeation of ions or that permeation of ions will not change the local electric field in the channel. However, the importance of this type of coupling will need to be evaluated quantitatively. Further progress in determining the mechanisms of gating and the permeation of single ion channels has resulted from combining the methods of membrane biophysics with those of molecular biology, as we shall describe in the next section.

6.7   Voltage-Gated Ion Channel Macromolecules

The molecular structures of voltage-gated ion channel macromolecules are under intense investigation. Of these channels, more is known about the structure of the sodium channel macromolecule than about any other. We shall describe this channel first and follow with a description of other ion channel macromolecules. More extensive descriptions are found elsewhere (Hille and Fambrough, 1987; Rydström, 1987; Agnew et al., 1988; Catterall, 1988; Jan and Jan, 1989; Numa, 1989; Guy, 1990; Guy and Conti, 1990; Patlak, 1991; Hille, 1992; Catterall, 1993).

6.7.1   Sodium Channel Macromolecule

The sodium channel macromolecule was the first voltage-gated ion channel to be isolated and then sequenced. Its structural and functional parts are under intense investigation.

6.7.1.1   *Isolation of the Sodium Channel Protein*

The technique for isolating and purifying the sodium channel macromolecule is similar to that for any other integral membrane protein. First, a substance must be used to mark the macromolecule during isolation procedures. Fortunately, there are a number of neurotoxins that are known to bind to the sodium channel macromolecule: STX, TTX, scorpion toxin, etc. These toxins have been prepared with markers (e.g., radioactive tracers or photoreactive groups) and used to identify membrane fractions containing the sodium channel macromolecule obtained from tissues known to be rich in sodium channels (e.g., eel electric organs). The markers are used in successive stages of purification, as indicated schematically in Figure 6.67. Cellular membranes are disrupted and solubilized by use of ultracentrifugation and detergents. The resulting membrane fragments contain many membrane macromolecules, including the sodium channel macromolecule. The membrane fragments are

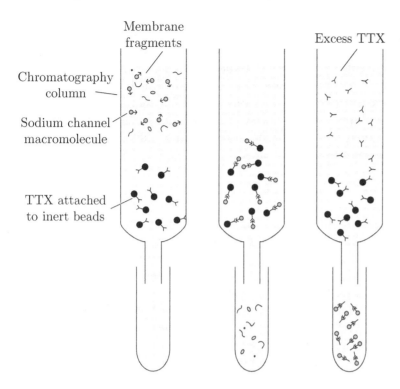

Figure 6.67 Schematic diagram illustrating the method of separating sodium channel macromolecules from membrane fragments obtained from detergent-treated cell membranes.

passed through a chromatography column that contains inert beads to which TTX has been attached. The sodium channel macromolecule binds to the TTX, while other membrane fragments pass through the column. A quantity of TTX in excess of the quantity bound to the beads is passed through the column. The free TTX competes for sodium channels with the TTX bound to beads. Because there is an excess of free TTX, most of the sodium channel macromolecules are bound to free TTX and then pass through the column and are collected.

Using methods similar to those outlined above, the sodium channel macromolecule was first isolated from the electric organs (or electroplax) of the electric eel *Electrophorus electricus* (Agnew et al., 1978), and later from a number of tissues including rat skeletal muscle, rabbit skeletal muscle, rat brain, pig brain, human brain, chick cardiac muscle, etc. The macromolecule can be reconstituted in artificial membranes and retains its ability to bind toxins and to transport sodium with kinetics similar to those of the native channel. The isolated sodium channel macromolecule is a glycoprotein with a ubiquitous large subunit called the α-subunit, which has a molecular weight of about

260 kD, and either zero, one, or two smaller β-subunits of molecular weights in the range of 33–38 kD. The number of different β-subunits varies in sodium channel macromolecules of different tissues and species. About 30% of the mass of all subunits is carbohydrate. The α subunit, which consists of about 2,000 amino acids, contains the structural elements required for the basic functions of the sodium channel, such as voltage gating, selective permeability to sodium, inactivation, and pharmacology (Numa, 1989). The functions of the β-subunits have not been determined.

6.7.1.2 Structure of the Sodium Channel Protein

Isolation of the sodium channel macromolecule enabled the use of the methods of molecular biology to determine the nucleotide sequence of the gene encoding the primary structure of the protein. Briefly, the purified channel protein was used to produce antibodies that were later used to identify the protein. Messenger RNA (mRNA) from eel electroplax were used to prepare complementary DNA (cDNA) libraries that were screened with the antibody to isolate the cDNA clones encoding the protein. These methods resulted in the determination of the amino acid sequence of the sodium channel protein initially in the eletroplax of the electric eel (Noda et al., 1984), but later in several tissue types and in several species, including both invertebrates and vertebrates. While there is a great deal of homology between the amino acid sequences obtained from all these different sources, there are also some differences. Therefore, there is a family of sodium channel macromolecules that share important structural and functional properties, but that show some differences.

The amino acid sequence defines the primary structure of the protein portion of the sodium channel macromolecule. The secondary, tertiary, and quaternary structures have not been determined directly. Nevertheless, as with other integral membrane proteins, an examination of the hydrophobicity of the amino acids (Weiss, 1996, Chapters 1, 4, and 6) reveals that the protein consists of segments that are predominantly hydrophobic and others that are predominantly hydrophilic, as shown in Figure 6.68. Some of the hydrophobic segments consist of amino acid sequences that can form α helices of about twenty amino acids, which is a number sufficient to span the membrane.

Such analyses have led to several proposals for the three-dimensional structure of the α-subunit of the sodium channel protein (Guy, 1990; Cat-

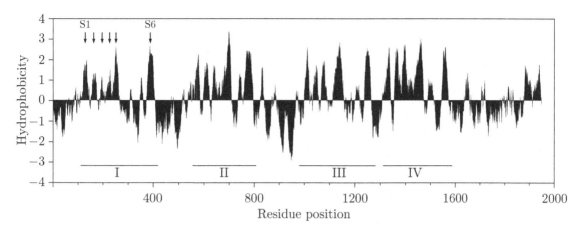

Figure 6.68 Hydrophobicity plot of the sodium channel protein of the electric organ of the electric eel *Electrophorus electricus*. The mRNA sequence (Noda et al., 1984) was obtained from the EMBL and GenBank database (accession no. X01119) and translated. The hydrophobicity of each amino acid was assigned (Kyte and Doolittle, 1982), and the resultant sequence of amino acids was averaged with a window that was 15 residues long to yield the averaged hydrophobicity shown. Domains I-IV are indicated. Six hydrophobic segments (S1-S6) are indicated with arrows in domain I. Similar segments can also be discerned in each of the other domains, although S4 and S5 cannot be resolved in some segments.

terall, 1993). These proposed structures have many commonalities, which we shall describe. The primary amino acid structure of the α-subunit consists of four homologous repeating sections called *domains*, which are labeled I to IV (Figure 6.68). Each of these domains consists of six to eight hydrophobic amino acid segments, six long segments and two shorter segments. These hydrophobic amino acid segments are separated by hydrophilic amino acid segments that stick out into the cytoplasmic and extracellular spaces on either side of the membrane. In particular, the segments on both sides of each domain are located on the cytoplasmic side of the membrane. The six long hydrophobic amino acid segments, labeled S1 to S6, are α helices that span the membrane. The fourth long segment in the third domain is referred to as III-S4. The short segments, labeled SS1 and SS2, occur between S5 and S6. Figure 6.69 shows this arrangement of the sodium channel protein. Proposed three-dimensional structures of the channel protein differ in the location and secondary structure of the short segments (Guy, 1990; Catterall, 1993).

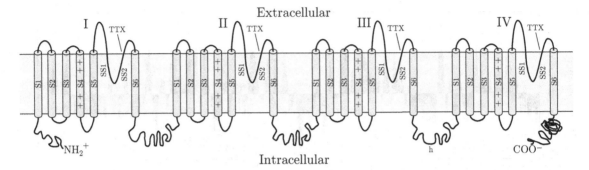

Figure 6.69 Putative two-dimensional organization of the sodium channel protein in the membrane (Guy, 1990; Catterall, 1993). The protein consists of four domains, I–IV, each of which consists of eight hydrophobic segments. Six long segments (S1-S6) are α helices that span the membrane (shown as cylinders), and two short segments (SS1 and SS2) are inserted in the membrane. Regions on the macromolecule are shown that involve tetrodotoxin (TTX) binding, channel gating (+), and inactivation (h).

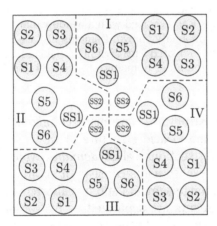

Figure 6.70 Putative three-dimensional organization of hydrophobic segments of the amino acid sequence of the sodium channel protein (adapted from Patlak, 1991, Figure 1).

The channel is formed as a tetramer where each of the domains I to IV forms a quarter of the channel, as shown in Figure 6.70. The homology of the four domains is indicated in Figure 6.71, which shows their amino acid sequences aligned.

6.7.1.3 Functional Portions of the Sodium Channel Protein

A variety of techniques have been used to determine the orientation of the macromolecule in the membrane. For example, it has been possible to identify

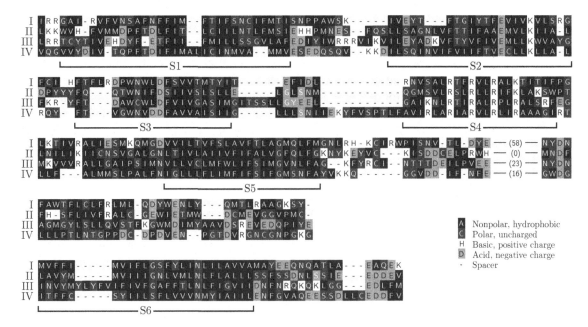

Figure 6.71 Amino acid sequences for domains I–IV of the α-subunit of the sodium channel protein from the electroplax of *Electrophorus electricus* (adapted from Noda et al., 1984, Figure 5). The protein consists of 1820 amino acids. The four white columns in every third column of S4 are either arginine (R) or lysine (K), each of which contains a positively charged residue.

the site of binding of antibodies introduced on the extracellular side of the membrane. These binding sites then identify portions of the macromolecule that are in contact with the extracellular space. Methodologies have also been developed to determine which portion of the sodium channel macromolecule is responsible for each particular function of the sodium channel. Especially useful have been studies in which modified mRNAs are produced with well-defined mutations, sometimes leading to a change in a single amino acid in the protein, and the mRNAs are then injected into *Xenopus* oocytes, which express the channel protein. Electrophysiological studies of these oocytes are used to assess the change in function caused by the mutation. Results of such studies have led to a number of important conclusions that are summarized in the schematic diagram shown in Figure 6.69 (Guy, 1990; Catterall, 1993).

Voltage-Gated Channel Activation

Initial determination of the amino acid sequence (Noda et al., 1984) revealed that segment S4 contains four amino acids with positively charged residues (arginines or lysines) separated by two hydrophobic amino acids (Figure 6.71). It was proposed that these were the gating charges responsible for sensing the membrane potential. Mutations of S4 that reduce the charged residues decrease the voltage sensitivity of channel activation and mutations to other residues in S4 modify the dependence of sodium activation on membrane potential (Stühmer et al., 1989; Auld et al., 1990). These studies clearly suggest that S4 is involved in voltage gating. However, the mechanism by which motions of the charges in S4 change the state of conduction of the channel is unknown.

Permeation Pathway

The pathway for ion permeation is presumed to be a pore in the center of the tetramer (Figure 6.70). Identification of the segments that line the pore might indicate the basis of the ion selectivity of the sodium channel. In addition, based on indirect evidence it has been generally assumed that toxins that block permeation of the sodium channel, such as tetrodotoxin (TTX) and saxitoxin (STX), bind at the extracellular mouth of the pore. Thus, determining the toxin-binding sites should give insight into the mechanism of toxin binding as well as into the basis of permeation blocking the action of toxin binding. Studies with mutated sodium channels have determined the key amino acids required for toxin binding. Replacement of negatively charged glutamate with uncharged glutamine at position 387 (Glu387 or E387), which occurs in domain I at segment SS2, reduces the binding affinity by more than 1,000 (Noda et al., 1989). Mutations of charged residues with uncharged residues at sites homologous to E387 in all four domains also reduce the binding affinity of TTX and STX to the membrane. Another set of amino acids located nearby is also involved in toxin binding. Finally, sodium channels that have a relatively low binding affinity for these toxins, such as those in cardiac muscle, show differences in amino acids in this region. Thus, the sodium channel macromolecule contains amino acids in a restricted region that are highly implicated as the binding sites for TTX and STX. Their location (Figure 6.69) is assumed to be at the extracellular mouth of the pore.

Mutations in the same region as those amino acids that reduce the binding affinity for toxins also affect the permeation of the channel for sodium. In fact, the mutation of two amino acid residues, Lys1422 (positively charged), located in segment SS2 of domain III, and Ala1714 (nonpolar), located in segment SS2 of domain IV, to negatively charged glutamate changes the ion selectivity of the sodium channel to that of the calcium channel (Heinemann et al., 1992). Thus, these residues must be involved in the ionic selectivity of the sodium channel.

Channel Inactivation

Proteolytic enzymes, such as pronase, remove sodium inactivation without affecting sodium activation much. Thus, it has been assumed that the structure involved in sodium inactivation is located on the cytoplasmic membrane surface. Both site-directed binding of antibodies and mutations to the cytoplasmic segment that links domains III and IV eliminate inactivation without affecting activation much (Vassilev et al., 1988; Vassilev et al., 1989; Patton et al., 1992; West et al., 1992). These studies suggest that the loop of amino acids that links domains III and IV can move to occlude the cytoplasmic portion of the pore.

6.7.2   Other Channel Macromolecules

The primary structures have been determined for sodium, calcium, and potassium voltage-gated ion channels (Catterall, 1993). Each consists of four homologous domains consisting of six hydrophobic α helical segments that form a pore through which ions permeate the membrane. Each of the four domains of the potassium channel is formed from a different gene, whereas, a single gene codes the α-subunit consisting of all four domains of the sodium and calcium channels. The S4 segments of potassium channels have also been modified and found to affect voltage activation. Blockers of the calcium and potassium channels act on segments SS1 and SS2, and these regions also affect ion selectivity of the channel. Thus, the different ion channels have similar three-dimensional structures. Therefore, the different voltage-gated ion channels have common structural/functional components, suggesting a common phylogenetic ancestry.

Appendix 6.1 Markov Process Models of Single Channels

Although the kinetic properties of single channels have been represented by more than one type of mathematical model (Sansom et al., 1989), Markov process models remain the most successful in providing a conceptual framework for a broad range of measurements on single channels. The development in this section is based on general treatments of Markov processes (Parzen, 1962; Cox and Miller, 1965) as well on those specialized for the analysis of single ion channels (Colquhoun and Hawkes, 1977; Neher and Stevens, 1977; Colquhoun and Hawkes, 1983; Horn, 1984). Some familiarity with probability theory, random process theory, and linear algebra is assumed in this appendix.

General Description

A Markov process is defined in terms of the properties of the probabilities of transitions between states. These probabilities characterize the process in the sense that all properties of a Markov process can be derived from these probabilities. In this section, we shall define these state transition probabilities and their relations to the state transition rates.

State Transition Probabilities

Assume that a channel can exist in M distinct kinetic states, and let $\tilde{s}(t)$ be the state occupied by a channel at time t. We shall assume that $\tilde{s}(t)$ can be described as a discrete-state, continuous-time Markov process. In such a process, the state of the process at some future time is dependent only on its present state and not on past states. That is, the state of a Markov process at some future time is independent of how the system reached its present state. Therefore, we can define the state transition probability $y_{ij}(u, t)$ that the system is in state j at time t given that it was in state i at a previous time u as

$$y_{ij}(u, t) = \text{Prob}\{\tilde{s}(t) = j | \tilde{s}(u) = i\}. \tag{6.55}$$

Note that this probability is independent of states prior to u. For a time-independent Markov process, $y_{ij}(u, t)$ depends only on $t - u$ and not on t and u individually. We shall consider only time-independent Markov processes unless otherwise stated. For these conditions, we have

$$y_{ij}(t) = \text{Prob}\{\tilde{s}(t + \tau) = j | \tilde{s}(\tau) = i\}. \tag{6.56}$$

We assume that at any time t, the states of the channel are mutually exclusive and collectively exhaustive. That is, the channel is in one of its states at every time and occupies only one state at a time. This assumption implies that $0 \leq y_{ij}(t) \leq 1$ and $\sum_j y_{ij}(t) = 1$. Therefore, $y_{ii} = 1 - \sum_{j \neq i} y_{ij}(t)$. It is convenient to represent the set of state transition probabilities as a matrix

$$\mathbf{y}(t) = \begin{bmatrix} y_{11}(t) & y_{12}(t) & \cdots & y_{1M}(t) \\ y_{21}(t) & y_{22}(t) & \cdots & y_{2M}(t) \\ \cdots & \cdots & \cdots & \cdots \\ y_{M1}(t) & y_{M2}(t) & \cdots & y_{MM}(t) \end{bmatrix}$$

in which the rows are the initial states and the columns are the final states, and we have assumed that the process has M discrete states. From the definition of the state transition probability, it follows that $y_{ij}(0) = \delta_{ij}$, where δ_{ij} is the Kronecker delta,

$$\delta_{ij} = \begin{cases} 0 & \text{for } i \neq j, \\ 1 & \text{for } i = j. \end{cases}$$

Therefore, $\mathbf{y}(0) = \mathbf{I}$, where $\mathbf{I}$ is the identity matrix whose off-diagonal terms are zero and whose diagonal terms are one.

In general, for any discrete-state, continuous-time stochastic process, we have

$$\text{Prob}\{\tilde{s}(t) = k | \tilde{s}(t - \tau) = i\}$$

$$= \sum_j (\text{Prob}\{\tilde{s}(t) = k | \tilde{s}(t - \tau) = i, \tilde{s}(t - u) = j\}$$

$$\times \text{Prob}\{\tilde{s}(t - u) = j | \tilde{s}(t - \tau) = i\}), \quad \text{for } \tau > u.$$

But for a Markov process,

$$\text{Prob}\{\tilde{s}(t) = k | \tilde{s}(t - \tau) = i, \tilde{s}(t - u) = j\} = \text{Prob}\{\tilde{s}(t) = k | \tilde{s}(t - u) = j\}$$

which, with the use of Equation 6.55, yields a fundamental relation between state transition probabilities of a Markov process known as the Chapman-Kolmogorov equation,

$$y_{ik}(t) = \sum_j y_{ij}(\tau) y_{jk}(t - \tau), \quad \text{for } t > \tau. \tag{6.57}$$

The Chapman-Kolmogorov equation can be written compactly in matrix form as

$$\mathbf{y}(t) = \mathbf{y}(\tau)\mathbf{y}(t - \tau). \tag{6.58}$$

State Transition Rates

What is the probability that a transition between two states will occur during a brief time interval Δt? We shall make the plausible assumption that for a sufficiently brief interval the probability of a transition in the interval will be proportional to the length of the interval, i.e.,

$$y_{ij}(\Delta t) = \alpha_{ij}\Delta t + o(\Delta t) \quad \text{for } i \neq j,$$

$$y_{ii}(\Delta t) = 1 + \alpha_{ii}\Delta t + o(\Delta t), \tag{6.59}$$

where $\lim_{\Delta t \to 0} o(\Delta t)/\Delta t = 0$, and that α_{ij} is the state transition rate from state i to state j. Therefore, the probability that no transition occurs from i to j in that time interval is $1 - \alpha_{ij}\Delta t - o(\Delta t)$, and the probability that no transition occurs to any state that differs from i is $1 - \sum_{j \neq i} \alpha_{ij}\Delta t - o(\Delta t)$. Thus, the state transition rate α_{ij} is defined as

$$\alpha_{ij} = \lim_{\Delta t \to 0} \frac{y_{ij}(\Delta t)}{\Delta t}, \quad \text{for } i \neq j,$$

$$\alpha_{ii} = \lim_{\Delta t \to 0} \frac{y_{ii}(\Delta t) - 1}{\Delta t}. \tag{6.60}$$

We can define a matrix of state transition rates as follows:

$$\alpha = \begin{bmatrix} \alpha_{11} & \alpha_{12} & \cdots & \alpha_{1M} \\ \alpha_{21} & \alpha_{22} & \cdots & \alpha_{2M} \\ \cdots & \cdots & \cdots & \cdots \\ \alpha_{M1} & \alpha_{M2} & \cdots & \alpha_{MM} \end{bmatrix}.$$

Because the rows of the transition rate matrix sum to zero, α is a singular matrix.

The Kolmogorov Equation

If the state transition rates are known, the state transition probabilities can be determined by solving a differential equation, the Kolmogorov equation, which $\mathbf{y}(t)$ satisfies. This equation can be determined by expressing the state transition probability at time $t + \Delta t$ in terms of the state transition probability at time t. The former can be expressed as the sum of two terms,

$$y_{ij}(t + \Delta t) = y_{ij}(t)(1 + \alpha_{jj}\Delta t) + \sum_{k \neq j} y_{ik}(t) \left(\alpha_{kj}\Delta t + o(\Delta t) \right).$$

The first term is the probability that $\tilde{s}(t) = j$ and no transition occurs in the interval $t + \Delta t$. The second term is the probability that $\tilde{s}(t) = k$ and there is a transition from k to j in the interval $t + \Delta t$. If we collect terms and divide by Δt, we obtain

$$\frac{y_{ij}(t + \Delta t) - y_{ij}(t)}{\Delta t} = y_{ij}\alpha_{jj} + \sum_{k \neq j} y_{ik} \left(\alpha_{kj} + \frac{o(\Delta t)}{\Delta t} \right).$$

If we take the limit as $\Delta t \to 0$ and combine terms, we obtain the Kolmogorov equation

$$\frac{dy_{ij}(t)}{dt} = \sum_k y_{ik}(t)\alpha_{kj}, \tag{6.61}$$

which can be written in matrix notation as

$$\frac{d\mathbf{y}(t)}{dt} = \mathbf{y}(t)\alpha. \tag{6.62}$$

The solution can be written in the form

$$\mathbf{y}(t) = \exp(\alpha t), \tag{6.63}$$

and $\mathbf{y}(t)$ can be computed directly from α by using power series expansion of the exponential

$$\mathbf{y}(t) = \mathbf{I} + \alpha t + (\alpha t)^2/2! + \cdots . \tag{6.64}$$

However, another method of solution is superior both computationally and conceptually. The matrix α can be expressed as

$$\alpha = \mathbf{E}\lambda\mathbf{E}^{-1}, \tag{6.65}$$

where λ is a diagonal matrix made up of the eigenvalues of α, $\mathbf{E}$ is an orthogonal matrix made up of the eigenvectors of α, and $\mathbf{E}\mathbf{E}^{-1} = \mathbf{I}$. The eigenvalues can be obtained by solving the equation

$$\det\{\alpha - \lambda\} = 0, \tag{6.66}$$

where $\det\{\mathbf{A}\}$ is the determinant of the matrix $\mathbf{A}$. Since the $M \times M$ matrix α is singular, one of the eigenvalues is zero. The remaining eigenvalues have negative real parts (Cox and Miller, 1965, page 184). In the remainder of this appendix, we shall assume for simplicity that the eigenvalues are distinct. The set of eigenvalues is denoted as $\{\lambda_1, \lambda_2, \cdots, \lambda_M\}$. Equation 6.63 can be expressed in terms of the eigenvalues as

$$\mathbf{y}(t) = \exp(\mathbf{E}\lambda t\mathbf{E}^{-1}) = \mathbf{E}\exp(\lambda t)\mathbf{E}^{-1}. \tag{6.67}$$

The equivalence of the two expressions of $\mathbf{y}(t)$ in Equation 6.67 can be verified by using Equation 6.64. Expansion of a typical term of $\mathbf{y}(t)$ in Equation 6.67 shows that

$$y_{ij}(t) = \sum_k w_{ijk}e^{\lambda_k t}, \tag{6.68}$$

i.e., the state transition probabilities are expressed as a linear superposition of exponential functions whose exponents are determined by the eigenvalues.

State Occupancy Probabilities

The state occupancy probability is the probability that the channel is in a particular state at a given time. That is, let $x_i(t)$ be the probability that the channel is in state i at time t. Given that the initial state occupancy probabilities are defined by $x_i(0)$, we can find the state occupancy probabilities at a later time as follows:

$$x_j(t) = \sum_i x_i(0) y_{ij}(t), \tag{6.69}$$

which can be written compactly in matrix notation if we first define the row matrix $\mathbf{x}(t)$ as the matrix of $x_i(t)$,

$$\mathbf{x}(t) = \mathbf{x}(0)\mathbf{y}(t). \tag{6.70}$$

Combining Equations 6.62 with 6.70 yields

$$\frac{d\mathbf{x}(t)}{dt} = \mathbf{x}(t)\boldsymbol{\alpha}. \tag{6.71}$$

Thus, the state occupancy probabilities satisfy the same first-order matrix differential equation as do the state transition probabilities. Given an initial set of state occupancies and the state transition rates, the state occupancies can be found as a function of time. They consist of a linear superposition of exponential time functions whose rate constants are the eigenvalues.

Single-Channel Variables

State Attributes
We associated certain attributes with each state of a channel. We denote by a_j the value of the attribute when the channel is in state j. It is convenient to define a column vector $\mathbf{a}$ of attributes. The attribute may be any state-related property of the channel. For example, the attribute may be the conductance of the channel, the current through the channel, or the gating charge (or polarization). This formulation allows us to compute the statistical properties of any random variable associated with the channel. As the channel jumps around from state to state, the variable also jumps from one of its attribute

values to another. Thus, we shall now compute the time course of the generic random variable $\tilde{v}(t)$.

Expected Value of Single-Channel Random Variable

The expected value of variable $\tilde{v}(t)$ is

$$v(t) = E[\tilde{v}(t)] = \sum_j a_j x_j(t) = \mathbf{x}(t)\mathbf{a}. \tag{6.72}$$

Thus, the expected value of the conductance $g(t) = E[\tilde{g}(t)] = \sum_j y_j x_j(t)$, where y_j is the conductance of the channel when it is in state j. The expected value of the gating charge is $q_g(t) = E[\tilde{q}_g(t)] = \sum_j \mathcal{Q}_j x_j(t)$, where $\mathcal{Q}_j$ is the gating charge of the channel when it is in state j. If the ionic current through the channel when the channel is in state j is $\mathcal{I}_j = y_j(V_m - V_{eq})$, the expected value of the current is $i(t) = E[\tilde{i}(t)] = \sum_j y_j(V_m - V_{eq})x_j(t)$, where V_{eq} is the reversal potential of the channel.

Variance of Single-Channel Random Variable

A convenient measure of the variation of variable $\tilde{v}(t)$ is its second central moment or variance, which is

$$\sigma_v(t) = E[\tilde{v}^2(t)] - v^2(t) = \sum_j a_j^2 x_j(t) - \left(\sum_j a_j x_j(t) \right)^2. \tag{6.73}$$

Autocovariance of Single-Channel Random Variable

The autocovariance function for variable $\tilde{v}(t)$ is defined as

$$c_v(\tau) = E[\tilde{v}(t + \tau)\tilde{v}(t)] = \sum_{i,j} \text{Prob}\{\tilde{v}(t + \tau) = a_j \ \& \ \tilde{v}(t) = a_i\},$$

which can be written as

$$c_v(\tau) = \sum_{i,j} \text{Prob}\{\tilde{v}(t + \tau) = a_j | \tilde{v}(t) = a_i\} \, \text{Prob}\{\tilde{v}(t) = a_i\}.$$

The steady-state value of this autocovariance can be written as

$$c_v(\tau) = \sum_{i,j} y_{ij}(\tau) x_i(\infty), \tag{6.74}$$

which is an even function of τ. If we take into account the form of y_{ij} from Equation 6.68, Equation 6.74 has the form

$$c_v(\tau) = \sum_{i,j} \left(\sum_k w_{ijk} e^{\lambda_k |\tau|} \right) x_i(\infty),$$

which can be written as

$$c_V(\tau) = \sum_k d_k e^{\lambda_k |\tau|}. \tag{6.75}$$

Thus, the autocovariance function is a weighted sum of exponential functions whose exponents are the eigenvalues of the state transition rate matrix.

Power Density Spectrum of Single-Channel Random Variable

The power density spectrum of the changes in variable $\tilde{v}(t)$, $C_V(f)$ is the Fourier transform of the autocovariance function, which is

$$C_V(f) = \int_{-\infty}^{\infty} c_V(\tau) e^{-j2\pi f \tau} \, d\tau, \tag{6.76}$$

which can be written as

$$C_V(f) = \int_{-\infty}^{\infty} \sum_k d_k e^{\lambda_k |\tau|} e^{-j2\pi f \tau} \, d\tau,$$

which can be evaluated term by term to give

$$C_V(f) = d_1 \delta(f) + \sum_{k \neq 1} d_k \frac{-2\lambda_k}{(2\pi f)^2 + \lambda_k^2}, \tag{6.77}$$

where λ_1 has been arbitrarily chosen as the eigenvalue that is zero, and $\delta(f)$ is the unit impulse function.

Dwell Times

The time that a channel spends in a particular state or in a population of states is called the *dwell time*. Probability distributions of dwell times are also related to the state transition rates. Hence, measurements of dwell times can be used to estimate these rates. The simplest dwell time distribution is the distribution of the time intervals that a channel spends in state j. Let

$$\mathcal{F}_j(t) = \text{Prob}\{\tilde{s}(u) = j \quad \text{for } 0 \leq u \leq t\}.$$

We can express

$$\mathcal{F}_j(t + \Delta t) = \text{Prob}\{\text{No transition in } (t, t + \Delta t) | \tilde{s}(u) = j \quad \text{for } 0 \leq u \leq t\} \mathcal{F}_j(t).$$

Therefore,

$$\mathcal{F}_j(t + \Delta t) = \left(1 - \sum_{i \neq j} \alpha_{ji} \Delta t - o(\Delta t)\right) \mathcal{F}_j(t),$$

or

$$\mathcal{F}_j(t + \Delta t) = \left(1 + \alpha_{jj} \Delta t - o(\Delta t)\right) \mathcal{F}_j(t).$$

Rearranging terms yields

$$\frac{\mathcal{F}_j(t + \Delta t) - \mathcal{F}_j(t)}{\Delta t} = \left(\alpha_{jj} + \frac{o(\Delta t)}{\Delta t}\right) \mathcal{F}_j(t).$$

Taking the limit of this equation as $\Delta t \to 0$ yields

$$\frac{d\mathcal{F}_j(t)}{dt} = \alpha_{jj} \mathcal{F}_j(t),$$

which has the solution

$$\mathcal{F}_j(t) = e^{\alpha_{jj}t},$$

where the coefficient of the exponential is one because the probability that the dwell time in state j exceeds a duration of zero is one. This equation states that the probability that the dwell time in state j exceeds t is exponential with the rate constant equal to the negative of the sum of rates of leaving state j. Therefore, the probability that the dwell time is less than t is

$$F_j(t) = 1 - \mathcal{F}(t) = 1 - e^{\alpha_{jj}t}.$$

Hence, the probability density function of dwell times is

$$f_j(t) = \frac{dF_j(t)}{dt} = -\alpha_{jj}e^{\alpha_{jj}t}.$$

From this density function we can compute the expected value of the dwell time in state j as

$$E[t_j] = \int_0^\infty t_j f_j(t_j)\, dt_j = -\int_0^\infty t_j \alpha_{jj}e^{\alpha_{jj}t}\, dt_j,$$

which can be integrated by parts to yield

$$E[t_j] = -\frac{1}{\alpha_{jj}}.$$

Specific Channel Kinetic Schemes

Two-State Model

To illustrate the use of the formalism in the simplest case, we shall examine the two-state model discussed in detail in Section 6.5, in which there are two states with kinetic diagram

$$C \underset{\beta}{\overset{\alpha}{\rightleftharpoons}} O.$$

Let us arbitrarily assign the closed state to state 1 and the open state to state 2. Then the state transition rate matrix is

$$\alpha = \begin{bmatrix} -\alpha & \alpha \\ \beta & -\beta \end{bmatrix}.$$

Note that the rows of the state transition rate matrix sum to zero. The eigenvalues of α can be obtained from

$$\begin{vmatrix} -\alpha - \lambda & \alpha \\ \beta & -\beta - \lambda \end{vmatrix} = 0,$$

which leads to the characteristic equation

$$\lambda \left(\lambda - (\alpha + \beta) \right) = 0.$$

Therefore, there are two eigenvalues: $\lambda_1 = 0$ and $\lambda_2 = -(\alpha + \beta)$. For $\lambda = 0$, the $\alpha - \lambda$ matrix is

$$\begin{bmatrix} -\alpha & \alpha \\ \beta & -\beta \end{bmatrix},$$

so that the eigenvector is the column vector with components $\{1, 1\}$ such that

$$\begin{bmatrix} -\alpha & \alpha \\ \beta & -\beta \end{bmatrix} \begin{bmatrix} 1 \\ 1 \end{bmatrix} = \begin{bmatrix} 0 \\ 0 \end{bmatrix}.$$

Similarly, for $\lambda = -(\alpha + \beta)$ the $\alpha - \lambda$ matrix is

$$\begin{bmatrix} \beta & \alpha \\ \beta & \alpha \end{bmatrix},$$

and the eigenvector has components $\{1/\beta, -1/\alpha\}$ such that

$$\begin{bmatrix} \beta & \alpha \\ \beta & \alpha \end{bmatrix} \begin{bmatrix} 1/\beta \\ -1/\alpha \end{bmatrix} = \begin{bmatrix} 0 \\ 0 \end{bmatrix}.$$

The two eigenvectors form the columns of matrix $\mathbf{E}$, and $\mathbf{E}^{-1}$ is the inverse of $\mathbf{E}$:

$$\mathbf{E} = \begin{bmatrix} 1 & 1/\beta \\ 1 & -1/\alpha \end{bmatrix} \quad \text{and} \quad \mathbf{E}^{-1} = \begin{bmatrix} \beta/(\alpha + \beta) & \alpha/(\alpha + \beta) \\ \alpha\beta/(\alpha + \beta) & -\alpha\beta/(\alpha + \beta) \end{bmatrix}.$$

By performing the matrix multiplication, it can be verified that these matrices satisfy Equation 6.65 as follows:

$$\begin{bmatrix} 1 & 1/\beta \\ 1 & -1/\alpha \end{bmatrix} \begin{bmatrix} 0 & 0 \\ 0 & -(\alpha + \beta) \end{bmatrix} \begin{bmatrix} \beta/(\alpha + \beta) & \alpha/(\alpha + \beta) \\ \alpha\beta/(\alpha + \beta) & -\alpha\beta/(\alpha + \beta) \end{bmatrix} = \begin{bmatrix} -\alpha & \alpha \\ \beta & -\beta \end{bmatrix}.$$

According to Equation 6.67, the state transition probabilities are

$$\begin{bmatrix} y_{11}(t) & y_{12}(t) \\ y_{21}(t) & y_{22}(t) \end{bmatrix} = \begin{bmatrix} 1 & 1/\beta \\ 1 & -1/\alpha \end{bmatrix} \begin{bmatrix} 1 & 0 \\ 0 & e^{-(\alpha+\beta)t} \end{bmatrix}$$

$$\times \begin{bmatrix} \beta/(\alpha + \beta) & \alpha/(\alpha + \beta) \\ \alpha\beta/(\alpha + \beta) & -\alpha\beta/(\alpha + \beta) \end{bmatrix},$$

which, when evaluated, yields

$$\begin{bmatrix} y_{11}(t) & y_{12}(t) \\ y_{21}(t) & y_{22}(t) \end{bmatrix} = \begin{bmatrix} \frac{\beta + \alpha e^{-(\alpha+\beta)t}}{\alpha+\beta} & \frac{\alpha - \alpha e^{-(\alpha+\beta)t}}{\alpha+\beta} \\ \frac{\beta - \beta e^{-(\alpha+\beta)t}}{\alpha+\beta} & \frac{\alpha + \beta e^{-(\alpha+\beta)t}}{\alpha+\beta} \end{bmatrix}.$$

Note that the rows of the state transition probability matrix add up to one.

According to Equation 6.70, the state occupancy probabilities are

$$\begin{bmatrix} x_1(t) & x_2(t) \end{bmatrix} = \begin{bmatrix} x_1(0) & x_2(0) \end{bmatrix} \begin{bmatrix} \frac{\beta + \alpha e^{-(\alpha+\beta)t}}{\alpha+\beta} & \frac{\alpha - \alpha e^{-(\alpha+\beta)t}}{\alpha+\beta} \\ \frac{\beta - \beta e^{-(\alpha+\beta)t}}{\alpha+\beta} & \frac{\alpha + \beta e^{-(\alpha+\beta)t}}{\alpha+\beta} \end{bmatrix}.$$

To relate these results to those obtained in Section 6.5, let the probability that the channel is open be $x_2(t) = x(t)$, and let the probability that it is closed be $x_1(t) = 1 - x(t)$. Evaluation of the matrix and rearrangement of terms yields

$$x(t) = \frac{\alpha}{\alpha + \beta} + \left(x(0) - \frac{\alpha}{\alpha + \beta} \right) e^{-(\alpha+\beta)t},$$

which is an exponential time function with an initial value of $x(0)$, a final value of $\alpha/(\alpha + \beta)$, and a time constant of $1/(\alpha + \beta)$. This result is identical with that given in Equation 6.22.

The autocovariance function for the conductance of the two-state model can be evaluated if we assign conductance zero to state 1 and y to state 2. Then Equation 6.74 can be evaluated as

$$c_g(\tau) = \gamma^2 \left(\frac{\alpha + \beta e^{-(\alpha+\beta)\tau}}{\alpha + \beta} \right) \left(\frac{\alpha}{\alpha + \beta} \right) \quad \text{for } \tau > 0.$$

Since the autocovariance function is an even function of τ, we can write it in the following form:

$$c_g(\tau) = \left(\gamma \frac{\alpha}{\alpha + \beta} \right)^2 + \gamma^2 \frac{\alpha\beta}{(\alpha + \beta)^2} e^{-(\alpha+\beta)|\tau|}. \tag{6.78}$$

The autocovariance function also yields both the expected value and the variance of the conductance fluctuations. $E^2[\tilde{g}] = \lim_{\tau\to\infty} c_g(\tau)$, so $E[\tilde{g}] = \gamma\alpha/(\alpha + \beta)$. This result is simply the conductance of the open channel times the probability that the channel is open. The value of $c_g(0)$ is the second moment of $\tilde{g}$, and we can recognize the first term in Equation 6.78 as the square of the expected value. Hence, $\sigma_g^2 = \gamma^2\alpha\beta/(\alpha + \beta)^2$.

The power density spectrum is obtained from the Fourier transform of the autocovariance function (Equation 6.78) and is

$$C_g(f) = \left(\gamma \frac{\alpha}{\alpha + \beta} \right)^2 \delta(f) + \gamma^2 \frac{2\alpha\beta}{(\alpha + \beta)^2} \left(\frac{1}{1 + (f/f_c)^2} \right), \tag{6.79}$$

the corner frequency $f_c = (\alpha + \beta)/(2\pi)$.

Three-State Model

Next we shall consider a simple three-state model with kinetic diagram

$$S_1 \underset{\alpha_{21}}{\overset{\alpha_{12}}{\rightleftharpoons}} S_2 \underset{\alpha_{32}}{\overset{\alpha_{23}}{\rightleftharpoons}} S_3.$$

Therefore, the transition matrix is

$$\alpha = \begin{bmatrix} -\alpha_{12} & \alpha_{12} & 0 \\ \alpha_{21} & -\alpha_{21} - \alpha_{23} & \alpha_{23} \\ 0 & \alpha_{32} & -\alpha_{32} \end{bmatrix},$$

from which the eigenvalues are

$$\lambda_1 = 0,$$

$$\lambda_2 = -\frac{A}{2} + \frac{\sqrt{B}}{2},$$

$$\lambda_3 = -\frac{A}{2} - \frac{\sqrt{B}}{2},$$

where

$$A = \alpha_{12} + \alpha_{21} + \alpha_{23} + \alpha_{32},$$

$$B = \alpha_{12}^2 + 2\alpha_{12}\alpha_{21} + \alpha_{21}^2 - 2\alpha_{12}\alpha_{23} + 2\alpha_{21}\alpha_{23} +$$
$$\alpha_{23}^2 - 2\alpha_{12}\alpha_{32} - 2\alpha_{21}\alpha_{32} + 2\alpha_{23}\alpha_{32} + \alpha_{32}^2.$$

This result illustrates the complexity of the relation between the eigenvalues and the transition rates for even this simple kinetic scheme.

Exercises

6.1 Describe the distinction between conduction current and displacement current.

6.2 Explain the origin of gating current.

6.3 For a channel with one two-state gate, what is the distinction among the *single-channel ionic current random variable*, the *single open-channel current*, and the *average single-channel current*.

6.4 State whether each of the following is true or false, and give a reason for your answer.

   a. Tetrodotoxin blocks the flow of potassium through the sodium channel.

   b. The macroscopic sodium current recorded by an electrode in a cell is a sum of the single-channel sodium currents that flow through single sodium channels.

   c. The macroscopic sodium current recorded by an electrode in a cell is the average of the single-channel sodium currents that flow through single sodium channels.

   d. Ionic and gating currents give identical information about channel kinetic properties.

6.5 Figure 6.72 shows two putative records of membrane currents recorded from two membrane patches, each of which contains a single channel, in response to a step of depolarizing membrane potential. Each of these channels has a linear voltage-current characteristic when the channel is open.

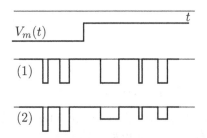

Figure 6.72 Two putative single-channel currents in response to a voltage step (Exercise 6.5).

  a. Which, if any, of these records could be from a single voltage-gated channel? Explain.

  b. Which, if any, of these records could be from a single channel that is not voltage gated? Explain.

6.6 Explain why the gating current is outward in response to a depolarization independent of the sign of the charge on the gate.

6.7 Figure 6.28 shows a schematic diagram of the recording arrangement used to record from a patch of membrane that is still attached to the cell. Find an equivalent electrical network that shows the relation between the source voltage, V, and the potential across the membrane of the patch, V_{mp}, in terms of two membrane patches, one across the patch of membrane in the micropipette and one for the rest of the cell membrane. Determine V_{mp} for the values of V shown in the traces in the right panel of Figure 6.28 under the assumption that the potential across the membrane not in the patch is $V_{mn} = -70$ mV.

6.8 Describe a prediction of the single-channel model with one two-state gate that is inconsistent with measurements of ionic currents through the potassium channels in squid giant axons.

6.9 Describe a prediction of the single-channel model consisting of four independent two-state gates that is inconsistent with measurements of ionic currents through the sodium channels in squid giant axons.

6.10 Describe four important functional differences between water channels (Weiss, 1996) and ion channels.

6.11 List four distinct properties shown by ionic currents measured from single voltage-gated ion channels.

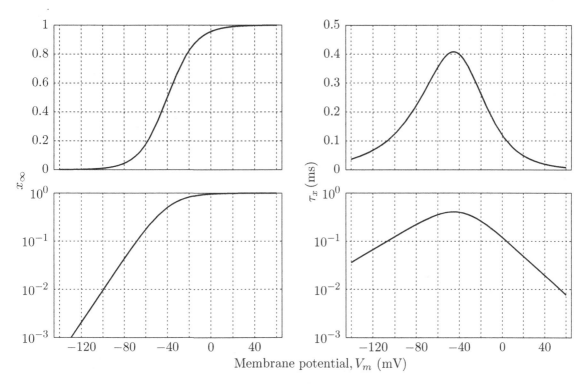

Figure 6.73 Dependence of x_∞ (left panels) and τ_x (right panels) versus V_m on linear (upper panels) and logarithmic (lower panels) ordinate scales (Problem 6.1).

Problems

6.1 The dependences of both x_∞ and τ_x on V_m are shown in Figure 6.73 for a single channel that contains one two-state gate such as the one shown in Figure 6.4.

    a. Is the gate in each channel an activation or an inactivation gate? Explain.

    b. Determine the parameters $V_{1/2}$, z, and ζ for this gate.

6.2 Let Q_∞ be the steady-state value of the gating charge density $Q_g(t)$, where $Q_g(t) = ze\mathcal{N}x(t)$ and $x(t)$ satisfies first-order kinetics, with rate constants determined by absolute reaction rate theory according to the model of a two-state molecular gate presented in Section 6.5.

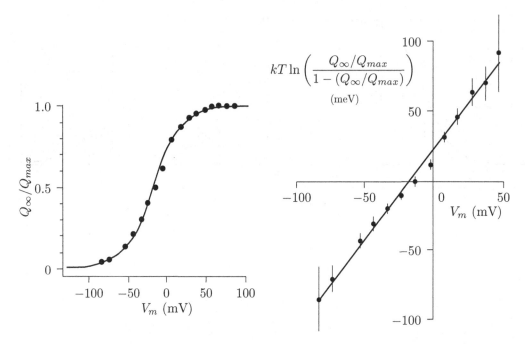

Figure 6.74 Measurements of gating charge versus membrane potential in squid giant axons (adapted from Keynes and Rojas, 1976, Figure 10) (Problem 6.2). The ordinate in the right-hand panel has units of energy and is expressed in milli-electron volts (meV).

a. Show that a plot of $kT \ln (Q_\infty/(Q_{max} - Q_\infty))$ versus V_m is a straight line, and find the equation of this line. $Q_\infty = Q_g(\infty)$, and $Q_{max} = \lim_{V_m \to \infty} Q_\infty$.

b. In Figure 6.74, measurements of Q_∞/Q_{max} versus V_m are shown. On the basis of these measurements, compute an estimate of the gate valence, z.

6.3 A single ion channel has an open-channel conductance of $\gamma = 20$ pS. You may assume that the channel is a cylindrical water-filled pore through the membrane (whose thickness is 75 Å) and that the resistivity of the solution in the pore is the same as physiological saline, i.e., 100 $\Omega \cdot$cm. What is the radius of this channel? In this calculation, you may ignore end effects due to current flow from the lumen of the cylinder to the cytoplasm on one side and to the extracellular space on the other.

6.4 The voltage across a membrane patch is stepped from V_m^o to V_m^f at $t = 0$, and single-channel ionic currents are recorded as a function of time. Typical records at six different values of V_m^f are shown in Figure 6.75.

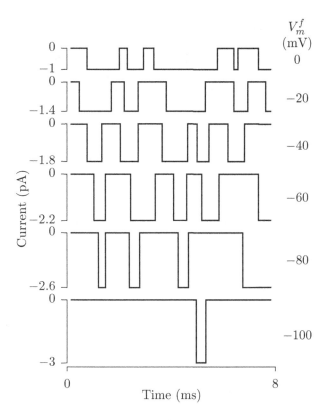

Figure 6.75 Single-channel currents (Problem 6.4).

a. Is the open-channel voltage-current characteristic of this channel linear or nonlinear?

b. What is the conductance of the open channel?

c. What is the equilibrium (reversal) potential of this channel?

d. It is proposed that this channel is the voltage-gated sodium channel responsible for sodium-activated action potentials. Discuss this suggestion.

6.5 A channel contains one two-state gate with states S_1 and S_2. The gate is governed by first-order kinetics as follows:

$$S_1 \underset{\beta}{\overset{\alpha}{\rightleftharpoons}} S_2.$$

The probability that the gate is in state S_2 is x, whose parameters are a function of membrane potential V_m as shown in Figure 6.73, where $x_\infty = \alpha/(\alpha + \beta)$ and $\tau_x = 1/(\alpha + \beta)$.

a. Determine the probability that the gate is in state S_2.

Let the conductance of the channel when the gate is in S_1 be zero and the conductance of the channel when the gate is in state S_2 be 40 pS. The membrane potential is held at -80 mV for a long time and then stepped to -20 mV at $t = 0$.

b. Determine $x(t)$.

c. Determine the average single-channel conductance $g(t)$ as a function of time.

d. Is this an activation or an inactivation gate? Explain.

Now let the conductance of the channel when the gate is in S_1 be 40 pS and the conductance of the channel when the gate is in state S_2 be 0. The membrane potential is held at -80 mV for a long time and then stepped to -20 mV at $t = 0$.

e. Determine $x(t)$.

f. Determine the average single-channel conductance $g(t)$ as a function of time.

g. Is this an activation or an inactivation gate? Explain.

6.6 Transport of an ion through a cell membrane can be represented by a population of voltage-gated channels where each channel contains one two-state gate. The two states are state S_0 and state S_1, and transitions between these states obey first-order kinetics with voltage-dependent rate constants. Therefore,

$$S_0 \underset{\beta(V_m)}{\overset{\alpha(V_m)}{\rightleftharpoons}} S_1.$$

In response to a step of voltage across the channel, the state occupancy of the channel, the single-channel current, and the probability that the channel occupies state S_1 are shown in Figure 6.76.

a. For which state is the channel nonconducting?

b. Determine both the equilibrium (reversal) potential for conduction through this channel and the conductance of the channel when the channel is conducting.

c. For $V_m = 20$ mV, determine the rate constants $\alpha(20)$ and $\beta(20)$ where the voltage is expressed in mV.

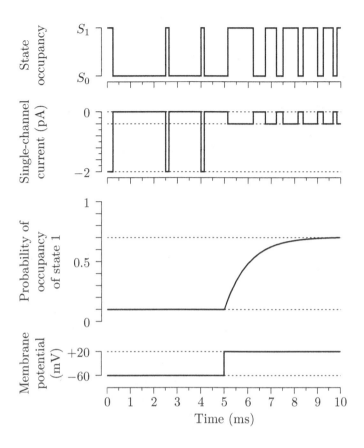

Figure 6.76 Variables associated with a channel with a two-state gate (Problem 6.6).

d. Sketch the probability that the channel occupies state S_0 as a function of time.

e. Briefly describe one experimental method that can provide an estimate of channel density. Be specific about which data you propose to use and how you propose to estimate the density from these data.

f. Measurements indicate that there are 1,000 channels per μm^2 in the membrane of this cell. Sketch the ionic current density, $J_m(t)$, that would be expected with the voltage step shown in the figure. Indicate relevant dimensions on the sketch.

6.7 A membrane contains a population of ionic channels, each one of which is gated by a charged macromolecule that has two states: open with energy E_O and closed with energy E_C. The barrier separating open and

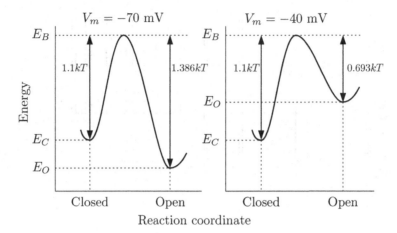

Figure 6.77 Dependence of conformational energy on membrane potential (Problem 6.7).

closed states has energy E_B. Transitions between open and closed states are governed by first-order kinetics, and both the opening and closing rate constants $= 3 \times 10^3/\text{sec}$ when $E_O - E_B = E_C - E_B = 0$. The energy E_O depends on the membrane potential. The membrane potential is held at -70 mV for a long time and is changed abruptly to -40 mV at $t = 0$. The energy diagrams for $V_m = -70$ and $V_m = -40$ mV are shown in Figure 6.77, where k is Boltzmann's constant and T is absolute temperature. Let $x(t)$ be the fraction of channels open at time t.

a. Find $x(0)$, the value of $x(t)$ for $t < 0$.

b. Find $x(\infty)$, the value of $x(t)$ for $t \to \infty$.

c. Find τ_x, the time constant of the transition from $x(0)$ to $x(\infty)$.

d. Sketch $x(t)$.

e. Could this mechanism represent sodium activation (m), sodium inactivation (h), or potassium activation (n)? Explain.

6.8   Nobel P. Seeker has done it again! He has discovered a new voltage-gated channel permeable to the monovalent cation X only. Each channel contains a single gate that has one open and one closed state, with transitions between states governed by first-order kinetics as shown schematically in Figure 6.4. The opening and closing rate constants depend upon the membrane potential, V_m, as follows:

$$\alpha = Ae^{z(F/RT)\left(\frac{1}{2}V_m + V_C\right)} \quad \text{and} \quad \beta = Ae^{z(F/RT)\left(-\frac{1}{2}V_m + V_O\right)}, \tag{6.80}$$

where A, V_C, and V_O are constants, R is the molar gas constant, and T is absolute temperature. The gate has a positive charge of $\mathcal{Q} = ze$, where z is the valence of the charge and $e = 1.6 \times 10^{-19}$ C is the charge of an electron. The charge $\mathcal{Q}$, which is located at the tip of the gate, swings from the inside surface to the outside surface of the membrane, i.e., through the whole potential difference across the membrane, as the gate opens in response to a depolarization.

Nobel has conducted two types of experiments (at a temperature of 300°K): whole-cell voltage-clamp experiments in which macroscopic membrane current densities, J_m (which consist of capacitative and ionic currents), are measured, and patch-clamp experiments, in which single-channel ionic currents are measured on patches of membrane obtained from the same cell.

Whole-Cell Experiments

Using whole-cell recording (Figure 6.78), Nobel immerses the cell in a bath having a concentration of X of 150 mmol/L. He steps the mem-

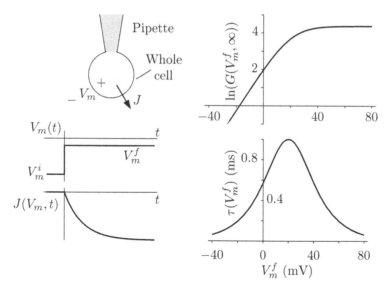

Figure 6.78 Whole-cell voltage-clamp currents (Problem 6.8). The method of obtaining the measurements is shown on the left, and the results are shown on the right.

brane potential from the resting potential to a final potential, V_m^f, and measures the membrane ionic current density for ion X, $J(V_m, t)$. The ionic current density is an exponential function of time,

$$J(V_m, t) = G(V_m^f, t)(V_m^f - V_X) = G(V_m^f, \infty)(1 - e^{-t/\tau(V_m^f)})(V_m^f - V_X), \quad (6.81)$$

where $G(V_m^f, t)$ is the conductance of the membrane for ion X per unit area and $G(V_m^f, \infty)$ is its steady-state value. V_X is the Nernst equilibrium potential for ion X, and $\tau(V_m^f)$ is the time constant of the conductance. Nobel performs a series of voltage-clamp experiments to measure the dependence on membrane potential of both the steady-state conductance and the time constant, and his results are shown in Figure 6.78. Nobel finds that

$$G(V_m^f, \infty) = \frac{80}{1 + e^{-3(V_m^f - 20)/26}}, \quad (6.82)$$

and

$$\tau(V_m^f) = \frac{1}{\cosh\left(3(V_m^f - 20)/52\right)}, \quad (6.83)$$

where the $G(V_m^f, \infty)$ has units of mS/cm^2, $\tau(V_m^f)$ has units of ms, and V_m^f has units of mV.

Single-Channel Experiments
In a tour de force, Nobel measures single-channel currents, i, under voltage-clamp conditions through outside-out membrane patches (Figure 6.79) obtained from the same cell that was used for the whole-cell recording. The solution in the pipette has the same ionic composition as the cell's cytoplasm, and the bath solution is the same as that used in the whole-cell measurements. Nobel measures the amplitude of the single-channel current, $\mathcal{I}$, as a function of the V_m^f, with the results shown in Figure 6.79.

a. Assume that these channels are distributed uniformly over the surface of the cell, and determine the number of channels per μm^2 of membrane surface area.

b. Determine the intracellular concentration of ion X.

c. Determine the valence, z, of the charge on the gate, $\mathcal{Q}$.

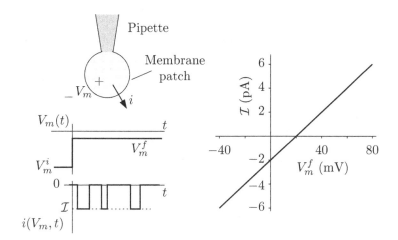

Figure 6.79 Single-channel measurements (Problem 6.8).

    d. In response to a step change in membrane potential from an initial potential of −80 mV to a final potential of +20 mV, determine the whole-cell gating current density, $J_g(t)$, and sketch your result.

6.9 A cell contains ionic channels that are permeable to sodium only and have linear voltage-current characteristics. When these channels are open, they have a single-channel conductance of 30 pS. The internal and external concentrations of sodium are 20 and 200 mmol/L, respectively, and the membrane potential is maintained at −20 mV.

    a. Determine the number of sodium ions per second that flow through an open channel.

    b. In which direction through the membrane do the ions flow?

6.10 This problem involves the five hypothetical voltage-gated ionic channels shown in Figure 6.80. These channels are found in membranes that separate ionic solutions whose compositions are shown in Table 6.6. Each of these channels is perfectly ion selective, so that only one ionic species permeates an open channel. Furthermore, conduction through an open channel is linear and given by the relation

$$\mathcal{I}_i = \gamma_i(V_m - V_i), \tag{6.84}$$

where $\mathcal{I}_i$ is the single open-channel current (the current through the channel when it is open), γ_i is the single-channel conductance, V_m is the membrane potential, and V_i is the Nernst equilibrium potential for ion i. In each part of this problem, a through d, measurements of cur-

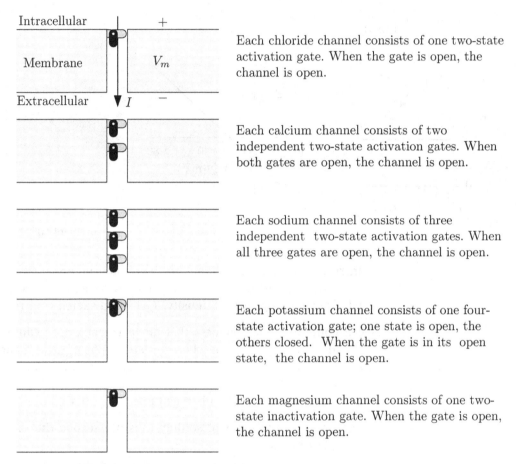

Each chloride channel consists of one two-state activation gate. When the gate is open, the channel is open.

Each calcium channel consists of two independent two-state activation gates. When both gates are open, the channel is open.

Each sodium channel consists of three independent two-state activation gates. When all three gates are open, the channel is open.

Each potassium channel consists of one four-state activation gate; one state is open, the others closed. When the gate is in its open state, the channel is open.

Each magnesium channel consists of one two-state inactivation gate. When the gate is open, the channel is open.

Figure 6.80 Schematic diagrams of five hypothetical membrane ionic channels (Problem 6.10). Each channel contains gates that change conformation in response to a change in membrane potential. All the gates are activation gates, i.e., they open when the membrane potential increases, except the gate in the magnesium channel, which is an inactivation gate, i.e., it closes when the membrane potential increases. Each channel has a single open state, indicated by the darkly filled gate(s), and one or more closed states. Reference directions for the membrane potential and for all current variables are shown for the chloride channel, but apply to all the channels.

Table 6.6 Concentrations of ions for inside and outside solutions (Problem 6.10). The rightmost column gives the magnitude of the logarithmic ratio shown.

	Concentration (mmol/L)		$\left| 59 \log_{10} \frac{c_i^o}{c_i^i} \right|$
Ion	Inside	Outside	(mV)
K^+	168	5	90
Na^+	13	137	60
Cl^-	37	120	30
Ca^{++}	0.002	5	200
Mg^{++}	5	5	0

rent variables under voltage clamp are given. Your task is to determine which of the results shown are consistent with each of the five channels. Assume that the four parts of this problem are independent of each other.

a. Macroscopic membrane *ionic* current densities (J_{i1}, J_{i2}, J_{i3}, J_{i4}, and J_{i5}) in response to a step voltage clamp from -60 to 0 mV are shown in Figure 6.81 (left panel). Assume that each of these ionic currents results from a population of identical channels. Determine *all* the channels, from among those shown in Figure 6.80, that could have been responsible for *each* of these currents. If none of the channels could have given rise to that current, answer, "None of the above." Indicate your answer with a check mark in the appropriate space in Table 6.7.

b. Macroscopic membrane gating current densities (J_{g1}, J_{g2}, J_{g3}, J_{g4}, and J_{g5}) in response to a step voltage clamp from -60 to 0 mV are shown in Figure 6.81 (right panel). Assume that each of these gating currents results from a population of identical channels. Determine *all* the channels, from among those shown in Figure 6.80, that could have been responsible for *each* of these currents. If none of the channels could have given rise to that current, answer, "None of the above." Indicate your answer with a check mark in the appropriate space in Table 6.8.

c. Single-channel currents ($\tilde{\imath}_{i1}$, $\tilde{\imath}_{i2}$, $\tilde{\imath}_{i3}$, $\tilde{\imath}_{i4}$, and $\tilde{\imath}_{i5}$) in response to a step voltage clamp from -60 to 0 mV are shown in Figure 6.82. Determine *all* the channels, from among those shown in Figure 6.80, that could have been responsible for *each* current. If none of the channels could have given rise to that current, answer "None of the above."

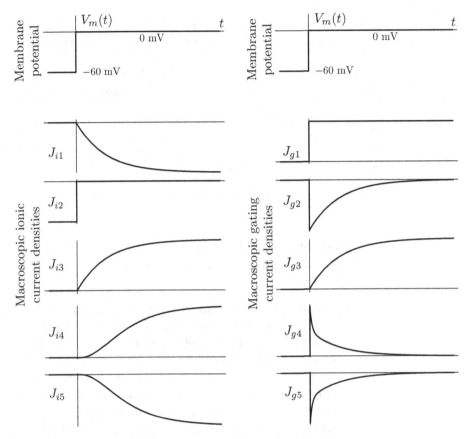

Figure 6.81 Macroscopic membrane currents (Problem 6.10). Macroscopic membrane ionic current densities (left panel) and macroscopic membrane gating current densities (right panel) in response to a step voltage clamp. The reference directions for membrane potential and currents are defined in Figure 6.80. The axes are drawn at zero potential and zero current. The curves shown have been computed and not drawn by hand, although the current scales have been normalized so that the maximum excursion is the same in each trace.

Table 6.7 Macroscopic membrane ionic current density (Problem 6.10).

Channel	J_{i1}	J_{i2}	J_{i3}	J_{i4}	J_{i5}
Chloride	___	___	___	___	___
Calcium	___	___	___	___	___
Sodium	___	___	___	___	___
Potassium	___	___	___	___	___
Magnesium	___	___	___	___	___
None of the above	___	___	___	___	___

Table 6.8 Macroscopic membrane gating current density (Problem 6.10).

Channel	J_{g1}	J_{g2}	J_{g3}	J_{g4}	J_{g5}
Chloride	___	___	___	___	___
Calcium	___	___	___	___	___
Sodium	___	___	___	___	___
Potassium	___	___	___	___	___
Magnesium	___	___	___	___	___
None of the above	___	___	___	___	___

Indicate your answer with a check mark in the appropriate space in Table 6.9.

d. The dependence of single open-channel currents ($\mathcal{I}_{i1}$, $\mathcal{I}_{i2}$, $\mathcal{I}_{i3}$, $\mathcal{I}_{i4}$, and $\mathcal{I}_{i5}$) on membrane potential are shown in Figure 6.83. Determine *all* the channels, from among those shown in Figure 6.80, that could have been responsible for each of these current-voltage relations. If none of the channels could have given rise to that relation, answer, "None of the above." Indicate your answer with a check mark in the appropriate space in Table 6.10.

6.11 This problem deals with the relation of current to voltage for single ion channels. Assume that conduction through an open ion channel is governed by the equation

$$\mathcal{I} = \gamma(V_m - V_e),$$

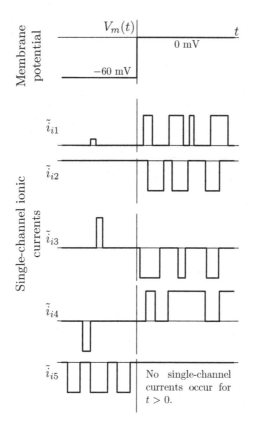

Figure 6.82 Single-channel currents (Problem 6.10). Single-channel currents in response to a step voltage clamp. Reference directions for membrane potential and currents are defined in Figure 6.80. The time axes are drawn at zero potential and zero current. These currents have been normalized so that the maximum current excursion is the same in each trace.

Table 6.9 Single-channel ionic currents (Problem 6.10).

Channel	$\tilde{\imath}_{i1}$	$\tilde{\imath}_{i2}$	$\tilde{\imath}_{i3}$	$\tilde{\imath}_{i4}$	$\tilde{\imath}_{i5}$
Chloride	___	___	___	___	___
Calcium	___	___	___	___	___
Sodium	___	___	___	___	___
Potassium	___	___	___	___	___
Magnesium	___	___	___	___	___
None of the above	___	___	___	___	___

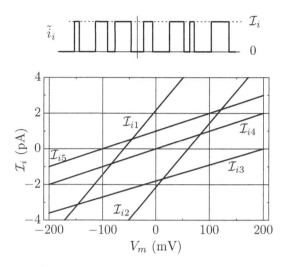

Figure 6.83 Single open-channel currents versus membrane potential (Problem 6.10).

Table 6.10 Single open-channel membrane ionic currents (Problem 6.10).

Channel	$\mathcal{I}_{i1}$	$\mathcal{I}_{i2}$	$\mathcal{I}_{i3}$	$\mathcal{I}_{i4}$	$\mathcal{I}_{i5}$
Chloride	———	———	———	———	———
Calcium	———	———	———	———	———
Sodium	———	———	———	———	———
Potassium	———	———	———	———	———
Magnesium	———	———	———	———	———
None of the above	———	———	———	———	———

where $\mathcal{I}$ is the current through a single open channel, y is the conductance of a single open channel, V_m is the membrane potential across the channel, and V_e is the reversal potential for the channel. For each of the channels in this problem, assume that $y = 25$ pS and $V_e = 20$ mV.

a. The membrane potential, V_m, and the *average* single-channel current, i, obtained from three different single channels (A, B, and C) are shown in Figure 6.84. Both the membrane potential and the current are plotted on a time scale such that the changes appear instantaneous and only the final values of these variables can be discerned in

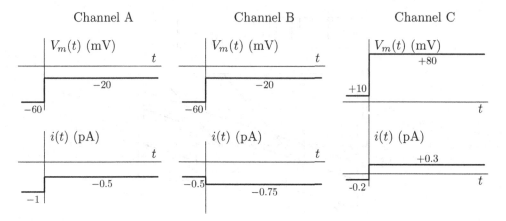

Figure 6.84 Average single-channel currents (Problem 6.11).

Table 6.11 Summary of channel density estimates (Problem 6.12).

Method	Density (channels/μm^2)
a. Ionic currents	
b. Gating currents	
c. Toxin binding	

the plots; i.e., the kinetics are not shown. For each of these channels, answer the following questions, and explain your answers:

i.   Is this channel voltage gated for the illustrated depolarization?

ii.   Is the channel activated (opened) or inactivated (closed) by the illustrated depolarization?

b. Assume that each voltage-gated channel contains one two-state gate, where τ is the time constant of transition between states. For each of the channels, sketch the time course of $i(t)$ on a normalized time scale t/τ. Clearly show the current near $t = 0$.

6.12 In parts a–c, which are completely independent, you will use three different sets of measurements to estimate the density of voltage-gated sodium channels in one type of axon; in part d, you will compare the results obtained with these three methods. Enter your estimates of the density of channels $\mathcal{N}$, in units of channels/μm^2, into a table such as Table 6.11.

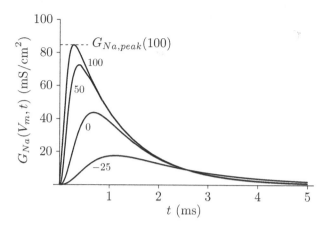

Figure 6.85 Macroscopic sodium conductance, $G_{Na}(V_m, t)$, shown as a function of time t for values of the membrane potential V_m of -25, 0, 50, and 100 mV (Problem 6.12). The definition of the peak sodium conductance, $G_{Na,peak}$, is illustrated for $V_m = 100$ mV.

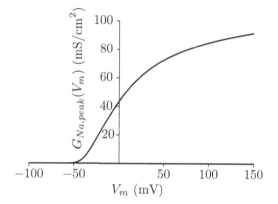

Figure 6.86 Dependence of the peak value of the macroscopic sodium conductance, $G_{Na,peak}$, on membrane potential, V_m (Problem 6.12).

a. The first method is based on a comparison of macroscopic and single-channel sodium currents measured under voltage clamp. Measurements of macroscopic sodium currents are used to measure the sodium conductance as a function of membrane potential and time during a step voltage clamp experiment (Figure 6.85). The peak value of the conductance, $G_{Na,peak}$, is determined for each value of the membrane potential, and the dependence of $G_{Na,peak}$ on the clamped membrane potential, V_m, is shown in Figure 6.86. Measurements of the dependence of single open-channel current, $\mathcal{I}$, on membrane potential, V_m, are summarized schematically in Figure 6.87. Based on the macroscopic and single-channel results, estimate the density of sodium channels in the axon membrane, and explain your method of estimation.

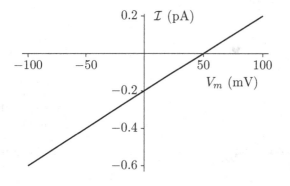

Figure 6.87 Dependence of single open-channel current, $\mathcal{I}$, on membrane potential, V_m (Problem 6.12).

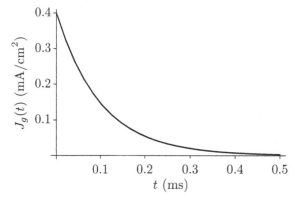

Figure 6.88 Gating current density as a function of time in response to a large step of depolarization that begins at $t = 0$ (Problem 6.12). The gating current is exponential with time constant 0.1 ms.

b. A second method is based on measurements of gating currents. Measurements of the sodium gating current density for a large step of depolarization of the membrane are shown in Figure 6.88. The depolarization is so large that you may assume that all the sodium gates are opened by this depolarization. The time course of the gating current is exponential, with the parameters shown in the figure. Assume that when the gate opens, the gating charge moves from one side of the membrane to the other and that the valence of the sodium gate is six. Using these results, estimate the density of sodium channels, and explain your method.

c. A third method of estimating sodium channel density is based on the binding of radioactive saxitoxin (STX) to sodium channels. The axon is bathed in a solution containing radioactive STX of concentration c, and the amount of STX removed from the bathing solution is measured. The density of bound STX, n, is determined from this measure-

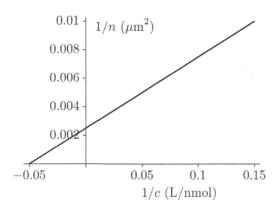

Figure 6.89 The reciprocal of the density of STX bound to an axon, $1/n$, as a function of the reciprocal of the concentration of STX in the bathing medium $1/c$ (Problem 6.12). Here n has units of (number of STX molecules)/$(\mu m)^2$.

ment and from an estimate of the surface area of the axon. When the density of bound STX molecules is plotted versus the concentration of STX in the bathing solution, it is found that the function is a straight line in double reciprocal coordinates as shown in Figure 6.89 (Weiss, 1996). Propose a model of STX binding that predicts the relation between the density of bound STX and the concentration of STX shown in Figure 6.89, and determine an estimate of the sodium channel density based on these results. Explain your method of estimation.

d. Briefly compare the results you obtained in parts a–c (which you should have summarized in Table 6.11). Briefly discuss the strengths and weaknesses of each of the three methods of estimation of the density of sodium channels. Are there reasons to expect the difference obtained between particular pairs of estimates? For example, does any one of the methods lead to an underestimate of the channel density? In your answer, please ignore the possibilities of instrumental errors, and focus on the inherent differences of the methods.

6.13 This problem deals with a comparison of measurements of ionic and gating currents for the sodium channel with predictions based on kinetic schemes consistent with the Hodgkin-Huxley model. Consider the model for activation of sodium shown in Figure 6.90, in which three identical but independent gating particles control a channel. Each gating particle has a probability m of being in the open conformation, and transitions between the open and closed states are governed by first-order kinetics:

$$C_g \underset{\beta}{\overset{\alpha}{\rightleftharpoons}} O_g,$$

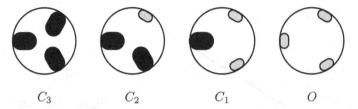

$$C_3 \qquad C_2 \qquad C_1 \qquad O$$

Figure 6.90 Activation states of a channel with 3 independent activation gates (Problem 6.13). The channel states are indicated by the states of the gates: C_3, C_2, C_1 are states for which the channel is closed (nonconducting) and O is the conducting state with all 3 gates in the open conformation.

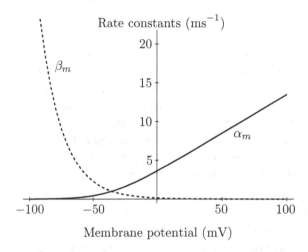

Figure 6.91 Rate constants of m as a function of membrane potential V_m (Problem 6.13).

where the transition from the closed (C_g) to the open (O_g) state of the gate has forward rate constant $\alpha(V_m)$ and reverse rate constant $\beta(V_m)$, both of which are dependent on membrane potential V_m as shown in Figure 6.91. Each gate has a positive charge of valence $+2$, and this charge moves from the inner to the outer surface of the membrane as the gate moves from its closed conformation to its open conformation.

The channel is open when all three gating particles are in the open conformation which has probability $x = m^3$. When the channel is open, the current through a single open channel is $\mathcal{I} = \gamma(V_m - V_{Na})$, where $\gamma = 20$ pS is the single open-channel conductance and $V_{Na} = 50$ mV is the Nernst equilibrium potential for sodium. The density of these channels is 200 channels/μm^2.

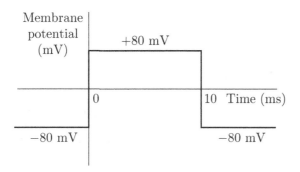

Figure 6.92 Membrane potential profile (Problem 6.13). The membrane potential is held at −80 mV for a long time and stepped to +80 mV at $t = 0$, then returned to −80 mV after 10 ms.

a. Determine and sketch the (ensemble) average macroscopic ionic current density as a function of time in response to the pulse of membrane potential shown in Figure 6.92.

b. Determine and sketch the (ensemble) average macroscopic gating current density as a function of time in response to the pulse of membrane potential shown in Figure 6.92.

c. In fifty words or less, describe similarities and differences in the kinetics of ionic and gating current densities predicted by the model. Consider the kinetics of both the onset and the offset of the potential.

6.14 Consider a two-state gate with the kinetic diagram

$$C_g \underset{\beta_n}{\overset{\alpha_n}{\rightleftharpoons}} O_g.$$

Determine a kinetic diagram for the Hodgkin-Huxley model of the potassium channel, which consists of four independent two-state activation gates. Determine all the states and the transition rates between states.

6.15 This problem compares predictions of the Hodgkin-Huxley model and measurements of sodium activation in squid giant axons. Parts a through d deal with predictions of the Hodgkin-Huxley model, and part e deals with a comparison to measurements. The Hodgkin-Huxley model for the voltage dependence of the equilibrium value and the time constant of activation factor m are shown in Figure 4.25. In the Hodgkin-Huxley model, the maximum sodium conductance $\bar{G}_{Na} = 120 \text{ mS/cm}^2$, and the Nernst equilibrium potential for sodium is $V_{Na} = +55$ mV. In this problem, you may assume that inactivation has been removed so that $h = 1$.

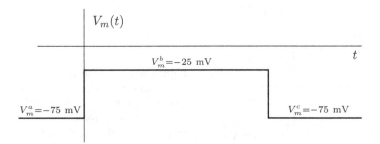

Figure 6.93 Pulse of membrane potential (Problem 6.15). The duration of the pulse is 5 ms.

a. Determine and sketch $m(t)$ as predicted by the Hodgkin-Huxley model for the membrane potential shown in Figure 6.93.

b. Determine and sketch the sodium conductance predicted by the Hodgkin-Huxley model in response to the membrane potential shown in Figure 6.93.

c. Determine and sketch the sodium current density predicted by the Hodgkin-Huxley model in response to the membrane potential shown in Figure 6.93.

d. Determine the time constant of the current density at the offset of the voltage pulse, τ_J, as predicted by the Hodgkin-Huxley model.

e. Figure 6.59 shows a comparison between the dependence of τ_m and that of τ_J on the membrane potential at the offset of a membrane potential. For the sake of this part, assume that variability in the measurements of as much as 50% can be attributed to variations in the time constants of different axons and to measurement errors. Briefly discuss the implications of these measurements.

6.16 Three three-state voltage-gated channels (channels a, b, and c) have the kinetic diagram and state occupancy probabilities shown in Figure 6.94. These channels have the same voltage-dependent rate constants and the same equilibrium potential, which is +40 mV. For the membrane potential shown, the channels are in state 1 with probability one for $t < 0$ and have the indicated rate constants for $t > 0$. The channels differ only in their state conductances and state gating charges, as shown in Figure 6.95. Denote the expected values of the single-channel random variables as follows: the conductances as $g_a(t)$, $g_b(t)$, and $g_c(t)$; the ionic currents as $i_a(t)$, $i_b(t)$, and $i_c(t)$; the gating charges as $q_a(t)$, $q_b(t)$, and $q_c(t)$; and the gating currents as $i_{ga}(t)$, $i_{gb}(t)$, and $i_{gc}(t)$.

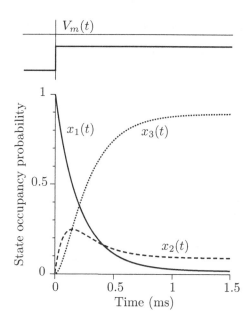

Figure 6.94 State diagram and occupancy probabilities for a three-state channel (Problem 6.16). The state occupancy probabilities for states S_1, S_2, and S_3 are $x_1(t)$, $x_2(t)$, and $x_3(t)$, respectively.

Channel a

$$\left(\begin{array}{c} S_1 \\ \gamma_1 = 10 \\ \mathcal{Q}_1 = -1 \end{array}\right) \underset{1}{\overset{5}{\rightleftharpoons}} \left(\begin{array}{c} S_2 \\ \gamma_2 = 0 \\ \mathcal{Q}_2 = 0 \end{array}\right) \underset{1}{\overset{10}{\rightleftharpoons}} \left(\begin{array}{c} S_3 \\ \gamma_3 = 0 \\ \mathcal{Q}_3 = 0 \end{array}\right)$$

Channel b

$$\left(\begin{array}{c} S_1 \\ \gamma_1 = 0 \\ \mathcal{Q}_1 = 0 \end{array}\right) \underset{1}{\overset{5}{\rightleftharpoons}} \left(\begin{array}{c} S_2 \\ \gamma_2 = 10 \\ \mathcal{Q}_2 = 0 \end{array}\right) \underset{1}{\overset{10}{\rightleftharpoons}} \left(\begin{array}{c} S_3 \\ \gamma_3 = 0 \\ \mathcal{Q}_3 = 1 \end{array}\right)$$

Figure 6.95 State diagrams of three three-state channel models (Problem 6.16). The models differ in state conductances and state gating charges, but not in rate constants.

Channel c

$$\left(\begin{array}{c} S_1 \\ \gamma_1 = 0 \\ \mathcal{Q}_1 = 1 \end{array}\right) \underset{1}{\overset{5}{\rightleftharpoons}} \left(\begin{array}{c} S_2 \\ \gamma_2 = 0 \\ \mathcal{Q}_2 = 0 \end{array}\right) \underset{1}{\overset{10}{\rightleftharpoons}} \left(\begin{array}{c} S_3 \\ \gamma_3 = 10 \\ \mathcal{Q}_3 = 0 \end{array}\right)$$

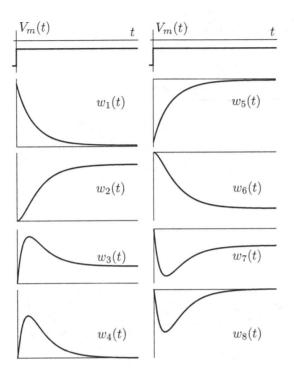

Figure 6.96 Waveforms of responses (Problem 6.16). The horizontal axis corresponds to $w(t) = 0$, and the vertical axis to $t = 0$.

a. Which of the waveforms shown in Figure 6.96 best represents $g_b(t)$? Explain.

b. Which of the waveforms shown in Figure 6.96 best represents $g_c(t)$? Explain.

c. Which of the waveforms shown in Figure 6.96 best represents $i_{ga}(t)$? Explain.

d. Which of the waveforms shown in Figure 6.96 best represents $i_{gc}(t)$? Explain.

e. Which of these channel models exhibits activation followed by inactivation of the ionic current? Explain.

f. Which of these channel models exhibits an ionic current that does not inactivate? Explain.

g. Which of these channel models represents a channel that closes on depolarization? Explain.

6.17 Figure 6.62 shows the state occupancy probabilities of three multiple-state voltage-gated channel models—a two-state model, a three-state model, and a four-state model—for a particular set of rate constants. In each of the three cases, the final values of the state occupancies approach the same value, but this equilibrium value is different for each of the models. Determine the equilibrium state occupancy probabilities for each of these channels.

6.18 Macroscopic sodium currents obtained under voltage-clamp conditions from the nodes of Ranvier of amphibians have been fit by the following kinetic equation:

$$I_{Na}(V_m, t) = \overline{G}_{Na} m^2(V_m, t) h(V_m, t)(V_m - V_{Na}),$$

where $I_{Na}(V_m, t)$ is the total sodium current through the node, $\overline{G}_{Na}$ is the maximum value of the sodium conductance, and m and h are the sodium activation and inactivation factors, which are solutions of first-order kinetic equations. Determine a multiple-state channel model that gives the right kinetics for the sodium current of the node. The channel model should consist of several independent two-state gates. Describe your model with a kinetic diagram that identifies the states and specifies the transition rates between states as quantitatively as possible.

6.19 The effects of certain pharmacological agents on the response of the squid giant axon to a voltage pulse can be represented by responses of the Hodgkin-Huxley model with suitable changes in key parameters. This problem refers to Problem 4.27. For each of the following questions, choose the appropriate waveform from Figure 4.108, and briefly justify your choice:

   a. $J_{ion}(V_m, t)$ after application of TTX to the outside of the axon.

   b. $J_{ion}(V_m, t)$ after application of TEA to the inside of the axon.

   c. $J_{Na}(V_m, t)$ after application of pronase to the inside of the axon.

   d. $J_{ion}(V_m, t)$ after application of ouabain to the outside of the axon.

6.20 The voltage across a membrane patch is stepped from V_m^i to V_m^f at time $t = 0$. Typical single-channel ionic currents and single-channel gating currents are shown in Figure 6.97 as a function of time for three different values of V_m^f. In each case, $V_m^i = 0$.

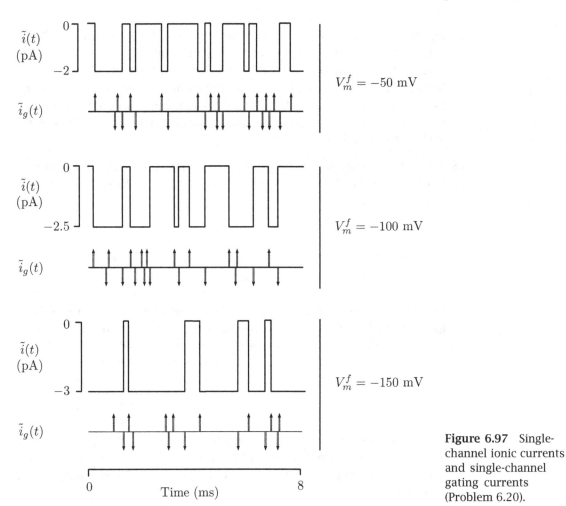

Figure 6.97 Single-channel ionic currents and single-channel gating currents (Problem 6.20).

a. Is the open-channel voltage-current characteristic of this channel linear or nonlinear? Explain.

b. Do the ionic currents show any evidence that the channel is voltage gated? If so, what is the evidence?

c. Compute the conductance of the open channel.

d. Compute the equilibrium (reversal) potential for this channel.

e. Estimate the probability that the channel is open when V_m^f is equal to -50, -100, and -150 mV.

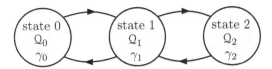

Figure 6.98 Three-state model (Problem 6.20).

    f. Are the measured gating currents consistent with a two-state channel model? Explain.

    g. It has been proposed that the three-state model shown in Figure 6.98 could account for both the ionic and the gating currents shown in Figure 6.97. Assume that $Q_2 = Q_0$ and that $Q_1 > Q_0$. Are there values of γ_0, γ_1, and γ_2 that are consistent with the measurements? If so, what are they? If not, why not?

6.21 A single channel contains one two-state activation gate whose kinetic diagram is

$$C \underset{\beta(V_m)}{\overset{\alpha(V_m)}{\rightleftharpoons}} O,$$

where C is the closed state and O is the open state and where $\alpha(V_m)$ and $\beta(V_m)$ are the voltage-dependent forward and reverse rate constants, respectively. The channel is permeable to sodium ions only. The current through the open channel is

$$\mathcal{I} = \gamma(V_m - V_{Na}),$$

where V_{Na} is the Nernst equilibrium potential for sodium.

    Figure 6.99 shows single-channel ionic current variables in response to a voltage step. Row 1 shows the membrane potential. In the remaining rows, the right panels show single-channel ionic current random variables and the left panels show average single-channel currents. Row 2 illustrates results for a default set of parameters. Rows A–E show results when one or two parameter values are changed. Determine which of rows A–E corresponds to each of the following changes, and give a brief reason for your choice.

    a. The single-open-channel conductance γ was increased.

    b. The final value of the membrane potential, V_m^f, was increased.

    c. The extracellular concentration of sodium was decreased.

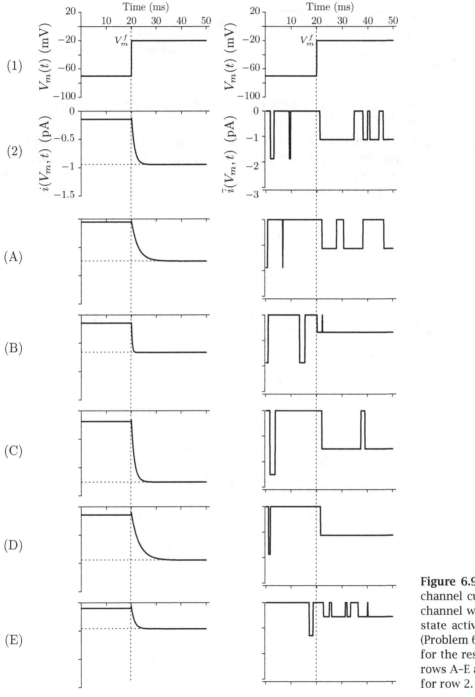

Figure 6.99 Single-channel currents of a channel with one two-state activation gate (Problem 6.21). The scales for the results shown in rows A–E are the same as for row 2.

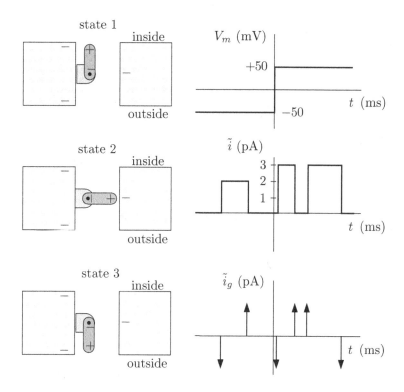

Figure 6.100 Channel with one three-state gate (Problem 6.22). The left panels illustrate the three states: states 1 and 3 are open states, and state 2 is a closed state. The right panels illustrate the responses of the channel to a step in membrane potential $V_m(t)$ at time $t = 0$ (top right), which gives rise to the ionic current $\tilde{\imath}(t)$ and the gating current $\tilde{\imath}_g(t)$ illustrated in the middle right and lower right panels, respectively.

    d. Both $\alpha(V_m)$ and $\beta(V_m)$ were decreased without changing the ratio $\alpha(V_m)/\beta(V_m)$.

    e. Only $\alpha(V_m)$ was decreased.

6.22 Figure 6.100 shows a model of a voltage-gated ion channel with one three-state gate plus representative single-channel ionic and gating current records.

    a. Assume that the voltage-current characteristic of the channel is the same for states 1 and 3 and is linear. Determine the open-channel conductance and the equilibrium (reversal) potential for this channel.

    b. The ionic current trace shown in Figure 6.100 has three nonzero segments. Determine which state the gate is in during each nonzero segment. Explain your reasoning.

    c. Figure 6.101 illustrates the dependence of the steady-state probability that the channel will be in each of its three states on the membrane potential. Let i_{ss} represent the average value of the ionic current that results after steady-state conditions are reached in a voltage-clamp

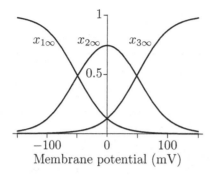

Figure 6.101 Steady-state probabilities for a channel with one three-state gate (Problem 6.22). Here, $x_{1\infty}$, $x_{2\infty}$, and $x_{3\infty}$ represent the steady-state probabilities of being in state 1, state 2, and state 3, respectively, as a function of membrane potential.

experiment in which V_m is held constant. Assume that the experiment is repeated for a number of different values of membrane potential V_m. Plot the relation between i_{ss} and V_m. Describe the important features of your plot.

References

Books and Reviews

Adams, D. J., Smith, S. J., and Thompson, S. H. (1980). Ionic currents in molluscan soma. *Ann. Rev. Neurosci.*, 3:141–167.

Adams, M. E. and Swanson, G. (1994). *TINS Neurotoxins Supplement 1994.* Trends Neurosciences, 1st edition.

Agnew, W. S., Claudio, T., and Sigworth, F. J., eds. (1988). *Molecular Biology of Ionic Channels*, vol. 33 of *Current Topics in Membranes and Transport*. Academic Press, New York.

Armstrong, C. M. (1975). Ionic pores, gates, and gating currents. *Quart. Rev. Biophys.*, 7:179–210.

Armstrong, C. M. (1981). Sodium channels and gating currents. *Physiol. Rev.*, 61:644–683.

Barchi, R. L. (1982). Biochemical studies of the excitable membrane sodium channel. *Int. Rev. Neurobiol.*, 23:69–101.

Baxter, D. A. and Byrne, J. H. (1991). Ionic conductance mechanisms contributing to the electrophysiological properties of neurons. *Curr. Opin. Neurobiol.*, 1:105–112.

Bezanilla, F. (1982). Gating charge movements and kinetics of excitable membrane proteins. In Haber, B., Perez-Polo, J. R., and Coulter, J. D., eds., *Proteins in the Nervous System: Structure and Function*, 3–16. Liss, New York.

Catterall, W. A. (1980). Neurotoxins that act on voltage-sensitive sodium channels in excitable membranes. *Ann. Rev. Pharmacol. Toxicol.*, 20:15–43.

Catterall, W. A. (1984). The molecular basis of neuronal excitability. *Science*, 223:653–661.

Catterall, W. A. (1988). Structure and function of voltage-sensitive ion channels. *Science*, 242:50–61.

Catterall, W. A. (1993). Structure and function of voltage-gated ion channels. *Trends Neurosci.*, 16:500–506.

Clapham, D. E. (1994). Direct G protein activation of ion channels? *Ann. Rev. Neurosci.*, 17:441–464.

Colquhoun, D. and Hawkes, A. G. (1983). The principles of the stochastic interpretation of ion-channel mechanisms. In Sakmann, B. and Neher, E., eds., *Single-Channel Recording*, 135–175. Plenum, New York.

Cox, D. R. and Miller, H. D. (1965). *The Theory of Stochastic Processes*. John Wiley & Sons, New York.

Debye, P. (1929). *Polar Molecules*. Dover, New York.

Dryer, S. E. (1994). Na^+-activated K^+ channels: A new family of large-conductance ion channels. *Trends Neurosci.*, 17:155–160.

Edwards, C. (1982). The selectivity of ion channels in nerve and muscle. *Neurosciences*, 7:1335–1366.

French, R. J. and Horn, R. (1983). Sodium channel gating: Models, mimics, and modifiers. *Ann. Rev. Biophys. Bioeng.*, 12:319–356.

Guggino, W. B., ed. (1994). *Current Topics in Membranes and Transport,* vol. 42, *Chloride Channels*. Academic Press, New York.

Guy, H. R. and Conti, F. (1990). Pursuing the structure and function of voltage-gated channels. *Trends Neurosci.*, 13:201–206.

Hagiwara, S. and Byerly, L. (1981). Calcium channel. *Ann. Rev. Neurosci.*, 4:69–125.

Hamill, O. P. (1983). Membrane ion channels. In Burgen, A. S. V. and Roberts, G. C. K., eds., *Topics in Molecular Pharmacology*. Elsevier, New York.

Haus, H. A. and Melcher, J. R. (1989). *Electromagnetic Fields and Energy*. Prentice Hall, Englewood Cliffs, NJ.

Hille, B. (1992). *Ionic Channels of Excitable Membranes*. Sinauer, Sunderland, MA.

Hille, B. and Fambrough, D. M., eds. (1987). *Proteins of Excitable Membranes*. Wiley, New York.

Hofmann, F., Biel, M., and Flockerzi, V. (1994). Molecular basis for Ca^{2+} channel diversity. *Ann. Rev. Neurosci.*, pages 399–418.

Horn, R. (1984). Gating of channels in nerve and muscle: A stochastic approach. In Bronner, F., ed., *Current Topics in Membrane Transport*, vol. 21, *Ionic Channels: Molecular and Physiological Aspects,* 53–97. Academic Press, New York.

Jan, L. Y. and Jan, Y. N. (1989). Voltage-sensitive ion channels. *Cell*, 56:13–25.

Jan, L. Y. and Jan, Y. N. (1992). Structural elements involved in specific K^+ channel functions. *Ann. Rev. Physiol.*, 54:537–555.

Jentsch, T. J. (1990). Molecular biology of voltage-gated chloride channels. In Guggino, W. E., ed., *Current Topics in Membranes and Transport*, vol. 42, *Chloride Channels*, 35–57. Academic Press, New York.

Johnston, D. and Wu, S. M. S. (1995). *Foundations of Cellular Neurophysiology*. MIT Press, Cambridge, MA.

Keynes, R. D. (1994b). The kinetics of voltage-gated ion channels. *Quart. Rev. Biophys.*, 27:339–434.

Koester, J. and Byrne, J. H., eds. (1980). *Molluscan Nerve Cells: From Biophysics to Behavior*. Cold Spring Harbor Laboratory, Cold Spring Harbor, NY.

Latorre, R., Oberhauser, A., Labarca, P., and Alvarez, O. (1989). Varieties of calcium-activated potassium channels. *Ann. Rev. Physiol.*, 51:385–399.

Levinson, S. R. (1981). The structure and function of the voltage-dependent sodium channel. In Singer and Ondarza, eds., *Molecular Basis of Drug Action*, 315–331. Elsevier, New York.

Marty, A. (1989). The physiological role of calcium-dependent channels. *Trends Neurosci.*, 12:420–424.

Montal, M., Darszon, A., and Schindler, H. (1981). Functional reassembly of membrane proteins in planar lipid bilayers. *Quart. Rev. Biophys.*, 14:1–79.

Neher, E. (1992). Ion channels for communication between and within cells. *Neuron*, 8:605–612.

Neher, E. and Stevens, C. F. (1977). Conductance fluctuations and ionic pores in membranes. *Ann. Rev. Biophys. Bioeng.*, 6:345–381.

Numa, S. (1989). A molecular view of neurotransmitter receptors and ionic channels. In *The Harvey Lectures,* ser. 83, 121–165. Alan R. Liss, New York.

Pallotta, B. S. and Wagoner, P. K. (1992). Voltage-dependent potassium channels since Hodgkin and Huxley. *Physiol. Rev.*, 72:S49–S67.

Parzen, E. (1962). *Stochastic Processes*. Holden-Day, San Francisco.

Patlak, J. (1991). Molecular kinetics of voltage-dependent Na^+ channels. *Physiol. Rev.*, 71:1047–1080.

Peracchia, C., ed. (1994). *Handbook of Membrane Channels*. Academic Press, New York.

Reuter, H. (1984). Ion channels in cardiac cell membranes. *Ann. Rev. Physiol.*, 46:473–484.

Ritchie, J. M. (1992). Voltage-gated ion channels in Schwann cells and glia. *Trends Neurosci*, 15:345–350.

Rogart, R. (1981). Sodium channels in nerve and muscle membrane. *Ann. Rev. Physiol.*, 43:711–725.

Rogawski, M. A. (1985). The A-current: How ubiquitous a feature of excitable cells is it? *Trends Neurosci.*, 8:214–219.

Rudy, B. (1988). Diversity and ubiquity of K channels. *Neurosciences*, 25:729–749.

Rydström, J., ed. (1987). *Membrane Proteins: Structure, Function, Assembly*. Cambridge University Press, New York.

Sakmann, B. and Neher, E., eds. (1983). *Single-Channel Recording*. Plenum, New York.

Salkoff, L., Baker, K., Butler, A., Covarrubias, M., Pak, M. D., and Wei, A. (1992). An essential 'set' of K^+ channels conserved in flies, mice and humans. *Trends Neurosci.*, 15:161–166.

Stevens, C. F. and Tsien, R. W., eds. (1979). *Ion Permeation through Membrane Channels*, vol. 3. Raven, New York.

Taylor, C. P. (1993). Na^+ currents that fail to inactivate. *Trends Neurosci.*, 16:455–460.

Tsien, R. W., Hess, P., McCleskey, E. W., and Rosenberg, R. L. (1987). Calcium channels: Mechanisms of selectivity, permeation, and block. *Ann. Rev. Biophys. Biophys. Chem.*, 16:265–290.

Watson, S. and Girdlestone, D. (1994). *1994 Receptor and Ion Channel Nomenclature Supplement*. Trends Pharmacol. Sci., 5th edition.

Weiss, T. F. (1996). *Cellular Biophysics,* vol. 1, *Transport.* MIT Press, Cambrdige, MA.

Yamamoto, D. (1985). The operation of the sodium channel in nerve and muscle. *Prog. Neurobiol.*, 24:257–291.

Original Articles

Agnew, W. S., Levinson, S. R., Brabson, J. S., and Raftery, M. A. (1978). Purification of the tetrodotoxin-binding component associated with the voltage-sensitive sodium channel from *Electrophorus electricus* electroplax membranes. *Proc. Natl. Acad. Sci. U.S.A.*, 75:2606–2610.

Ahmed, C. M. I., Ware, D. H., Lee, S. C., Patten, C. D., Ferrer-Montiel, A. V., Schinder, A. F., McPherson, J. D., Wagner-McPherson, C., Wasmuth, J. J., Evans, G. A., and Montal, M. (1992). Primary structure, chromosomal localization, and functional expression of a voltage-gated sodium channel from human brain. *Proc. Natl. Acad. Sci. U.S.A.*, 89:8220–8224.

Aldrich, R. W., Corey, D. P., and Stevens, C. F. (1983). A reinterpretation of mammalian sodium channel gating based on single channel recording. *Nature*, 306:436–441.

Arispe, N., Pollard, H. B., and Rojas, E. (1992). Calcium-independent K^+-selective channel from chromaffin granule membranes. *J. Membr. Biol.*, 130:191–202.

Armstrong, C. M. and Binstock, L. (1965). Anomalous rectification in the squid giant axon injected with tetraethylammonium chloride. *J. Gen. Physiol.*, 48:859–872.

Armstrong, C. M. and Cota, G. (1991). Calcium ion as a cofactor in Na channel gating. *Proc. Natl. Acad. Sci. U.S.A.*, 88:6528–6531.

Auld, V. J., Goldin, A. L., Krafte, D. S., Catterall, W. A., Lester, H. A., Davidson, N., and Dunn, R. J. (1990). A neutral amino acid change in segment IIS4 dramatically alters the gating properties of the voltage-dependent sodium channel. *Proc. Natl. Acad. Sci. U.S.A.*, 87:323–327.

Bauer, R. J., Bowman, B. F., and Kenyon, J. L. (1987). Theory of the kinetic analysis of patch-clamp data. *Biophys. J.*, 52:961–978.

Becker, J. D., Honerkamp, J., Hirsch, J., Fröbe, U., Schlatter, E., and Gregor, R. (1994). Analysing ion channels with hidden Markov models. *Pflügers Arch.*, 426:328–332.

Bekkers, J. M., Forster, I. C., and Greeff, N. G. (1990). Gating current associated with inactivated states of the squid axon sodium channel. *Proc. Natl. Acad. Sci. U.S.A.*, 87:8311–8315.

Benndorf, K., Koopmann, R., Lorra, C., and Pongs, O. (1994). Gating and conductance properties of a human delayed rectifier K^+ channel expressed in frog oocytes. *J. Physiol.*, 477.1:1–14.

Bezanilla, F. (1985). Gating of sodium and potassium channels. *J. Membr. Biol.*, 88:97–111.

Bezanilla, F. and Armstrong, C. M. (1975a). Kinetic properties and inactivation of the gating currents of sodium channels in squid axon. *Philos. Trans. R. Soc. London, Ser. B*, 270:449–458.

Bezanilla, F. and Armstrong, C. M. (1975b). Properties of the sodium channel gating current. In *Cold Spring Harbor Symposia on Quantitative Biology*, vol. 40, 297–304. Cold Spring Harbor Laboratory, Cold Spring Harbor, NY.

Bezanilla, F., Perozo, E., Papazian, D. M., and Stefani, E. (1991). Molecular basis of gating charge immobilization in *Shaker* potassium channels. *Science*, 254:679–683.

Bezanilla, F., Perozo, E., and Stefani, E. (1994). Gating of *Shaker* K^+ channels: II. The components of gating currents and a model of channel activation. *Biophys. J.*, 66:1011–1021.

Bezanilla, F., Taylor, R. E., and Fernàndez, J. M. (1982). Distribution and kinetics of membrane dielectric polarization: I. Long-term inactivation of gating currents. *J. Gen. Physiol.*, 79:21–40.

Březina, V., Evans, C. G., and Weiss, K. R. (1994a). Characterization of the membrane ion currents of a model molluscan muscle, the accessory radula closer muscle of *Aplysia californica*: I. Hyperpolarization-activated currents. *J. Neurophys.*, 71:2093–2112.

Březina, V., Evans, C. G., and Weiss, K. R. (1994b). Characterization of the membrane ion currents of a model molluscan muscle, the accessory radula closer muscle of *Aplysia californica*: II. Depolarization-actived K currents. *J. Neurophys.*, 71:2113–2125.

Březina, V., Evans, C. G., and Weiss, K. R. (1994c). Characterization of the membrane ion currents of a model molluscan muscle, the accessory radula closer muscle of *Aplysia californica*: III. Depolarization-actived Ca current. *J. Neurophys.*, 71:2126–2138.

Buchholtz, F., Golowasch, J., Epstein, I. R., and Marder, E. (1992). Mathematical model of an identified stomatogastric ganglion neuron. *J. Neurophys.*, 67:332–340.

Chiu, S. Y., Ritchie, J. M., Rogart, R. B., and Stagg, D. (1979). A quantitative description of membrane currents in rabbit myelinated nerve. *J. Physiol.*, 292:149–166.

Colquhoun, D. and Hawkes, A. G. (1977). Relaxation and fluctuations of membrane currents that flow through drug-operated channels. *Proc. R. Soc. London, Ser. B*, 199:231–262.

Colquhoun, D. and Hawkes, A. G. (1981). On the stochastic properties of single ion channels. *Proc. R. Soc. London, Ser. B*, 211:205–235.

Colquhoun, D. and Hawkes, A. G. (1982). On the stochastic properties of bursts of single ion channel openings and of clusters of bursts. *Philos. Trans. R. Soc. London, Ser. B*, 300:1–59.

Colquhoun, D. and Hawkes, A. G. (1990). Stochastic properties of ion channel openings and bursts in a membrane patch that contains two channels: Evidence concerning the number of channels present when a record containing only single openings is observed. *Proc. R. Soc. London, Ser. B*, 240:453–477.

Conti, F., Fioravanti, R., Segal, J. R., and Stühmer, W. (1982). Pressure dependence of the sodium currents of squid giant axon. *J. Membr. Biol.*, 69:23–34.

Cooper, K., Jakobsson, E., and Wolynes, P. (1985). The theory of ion transport through membrane channels. *Prog. Biophys. Mol. Biol.*, 46:51–96.

Covarrubias, M., Wei, A., and Salkoff, L. (1991). *Shaker, Shal*, and *Shaw* express independent K^+ current systems. *Neuron*, 7:763–773.

Crouzy, S. C. and Sigworth, F. J. (1993). Fluctuations in ion channel gating currents: Analysis of nonstationary shot noise. *Biophys. J.*, 64:68–76.

DeFelice, L. J. (1993a). Gating currents: Machinery behind the molecule. *Biophys. J.*, 64:5–6.

DeFelice, L. J. (1993b). Molecular and biophysical view of the Ca channel: A hypothesis regarding oligomeric structure, channel clustering, and macroscopic current. *J. Membr. Biol.*, 133:191–202.

Dodge, F. A. and Frankenhaeuser, B. (1958). Membrane currents in isolated frog nerve fibre under voltage clamp conditions. *J. Physiol.*, 143:76–90.

Dodge, F. A. and Frankenhaeuser, B. (1959). Sodium currents in the myelinated nerve fibre of *Xenopus laevis* investigated with the voltage clamp technique. *J. Physiol.*, 148:188-200.

Dubois, J. M. and Schneider, M. F. (1982). Kinetics of intramembrane charge movement and sodium current in frog node of Ranvier. *J. Gen. Physiol.*, 79:571-602.

Fenwick, E. M., Marty, A., and Neher, E. (1982a). A patch-clamp study of bovine chromaffin cells and of their sensitivity to acetylcholine. *J. Physiol.*, 331:577-597.

Fenwick, E. M., Marty, A., and Neher, E. (1982b). Sodium and calcium channels in bovine chromaffin cells. *J. Physiol.*, 331:599-635.

Fernàndez, J. M., Bezanilla, F., and Taylor, R. E. (1982). Distribution and kinetics of membrane dielectric polarization: II. Frequency domain studies of gating currents. *J. Gen. Physiol.*, 79:41-67.

Fox, A. P., Nowycky, M. C., and Tsien, R. W. (1987a). Kinetic and pharmacological properties distinguishing three types of calcium currents in chick sensory neurones. *J. Physiol.*, 394:149-172.

Fox, A. P., Nowycky, M. C., and Tsien, R. W. (1987b). Single-channel recordings of three types of calcium currents in chick sensory neurones. *J. Physiol.*, 394:173-200.

Fredkin, D. R. and Rice, J. A. (1992). Maximum likelihood estimation and identification directly from single-channel recordings. *Proc. R. Soc. London, Ser. B*, 249:125-132.

Fujii, S., Ayer Jr., R. K., and DeHaan, R. L. (1988). Development of the fast sodium current in early embryonic chick heart cells. *J. Membr. Biol.*, 101:209-223.

Fukushima, Y. (1981). Identification and kinetic properties of the current through a single Na^+ channel. *Proc. Natl. Acad. Sci. U.S.A.*, 78:1274-1277.

Fukushima, Y. (1982). Blocking kinetics of the anomalous potassium rectifier of tunicate egg studied by single channel recording. *J. Physiol.*, 331:311-331.

Goldman, L. and Schauf, C. L. (1973). Quantitative description of sodium and potassium currents and computed action potentials in *Myxicola* giant axons. *J. Gen. Physiol.*, 61:361-384.

Golowasch, J., Buchholtz, F., Epstein, I. R., and Marder, E. (1992). Contribution of individual ionic currents to activity of a model stomatogastric ganglion neuron. *J. Neurophys.*, 67:341-349.

Golowasch, J. and Marder, E. (1992). Ionic currents of the lateral pyloric neuron of the stomatogastric ganglion of the crab. *J. Neurophys.*, 67:318-331.

Gonoi, T. and Hille, B. (1987). Inactivation modifiers discriminate among models. *J. Gen. Physiol.*, 89:253-274.

Gordon, D., Merrick, D., Auld, V., Dunn, R., Goldin, A., Davidson, N., and Catterall, W. (1987). Tissue-specific expression of the R_I and R_{II} sodium channel subtypes. *Proc. Natl. Acad. Sci. U.S.A.*, 84:8682-8686.

Greeff, N. G., Keynes, R. D., and Van Helden, D. F. (1982). Fractionation of the asymmetry current in the squid giant axon into inactivating and non-inactivating components. *Proc. R. Soc. London, Ser. B*, 215:375-389.

Grissmer, S., Nguyen, A. N., and Cahalan, M. D. (1993). Calcium-activated potassium channels in resting and activated human T lymphocytes. *J. Gen. Physiol.*, 102:601-630.

Guy, H. R. (1990). Models of voltage- and transmitter-activated membrane channels based on their amino acid sequences. In Pasternak, C. A., ed., *Cations in Biological Systems*, 31-58. CRC Press, Boca Raton, FL.

Hagiwara, S. and Saito, N. (1959). Voltage-current relations in nerve cell membrane of *Onchidium verruculatum. J. Physiol.*, 148:161–179.

Hamill, O. P., Huguenard, J. R., and Prince, D. A. (1991). Patch-clamp studies of voltage-gated current in identified neurons of the rat cerebral cortex. *Cerebral Cortex*, 1:48–61.

Hamill, O. P., Marty, A., Neher, E., Sakmann, B., and Sigworth, F. J. (1981). Improved patch-clamp techniques for high-resolution current recording from cells and cell-free membrane patches. *Pflügers Arch.*, 391:85–100.

Hamill, O. P. and McBride, D. (1993). Molecular clues to mechanosensitivity. *Biophys. J.*, 65:17–18.

Hartline, D. K., Gassie, D. V., and Jones, B. R. (1993). Effects of soma isolation on outward currents measured under voltage clamp in spiny lobster stomatogastric motor neurons. *J. Neurophys.*, 69:2056–2071.

Heinemann, S. H., Terlau, H., Stühmer, W., Imoto, K., and Numa, S. (1992). Calcium channel characteristics conferred on the sodium channel by single mutations. *Nature*, 356:441–443.

Hille, B. (1966). Common mode of action of three agents that decrease the transient change in sodium permeability in nerves. *Nature*, 210:1220–1222.

Hille, B. (1967a). *A Pharmacological Analysis of the Ionic Channels of Nerve.* PhD thesis, Rockefeller University, New York.

Hille, B. (1967b). The selective inhibition of delayed potassium currents in nerve by tetraethylammonium ion. *J. Gen. Physiol.*, 50:1287–1302.

Hille, B. (1968). Pharmacological modifications of the sodium channels of frog nerve. *J. Gen. Physiol.*, 51:199–219.

Hille, B. (1971). The permeability of the sodium channel to organic cations in myelinated nerve. *J. Gen. Physiol.*, 58:599–619.

Hille, B. (1972). The permeability of the sodium channel to metal cations in myelinated nerve. *J. Gen. Physiol.*, 59:637–658.

Hille, B. (1973). Potassium channels in myelinated nerve: Selective permeability to small cations. *J. Gen. Physiol.*, 61:669–686.

Hille, B. (1988). Ionic channels: Molecular pores of excitable membranes. In *The Harvey Lectures,* series 82, 47–69. Alan R. Liss, New York.

Hille, B. (1989). Ionic channels: Evolutionary origins and modern roles. *Quart. J. Exp. Physiol.*, 61:785–804.

Horn, R. (1991). Estimating the number of channels in patch recordings. *Biophys. J.*, 60:433–439.

Horn, R. and Marty, A. (1988). Muscarinic activation of ionic currents measured by a new whole-cell recording method. *J. Gen. Physiol.*, 92:145–159.

Horn, R., Patlak, J., and Stevens, C. F. (1981). Sodium channels need not open before they inactivate. *Nature*, 291:426–427.

Isacoff, E. Y., Jan, Y. N., and Jan, L. Y. (1990). Evidence for the formation of heteromultimeric potassium channels in *Xenopus* oocytes. *Nature*, 345:530–534.

Julian, F. J., Moore, J. W., and Goldman, D. E. (1962). Membrane potentials of the lobster giant axon obtained by use of the sucrose-gap technique. *J. Gen. Physiol.*, 45:1195–1216.

Kallen, R. G., Sheng, Z. H., Yang, J., Chen, L., Rogart, R. B., and Barchi, R. L. (1990). Primary structure and expression of a sodium channel characteristic of denervated and immature rat skeletal muscle. *Neuron*, 4:233-242.

Katz, B. and Miledi, R. (1972). The statistical nature of the acetylcholine potential and its molecular components. *J. Physiol.*, 224:665-699.

Kayano, T., Noda, M., Flockerzi, V., Takahashi, H., and Numa, S. (1988). Primary structure of rat brain sodium channel III deduced from the cDNA sequence. *FEBS*, 228:187-194.

Keller, B. U., Hartshorne, R. P., Talvenheimo, J. A., Catterall, W. A., and Montal, M. (1986). Channel gating kinetics of purified sodium channels modified by batrachotoxin. *J. Gen. Physiol.*, 88:1-23.

Keynes, R. D. (1983). Voltage-gated ion channels in the nerve membrane. *Proc. R. Soc. London, Ser. B*, 220:1-30.

Keynes, R. D. (1990). A series-parallel model of the voltage-gated sodium channel. *Proc. R. Soc. London, Ser. B*, 240:425-432.

Keynes, R. D. (1991). On the voltage dependence of inactivation in the sodium channel of the squid giant axon. *Proc. R. Soc. London, Ser. B*, 243:47-53.

Keynes, R. D. (1994a). Bimodel gating of the Na^+ channel. *Trends Neurosci.*, 17:58-61.

Keynes, R. D., Greeff, N. G., and Forster, I. C. (1990). Kinetic analysis of the sodium gating current in the squid giant axon. *Proc. R. Soc. London, Ser. B*, 240:411-423.

Keynes, R. D., Greeff, N. G., and Van Helden, D. F. (1982). The relationship between the inactivating fraction of the asymmetry current and gating of the sodium channel in the squid giant axon. *Proc. R. Soc. London, Ser. B*, 215:391-404.

Keynes, R. D. and Kimura, J. E. (1983). Kinetics of activation of the sodium conductance in the squid giant axon. *J. Physiol.*, 336:621-634.

Keynes, R. D. and Rojas, E. (1974). Kinetics and steady-state properties of the charged system controlling sodium conductance in the squid giant axon. *J. Physiol.*, 239:393-434.

Keynes, R. D. and Rojas, E. (1976). The temporal and steady-state relationships between activation of the sodium conductance and movement of the gating particles in the squid giant axon. *J. Physiol.*, 255:157-189.

Kirsch, G. E. and Brown, A. M. (1989). Kinetic properties of single sodium channels in rat heart and rat brain. *J. Gen. Physiol.*, 93:85-99.

Koh, D., Jonas, P., Bräu, M. E., and Vogel, W. (1992). A TEA-insensitive flickering potassium channel active around the resting potential in myelinated nerve. *J. Membr. Biol.*, 130:149-162.

Kubo, Y., Baldwin, T. J., Jan, Y. N., and Jan, L. Y. (1993). Primary structure and functional expression of a mouse inward rectifier potassium channel. *Nature*, 362:127-133.

Kuo, C. and Bean, B. P. (1993). G-protein modulation of ion permeation through N-type calcium channels. *Nature*, 365:258-262.

Kuo, C. and Bean, B. P. (1994). Na^+ channels must deactivate to recover from inactivation. *Neuron*, 12:819-829.

Kyte, J. and Doolittle, R. F. (1982). A simple method for displaying the hydropathic character of a protein. *J. Mol. Biol.*, 157:105-132.

Leinders, T., van Kleef, R. G. D. M., and Vijverberg, H. P. M. (1992a). Distinct metal ion binding sites on Ca^{2+}-activated K^+ channels in inside-out patches of human erythrocytes. *Biochim. Biophys. Acta*, 1112:75-82.

Leinders, T., van Kleef, R. G. D. M., and Vijverberg, H. P. M. (1992b). Single Ca^{2+} activated K^+ channels in human erythrocytes: Ca^{2+} dependence of opening frequency but not of open lifetimes. *Biochim. Biophys. Acta*, 1112:67–74.

Leuchtag, H. R. (1994). Long-range interactions, voltage sensitivity, and ion conduction in S4 segments of excitable channels. *Biophys. J.*, 66:217–224.

Levis, R. A. and Rae, J. L. (1993). The use of quartz patch pipettes for low noise single channel recording. *Biophys. J.*, 65:1666–1677.

Levitt, D. G. (1986). Interpretation of biological ion channel flux data: Reaction-rate versus continuum theory. *Ann. Rev. Biophys. Biophys. Chem.*, 15:29–57.

Liman, E. R., Tytgat, J., and Hess, P. (1992). Subunit stoichiometry of a mammalian K^+ channel determined by construction of multimeric cDNAs. *Neuron*, 9:861–871.

Lipkind, G. M. and Fozzard, H. A. (1994). A structural model of the tetrodotoxin and saxitoxin binding site of the Na^+ channel. *Biophys. J.*, 66:1–13.

Lopez, G. A., Jan, Y. N., and Jan, L. Y. (1994). Evidence that the S6 segment of the Shaker voltage-gated K^+ channel comprises part of the pore. *Nature*, 367:179–182.

Meech, R. W. and Mackie, G. O. (1993a). Ionic currents in giant motor axons of the jellyfish, *Aglantha digitale*. *J. Neurophys.*, 69:884–893.

Meech, R. W. and Mackie, G. O. (1993b). Potassium channel family in giant motor axons of *Aglantha digitale*. *J. Neurophys.*, 69:894–901.

Mika, Y. H. and Palti, Y. (1994). Charge displacement in a single potassium ion channel macromolecule during gating. *Biophys. J.*, 67:1455–1463.

Montal, M. (1987). Reconstitution of channel proteins from excitable cells in planar lipid bilayer membranes. *J. Membr. Biol.*, 98:101–115.

Moore, H., Fritz, L., Raftery, M., and Brockes, J. (1982). Isolation and characterization of a monoclonal antibody against the saxitoxin-binding component from the electric organ of the eel *Electrophorus electricus*. *Proc. Natl. Acad. Sci. U.S.A.*, 79:1673–1677.

Nagy, K., Kiss, R., and Hof, D. (1983). Single Na channels in mouse neuroblastoma cell membrane; indications for two open states. *Pflügers Arch.*, 399:302–308.

Nakamura, Y., Nakajima, S., and Grundfest, H. (1965). Analysis of spike electrogenesis and depolarizing K inactivation in electroplaques of *Electrophorus electricus*. *J. Gen. Physiol.*, 49:321–349.

Narahashi, T., Moore, J. W., and Scott, W. R. (1964). Tetrodotoxin blockage of sodium conductance increase in lobster giant axons. *J. Gen. Physiol.*, 47:965–974.

Nealey, T., Spires, S., Eatock, R. A., and Begenisich, T. (1993). Potassium channels in squid neuron cell bodies: Comparison to axonal channels. *J. Membr. Biol.*, 132:13–25.

Neher, E. (1971). Two fast transient current components during voltage clamp on snail neurons. *J. Gen. Physiol.*, 58:36–53.

Neher, E. and Sakmann, B. (1976). Single-channel currents recorded from membrane of denervated frog muscle fibres. *Nature*, 260:799–802.

Noda, M., Ikeda, T., Kayano, T., Suzuki, H., Takeshima, H., Kurasaki, M., Takahashi, H., and Numa, S. (1986a). Existence of distinct sodium channel messenger RNAs in rat brain. *Nature*, 320:188–192.

Noda, M., Ikeda, T., Suzuki, H., Takeshima, H., Takahashi, T., Kuno, M., and Numa, S. (1986b). Expression of functional sodium channels from cloned cDNA. *Nature*, 322:826–828.

Noda, M., Shimizu, S., Tanabe, T., Takai, T., Kayano, T., Ikeda, T., Takahashi, H., Nakayama, H., Kanaoka, Y., Minamino, N., Kangawa, K., Matsuo, H., Raftery, M., Hirose, T., Inayama, S., Hayashida, H., Miyata, T., and Numa, S. (1984). Primary structure of *electrophorus electricus* sodium channel deduced from cDNA sequence. *Nature*, 312:121–127.

Noda, M., Suzuki, H., Numa, S., and Stühmer, W. (1989). A single point mutation confers tetrodotoxin and saxitoxin insensitivity on the sodium channel II. *Fed. Eur. Biochem. Soc.*, 259:213–216.

Oxford, G. S. (1981). Some kinetic and steady-state properties of sodium channels after removal of inactivation. *J. Gen. Physiol.*, 77:1–22.

Pantoja, O., Gelli, A., and Blumwald, E. (1992). Voltage-dependent calcium channels in plant vacuoles. *Science*, 255:1567–1570.

Patlak, J. and Horn, R. (1982). Effect of *N-bromoacetamide* on single sodium channel currents in excised membrane patches. *J. Gen. Physiol.*, 79:333–351.

Patlak, J., Ortiz, M., and Horn, R. (1986). Opentime heterogeneity during bursting of sodium channels in frog skeletal muscle. *Biophys. J.*, 49:773–777.

Patton, D. E., West, J. W., Catterall, W. A., and Goldin, A. L. (1992). Amino acid residues required for fast Na^+-channel inactivation: Charge neutralization and deletions in III-IV linker. *Proc. Natl. Acad. Sci. U.S.A.*, 89:10905–10909.

Perozo, E., Santacruz-Toloza, L., Stefani, E., Bezanilla, F., and Papazian, D. M. (1994). S4 mutations alter gating currents of Shaker K channels. *Biophys. J.*, 66:345–354.

Pichon, Y. and Boistel, J. (1967). Current-voltage relations in the isolated giant axon of the cockroach under voltage-clamp conditions. *J. Exp. Biol.*, 47:343–355.

Pongs, O. (1993). Structure-function studies on the pore of potassium channels. *J. Membr. Biol.*, 136:1–8.

Pusch, M. and Jentsch, T. J. (1994). Molecular physiology of voltage-gated chloride channels. *Physiol. Rev.*, 74:813–827.

Pusch, M. and Neher, E. (1988). Rates of diffusional exchange between small cells and a measuring patch pipette. *Pflügers Arch.*, 411:204–211.

Ritchie, J. M., Black, J. A., Waxman, S. G., and Angelides, K. J. (1990). Sodium channels in the cytoplasm of Schwann cells. *Proc. Natl. Acad. Sci. U.S.A.*, 87:9290–9294.

Ritchie, J. M. and Rogart, R. B. (1977). The binding of saxitoxin and tetrodotoxin to excitable tissue. *Rev. Physiol. Biochem. Pharmacol.*, 79:1–50.

Rogart, R. B., Cribbs, L. L., Muglia, L. K., Kephart, D. D., and Kaiser, M. W. (1989). Molecular cloning of a putative tetrodotoxin-resistant rat heart Na^+ channel isoform. *Proc. Natl. Acad. Sci. U.S.A.*, 86:8170–8174.

Rojas, E. (1975). Gating mechanism for the activation of the sodium conductance in nerve membranes. In *Cold Spring Harbor Symposia on Quantitative Biology*, vol. 40, 305–320. Cold Spring Harbor Laboratory, Cold Spring Harbor, NY.

Rosenthal, J. J. C. and Gilly, W. F. (1993). Amino acid sequence of a putative sodium channel expressed in the giant axon of the squid *Loligo opalescens*. *Proc. Natl. Acad. Sci. U.S.A.*, 90:10026–10030.

Rudy, B., Kentros, C., and Vega-Saenz de Miera, E. (1991). Families of potassium channel genes in mammals: Toward an understanding of the molecular basis of potassium channel diversity. *Mol. Cell. Neurosci.*, 2:89–102.

Ruppersberg, J. P., Schröter, K. H., Sakmann, B., Stocker, M., Sewing, S., and Pongs, O. (1990). Heteromultimeric channels formed by rat brain potassium-channel proteins. *Nature*, 345:535-537.

Saint, D. A., Pugsley, M. K., and Chung, S. H. (1994). An analysis of cardiac sodium channel properties using digital signal processing techniques. *Biochim. Biophys. Acta*, 1196:131-138.

Sansom, M. S. P., Ball, F. G., Kerry, C. J., McGee, R., Ramsey, R. L., and Usherwood, P. N. R. (1989). Markov, fractal, diffusion, and related models of ion channel gating: A comparison with experimental data from two ion channels. *Biophys. J.*, 56:1229-1243.

Shrager, P. (1974). Ionic conductance change in voltage clamped crayfish axons at low pH. *J. Gen. Physiol.*, 64:666-690.

Siemen, D. (1993). Nonselective cation channels. In *Nonselective Cation Channels: Pharmacology, Physiology and Biophysics*. Birkhäuser Verlag, Switzerland.

Sigworth, F. J. and Neher, E. (1980). Single Na^+ channel currents observed in cultured rat muscle cells. *Nature*, 287:447-449.

Smith, S. J. and Thompson, S. H. (1987). Slow membrane currents in bursting pacemaker neurones of *tritionia*. *J. Physiol.*, 382:425-448.

Sontheimer, H., Black, J. A., Ransom, B. R., and Waxman, S. G. (1992). Ion channels in spinal cord astrocytes *in vitro*: I. Transient expression of high levels of Na^+ and K^+ channels. *J. Neurophys.*, 68:985-1000.

Sontheimer, H. and Waxman, S. G. (1992). Ion channels in spinal cord astrocytes *in vitro*: II. Biophysical and pharmacological analysis of two Na^+ current types. *J. Neurophys.*, 68:1001-1011.

Spires, S. and Begenisich, T. (1989). Pharmacological and kinetic analysis of K channel gating currents. *J. Gen. Physiol.*, 93:263-283.

Spires, S. and Begenisich, T. (1992). Chemical properties of the divalent cation binding site on potassium channels. *J. Gen. Physiol.*, 100:181-193.

Stampe, P. and Vestergaard-Bogind, B. (1988). Ca^{2+}-activated K^+ conductance of human red cell membranes exhibits two different types of voltage dependence. *J. Membr. Biol.*, 101:165-172.

Stefani, E., Toro, L., Perozo, E., and Bezanilla, F. (1994). Gating of Shaker K^+ channels: I. Ionic and gating currents. *Biophys. J.*, 66:996-1010.

Stimers, J. R., Bezanilla, F., and Taylor, R. E. (1985). Sodium channel activation in the squid giant axon. *J. Gen. Physiol.*, 85:65-82.

Stühmer, W., Conti, F., Suzuki, H., Wang, X., Noda, M., Yahagi, N., Kubo, H., and Numa, S. (1989). Structural parts involved in activation and inactivation of the sodium channel. *Nature*, 339:597-603.

Suzuki, H., Beckh, S., Kubo, H., Yahagi, N., Ishida, H., Kayano, T., Noda, M., and Numa, S. (1988). Functional expression of cloned cDNA encoding sodium channel III. *FEBS*, 228:195-200.

Tasaki, I. and Hagiwara, S. (1957). Demonstration of two stable potential states in the squid giant axon under tetraethylammonium chloride. *J. Gen. Physiol.*, 40:859-885.

Taylor, R. E. and Bezanilla, F. (1983). Sodium and gating current time shifts resulting from changes in initial conditions. *J. Gen. Physiol.*, 81:773-784.

Tempel, B. L., Jan, Y. N., and Jan, L. Y. (1988). Cloning of a probable potassium channel gene from mouse brain. *Nature*, 332:837-839.

Thompson, S. H. (1977). Three pharmacologically distinct potassium channels in molluscan neurones. *J. Physiol.*, 265:465–488.

Thompson, S. H. (1982). Aminopyridine block of transient potassium current. *J. Gen. Physiol.*, 80:1–18.

Tosteson, M. T., Auld, D. S., and Tosteson, D. C. (1989). Voltage-gated channels formed in lipid bilayers by a positively charged segment of the Na-channel polypeptide. *Proc. Natl. Acad. Sci. U.S.A.*, 86:707–710.

Trimmer, J. S., Cooperman, S. S., Tomiko, S. A., Zhou, J., Crean, S. M., Boyle, M. B., Kallen, R. G., Sheng, Z., Barchi, R. L., Sigworth, F. J., Goodman, R. H., Agnew, W. S., and Mandel, G. (1989). Primary structure and functional expression of a mammalian skeletal muscle sodium channel. *Neuron*, 3:33–49.

Vassilev, P. M., Scheuer, T., and Catterall, W. A. (1988). Identification of an intracellular peptide segment involved in sodium channel inactivation. *Science*, 241:1658–1661.

Vassilev, P. M., Scheuer, T., and Catterall, W. A. (1989). Inhibition of inactivation of single sodium channels by a site-directed antibody. *Proc. Natl. Acad. Sci. U.S.A.*, 86:8147–8151.

Weigele, J. B. and Barchi, R. L. (1982). Functional reconstitution of the purified sodium channel protein from rat sarcolemma. *Proc. Natl. Acad. Sci. U.S.A.*, 79:3651–3655.

Weiss, D. S. and Magleby, K. L. (1990). Voltage dependence and stability of the gating kinetics of the fast chloride channel from rat skeletal muscle. *J. Physiol.*, 426:145–176.

West, J. W., Numann, R., Murphy, B. J., Scheuer, T., and Catterall, W. A. (1991). A phosphorylation site in the Na^+ channel required for modulation by protein kinase C. *Science*, 254:866–868.

West, J. W., Patton, D. E., Scheuer, T., Wang, Y., Goldin, A. L., and Catterall, W. A. (1992). A cluster of hydrophobic amino acid residues required for fast Na^+-channel inactivation. *Proc. Natl. Acad. Sci. U.S.A.*, 89:10910–10914.

Wollmuth, L. P. and Hille, B. (1992). Ionic selectivity of I_h channels of rod photoreceptors in tiger salamanders. *J. Gen. Physiol.*, 100:749–765.

Yang, J., Ellinor, P. T., Sather, W. A., Zhang, J., and Tsien, R. W. (1993). Molecular determinants of Ca^{2+} selectivity and ion permeation in L-type Ca^{2+} channels. *Nature*, 366:158–161.

Yue, D. T. (1993). Bridging the gap between anomalous sodium channel molecules and aberrant physiology. *Biophys. J.*, 65:13–14.

Zamponi, G. W. and French, R. J. (1994a). Amine blockers of the cytoplasmic mouth of sodium channels: A small structural change can abolish voltage dependence. *Biophys. J.*, 67:1015–1027.

Zamponi, G. W. and French, R. J. (1994b). Open-channel block by internally applied amines inhibits activation gate closure in batrachotoxin-activated sodium channels. *Biophys. J.*, 67:1040–1051.

Zamponi, G. W. and French, R. J. (1994c). Transcainide causes two modes of open-channel block with different voltage sensitivities in batrachotoxin-activated sodium channels. *Biophys. J.*, 67:1028–1039.

List of Figures

1 Introduction to Electrical Properties of Cells

2   Lumped-Parameter and Distributed-Parameter Models of Cells

3   Linear Electrical Properties of Cells

4   The Hodgkin-Huxley Model

5   Saltatory Conduction in Myelinated Nerve Fibers

6   Voltage-Gated Ion Channels

List of Tables

5   Saltatory Conduction in Myelinated Nerve Fibers

6   Voltage-Gated Ion Channels

Contents of Volume 1

Index

Page numbers in italic indicate figures or tables; *n* indicates footnotes.

Printed in the United States
By Bookmasters